The Role of Fibrinolytic System in Health and Disease

The Role of Fibrinolytic System in Health and Disease

Editor

Hau C. Kwaan

MDPI • Basel • Beijing • Wuhan • Barcelona • Belgrade • Manchester • Tokyo • Cluj • Tianjin

Editor
Hau C. Kwaan
Northwestern University
USA

Editorial Office
MDPI
St. Alban-Anlage 66
4052 Basel, Switzerland

This is a reprint of articles from the Special Issue published online in the open access journal *International Journal of Molecular Sciences* (ISSN 1422-0067) (available at: https://www.mdpi.com/journal/ijms/special_issues/_Fibrinolysis).

For citation purposes, cite each article independently as indicated on the article page online and as indicated below:

LastName, A.A.; LastName, B.B.; LastName, C.C. Article Title. *Journal Name* **Year**, *Volume Number*, Page Range.

ISBN 978-3-0365-4619-3 (Hbk)
ISBN 978-3-0365-4620-9 (PDF)

Contents

About the Editor

Hau C. Kwaan

Hau C. Kwaan, M.D., Ph.D., F.R.C.P. (London), F.R.C.P. (Edinburgh), F.A.C.P., Professor of Medicine, and Marjorie C. Barnett, Professor of Hematology and Oncology, Division of Hematology and Medical Oncology, Feinberg School of Medicine, Northwestern University, Chicago, IL. Attending physician at Northwestern Memorial Hospital, Chicago, IL. Qualified MB, BS, University of Hong Kong in 1952. Research Fellow in Pharmacology at Columbia University College of Physicians and Surgeons, New York 1957–1958. Member of the Royal College of Physicians of Edinburgh, Scotland, in 1958. Certified by the American Board of Internal Medicine in 1969, and recertified in 1977. Certified by the Hematology Board in 1974 and the Oncology Board in 1979. 2012–present - Senior Editor, Seminars in Thrombosis and Hemostasis Research. Accomplishments include his demonstration that a blood clot-dissolving enzyme (vascular plasminogen activator) is pre¬sent in blood vessel walls and can be released to dissolve blood clots when they are formed in the blood vessels.

International Journal of
Molecular Sciences

Editorial

The Role of Fibrinolytic System in Health and Disease

Hau C. Kwaan

Division of Hematology-Oncology, Feinberg School of Medicine, Northwestern University, Chicago, IL 60611, USA; h-kwaan@northwestern.edu

Abstract: The fibrinolytic system is composed of the protease plasmin, its precursor plasminogen and their respective activators, tissue-type plasminogen activator (tPA) and urokinase-type plasminogen activator (uPA), counteracted by their inhibitors, plasminogen activator inhibitor type 1 (PAI-1), plasminogen activator inhibitor type 2 (PAI-2), protein C inhibitor (PCI), thrombin activable fibrinolysis inhibitor (TAFI), protease nexin 1 (PN-1) and neuroserpin. The action of plasmin is counteracted by α2-antiplasmin, α2-macroglobulin, TAFI, and other serine protease inhibitors (antithrombin and α2-antitrypsin) and PN-1 (protease nexin 1). These components are essential regulators of many physiologic processes. They are also involved in the pathogenesis of many disorders. Recent advancements in our understanding of these processes enable the opportunity of drug development in treating many of these disorders.

Keywords: fibrinolysis; plasmin; plasminogen activator; PAI-1; PAI-2; antiplasmin

Citation: Kwaan, H.C. The Role of Fibrinolytic System in Health and Disease. *Int. J. Mol. Sci.* **2022**, 23, 5262. https://doi.org/10.3390/ijms23095262

Received: 6 April 2022
Accepted: 5 May 2022
Published: 9 May 2022

Publisher's Note: MDPI stays neutral with regard to jurisdictional claims in published maps and institutional affiliations.

The fibrinolytic system, also known as the plasminogen–plasmin system, is composed of a proteolytic enzyme plasmin (Pm) with its precursor plasminogen (Pg) (Figure 1) [1–7]. There are two naturally occurring activators of plasminogen, tissue-type plasminogen activator (tPA) and urokinase-type plasminogen activator (uPA). In addition, the conversion from Pg to Pm can be accomplished by other proteases such as streptokinase, staphylokinase and plasmin. The actions of activators are counteracted by inhibitors, plasminogen activator type 1 (PAI-1), plasminogen activator type 2 (PAI-2), protein C inhibitor (PCI), thrombin activable fibrinolysis inhibitor (TAFI), protease nexin 1 (PN-1) and neuroserpin. Pm, on the other hand, are inhibited by α2-antiplasmin, α2-macroglobulin, TAFI and serine protease inhibitors, including antithrombin and α2-antitrypsin, and by PN-1. In addition, there are receptors for plasminogen [8] and for tPA in the form of annexin II [9–11], which is co-localized on the cell surface S-100A10 [12], as well as a receptor for uPA, uPAR [13].

When first discovered, the fibrinolytic system was thought to primarily function as a regulator of fibrin formation and breakdown. Soon it was found that it is involved in many physiological and pathological functions (Tables 1 and 2). A complete review is beyond the scope of this article, but a few examples are shown below.

Table 1. Physiological functions of the fibrinolytic system.

Embryogenesis Ovulation, menstruation
Pregnancy
Neuron growth
Brain function
Regulation of blood–brain barrier
Immunity
Wound healing
Senescence
Fibrosis

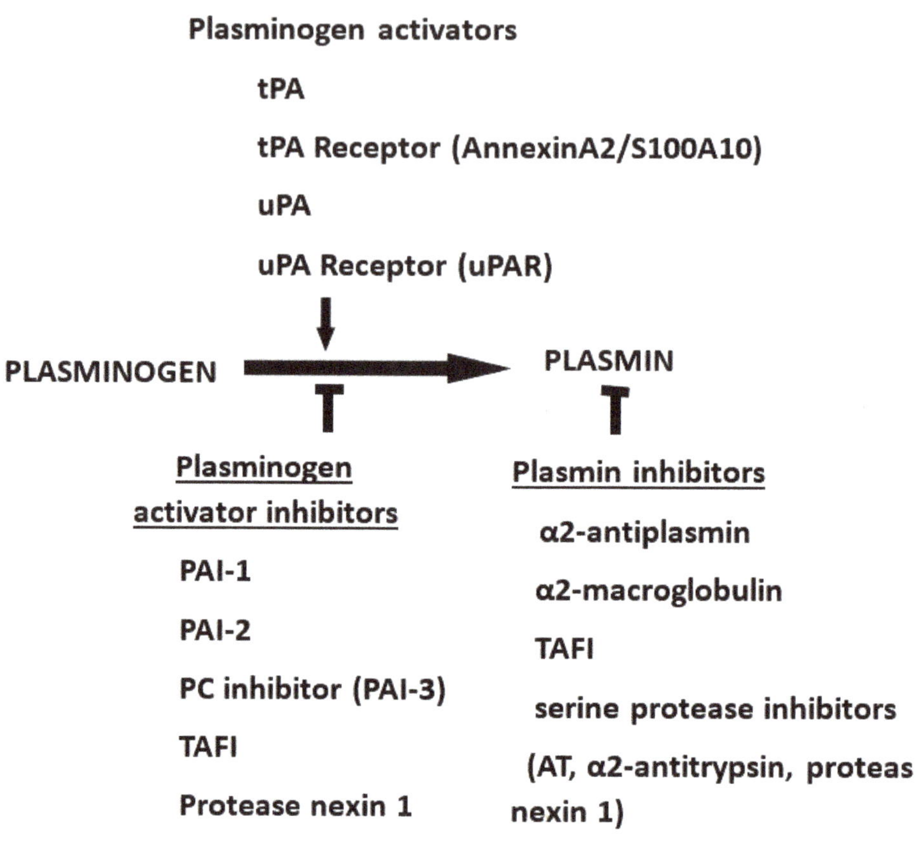

Figure 1. The fibrinolytic (plasminogen–plasmin) system.

Table 2. Role of the fibrinolytic system in multiple disorders.

Neurologic disorders
Stroke/Hemorrhagic transformation
Degenerative disorders
Cancer proliferation, invasion/metastasis, angiogenesis
Vascular diseases
Atherosclerosis, myocardial infarction
Metabolic syndrome
Trauma
Fibrosis

Components of the Pg–Pm system are involved in the regulation of menstruation and pregnancy [14], with interactions between the gonadotrophins and tPA, uPA and uPAR. tPA is involved in neuronal growth and learning [15], the regulation of the blood–brain barrier [16–19], and the regulation of glucose metabolism in the brain [20]. Through multiple pathways, fibrinolytic components can modulate host immunity [21–23]. PAI-1 is involved in cell senescence [24] and physiological aging [25]. uPA/uPAR and PAI-1 regulate cell motility and migration and thus are important in wound healing [26,27].

In many types of cancer, there is evidence that there is a correlation between uPA, uPAR and PAI-1 and the aggressiveness and metastatic potential in both tumor cell cultures and tumor tissues [28–33]. In carcinoma of the breast, elevated levels of uPA and PAI-1 were found to be associated with a worse prognosis [34]. This association was used in the management of the tumor [35,36]. In carcinoma of the pancreas, the postoperative survival of those with high uPA and PAI-1 was found to be 9 months, while those without these markers was 18 months [37]. A poor response to chemotherapy in small-cell carcinoma of the lung was observed in those with high uPAR [38]. In an athymic mouse model, the transfection of PAI-1 to prostate cancer cells (PC-3) was found to inhibit growth and metastasis [39].

Fibrinolysis is a major component of trauma-induced coagulopathy [40]. As hemorrhage is the major cause of death, excessive fibrinolysis has been observed early after injury and showed a negative predictive value of outcome [41–46]. However, the status of fibrinolysis rapidly changes to a hypofibrinolytic phase, often referred to as "fibrinolytic shutdown". Such a temporal change is part of the body's response to injury. Persistent low fibrinolytic activity is, however, associated with poor outcomes with multi-organ failure. In one study, patients with low fibrinolytic activity for 7 days had an eightfold higher mortality rate than those whose fibrinolytic activity recovered [47]. In another study, a threefold higher mortality was seen in those with persistent fibrinolytic shutdown at 24 h after injury [48]. Furthermore, the fibrinolytic components play a major role in the pathogenesis of intracranial hemorrhage in trauma patients [40]. Notably, hyperfibrinolysis carries with it a poor prognosis. This is due in part to a breakdown of the blood–brain barrier [16,18,40], as discussed below.

In acute and chronic stress, the hemostatic balance, including endothelial activation, the activation of coagulation and altered fibrinolytic balance, are altered [49]. These changes are pro-thrombotic and hypofibrinolytic with an increase in PAI-1.

Impaired fibrinolysis has been observed in depression with an increase in PAI-1 [50]. Fibrinolysis is an important factor in brain remodeling [51]. Biomarkers for depression are correlated with hypofibrinolysis.

Neurologic functions are another area where tPA and PAI-1 are involved. A wide range of these functions includes ovulation [52], embryogenesis [53], neuronal migration [54], learning [55], the degradation of amyloid [16], stress/fear response [56], and the regulation of the blood–brain barrier [10,12,34]. Notably, tPA increases the permeability of the blood–brain barrier in both a plasmin-dependent and a plasmin-independent pathway. Plasmin activates metalloproteinase directly. Alternatively, tPA can directly activate latent platelet-derived growth factor CC (PDGF-CC), and the tPA–PAI-1 complex activates PDGF receptor alpha (PDGFRa), thus signaling intracellular PI 3K, Ras MAPK, p38 MAPK and PLc-g. These signal an increase in vascular permeability and open the blood–brain barrier [16,18,57,58].

Clinically, these characteristics of tPA are seen as adverse effects. In the treatment of acute ischemic stroke, tPA can increase the risk of hemorrhagic conversion. In nine clinical trials [59,60], hemorrhagic conversion resulted in severe intracranial hemorrhage within 24–36 h in 6.8% of patients, while this figure is 1% in those not receiving tPA, an increase of over fivefold. Fatal intercranial hemorrhage within 7 days occurred in 2.7% of the tPA-treated patients versus 0.4% in those not treated, an over sixfold increase.

In the cardiovascular system, PAI-1 is elevated in cardiovascular diseases [61–63] and in metabolic syndrome [64]. In a clinical trial assessing the dietary intake of saturated fatty acids, PAI-1 concentrations were twofold higher in participants at increased risk for cardiometabolic diseases compared with healthy participants. Patients with acute myocardial infarction and acute ischemic strokes were found to have higher PAI-1 levels [63]. In patients with metabolic syndrome, dietary restriction results in the lowering of the PAI-1 level [61].

During response to injury, plasmin and PAI-1 are involved in the wound healing process. Plasmin activates many latent growth factors and proteases, including metalloproteinases (MMP). The latter are responsible for the breakdown of the intercellular matrix. The failure of this process results in delayed healing and chronic inflammation with fibrosis [65]. PAI-1 enhances fibrosis in chronic inflammation in many organs [66–68].

In experimental animals, for example, transgenic mouse with the PAI-1 $-/-$ genotype, fibrosis does not occur following injury, such as the bleomycin injury model, to the lung.

In COVID-19, both tPA and PAI-1 are involved in the complex pathway in which the spike protein of SARS-Co-2 attaches to a component of the renin–aldosterone–angiotensin system, ACE 2, during the invasion of the host cells [69–71]. First, plasmin and with other proteases, trypsin and transmembrane proteases (TMPRSS 2), facilitate the binding of the spike protein to ACE 2. Following binding, ACE is internalized and unable to process the breakdown of angiotensin II, leading to its excess. The excess of angiotensin II leads to an increase in PAI-1 [72]. This contributes to the hypercoagulable state seen in COVID-19. In addition, the excess of angiotensin II binds to its receptor angiotensin II receptor 1a, causing lung injury and leading to pulmonary edema with the formation of a hyaline membrane with fibrin in the alveoli [73,74]. Here, again, plasmin is involved in clearing fibrin. Furthermore, the diffuse alveolar damage with damaged type II alveolar cells leads to decreased surfactant, which results in the induction of the p53 pathway and increased PAI-1 [75].

In summary, the role of the fibrinolytic system is not limited to the resolution of fibrin and thrombi, but is involved a wide range of physiologic conditions and pathologic disorders. The intensive research carried out in recent years has revealed many new findings. This knowledge offers an opportunity for therapeutic development, particularly in the mitigation of the adverse effects of PAI-1. Greater understanding of these functions is essential for the management of many disorders.

Funding: This research received no external funding.

Conflicts of Interest: The authors declare no conflict of interest.

References

1. Astrup, T. Blood coagulation and fibrinolysis in tissue culture and tissue repair. *Biochem. Pharmacol.* **1968**, *17*, 241–257. [CrossRef]
2. Sherry, S. Fibrinolysis and Clinical Medicine. *Triangle* **1964**, *7*, 294–300. Available online: https://www.ncbi.nlm.nih.gov/pubmed/14251104 (accessed on 1 May 2022).
3. Sherry, S. Fibrinolysis in health and disease. *Med. Times* **1967**, *95*, 945–956. Available online: https://www.ncbi.nlm.nih.gov/pubmed/6074071 (accessed on 1 May 2022).
4. Sherry, S. Fibrinolysis. *Annu. Rev. Med.* **1968**, *19*, 247–268. [CrossRef]
5. Kwaan, H.C. Disorders of fibrinolysis. *Med. Clin. N. Am.* **1972**, *56*, 163–176. [CrossRef]
6. Kwaan, H.C. Fibrinolysis–A perspective. *Prog. Cardiovasc. Dis.* **1979**, *21*, 397–403. [CrossRef]
7. Kwaan, H.C. The role of fibrinolysis in disease processes. *Semin. Thromb. Hemost.* **1984**, *10*, 71–79. [CrossRef]
8. Miles, L.A.; Parmer, R.J. Plasminogen receptors: The first quarter century. *Semin. Thromb. Hemost.* **2013**, *39*, 329–337. [CrossRef]
9. Cesarman, G.M.; Guevara, C.A.; Hajjar, K.A. An endothelial cell receptor for plasminogen/tissue plasminogen activator (t-PA). II. Annexin II-mediated enhancement of t-PA-dependent plasminogen activation. *J. Biol. Chem.* **1994**, *269*, 21198–21203. Available online: https://www.ncbi.nlm.nih.gov/pubmed/8063741 (accessed on 1 May 2022). [CrossRef]
10. Hajjar, K.A.; Jacovina, A.T.; Chacko, J. An endothelial cell receptor for plasminogen/tissue plasminogen activator. I. Identity with annexin II. *J. Biol. Chem.* **1994**, *269*, 21191–21197. Available online: https://www.ncbi.nlm.nih.gov/pubmed/8063740 (accessed on 1 May 2022). [CrossRef]
11. Hajjar, K.A.; Menell, J.S. Annexin II: A novel mediator of cell surface plasmin generation. *Ann. N. Y. Acad. Sci.* **1997**, *811*, 337–349. [CrossRef] [PubMed]
12. O'Connell, P.A.; Madureira, P.A.; Berman, J.N.; Liwski, R.S.; Waisman, D.M. Regulation of S100A10 by the PML-RAR-alpha oncoprotein. *Blood* **2011**, *117*, 4095–4105. [CrossRef] [PubMed]
13. Ploug, M.; Ronne, E.; Behrendt, N.; Jensen, A.L.; Blasi, F.; Dano, K. Cellular receptor for urokinase plasminogen activator. Carboxyl-terminal processing and membrane anchoring by glycosyl-phosphatidylinositol. *J. Biol. Chem.* **1991**, *266*, 1926–1933. Available online: https://www.ncbi.nlm.nih.gov/pubmed/1846368 (accessed on 1 May 2022). [CrossRef]
14. Ny, T.; Liu, Y.X.; Ohlsson, M.; Jones, P.B.; Hsueh, A.J. Regulation of tissue-type plasminogen activator activity and messenger RNA levels by gonadotropin-releasing hormone in cultured rat granulosa cells and cumulus-oocyte complexes. *J. Biol. Chem.* **1987**, *262*, 11790–11793. Available online: https://www.ncbi.nlm.nih.gov/pubmed/3114254 (accessed on 1 May 2022). [CrossRef]
15. Seeds, N.W.; Williams, B.L.; Bickford, P.C. Tissue plasminogen activator induction in Purkinje neurons after cerebellar motor learning. *Science* **1995**, *270*, 1992–1994. [CrossRef]
16. Medcalf, R.L. Fibrinolysis: From blood to the brain. *J. Thromb. Haemost.* **2017**, *15*, 2089–2098. [CrossRef]

17. Niego, B.; Lee, N.; Larsson, P.; De Silva, T.M.; Au, A.E.; McCutcheon, F.; Medcalf, R.L. Selective inhibition of brain endothelial Rho-kinase-2 provides optimal protection of an in vitro blood-brain barrier from tissue-type plasminogen activator and plasmin. *PLoS ONE* **2017**, *12*, e0177332. [CrossRef]

18. Yepes, M.; Sandkvist, M.; Moore, E.G.; Bugge, T.H.; Strickland, D.K.; Lawrence, D.A. Tissue-type plasminogen activator induces opening of the blood-brain barrier via the LDL receptor-related protein. *J. Clin. Investig.* **2003**, *112*, 1533–1540. [CrossRef]

19. Fredriksson, L.; Lawrence, D.A.; Medcalf, R.L. tPA Modulation of the Blood-Brain Barrier: A Unifying Explanation for the Pleiotropic Effects of tPA in the CNS. *Semin. Thromb. Hemost.* **2017**, *43*, 154–168. [CrossRef]

20. Parcq, J.; Bertrand, T.; Montagne, A.; Baron, A.F.; Macrez, R.; Billard, J.M.; Briens, A.; Hommet, Y.; Wu, J.; Yepes, M.; et al. Unveiling an exceptional zymogen: The single-chain form of tPA is a selective activator of NMDA receptor-dependent signaling and neurotoxicity. *Cell Death Differ.* **2012**, *19*, 1983–1991. [CrossRef]

21. Hastings, S.; Myles, P.S.; Medcalf, R.L. Plasmin, Immunity, and Surgical Site Infection. *J. Clin. Med.* **2021**, *10*, 2070. [CrossRef] [PubMed]

22. Keragala, C.B.; Draxler, D.F.; McQuilten, Z.K.; Medcalf, R.L. Haemostasis and innate immunity-a complementary relationship: A review of the intricate relationship between coagulation and complement pathways. *Br. J. Haematol.* **2018**, *180*, 782–798. [CrossRef] [PubMed]

23. Kolev, K.; Medcalf, R.L. Editorial: Fibrinolysis in Immunity. *Front. Immunol.* **2020**, *11*, 582. [CrossRef] [PubMed]

24. Eren, M.; Boe, A.E.; Klyachko, E.A.; Vaughan, D.E. Role of plasminogen activator inhibitor-1 in senescence and aging. *Semin. Thromb. Hemost.* **2014**, *40*, 645–651. [CrossRef] [PubMed]

25. Aillaud, M.F.; Pignol, F.; Alessi, M.C.; Harle, J.R.; Escande, M.; Mongin, M.; Juhan-Vague, I. Increase in plasma concentration of plasminogen activator inhibitor, fibrinogen, von Willebrand factor, factor VIII:C and in erythrocyte sedimentation rate with age. *Thromb. Haemost.* **1986**, *55*, 330–332. Available online: https://www.ncbi.nlm.nih.gov/pubmed/3092390 (accessed on 1 May 2022). [PubMed]

26. Romer, J.; Bugge, T.H.; Pyke, C.; Lund, L.R.; Flick, M.J.; Degen, J.L.; Dano, K. Plasminogen and wound healing. *Nat. Med.* **1996**, *2*, 725. [CrossRef]

27. Creemers, E.; Cleutjens, J.; Smits, J.; Heymans, S.; Moons, L.; Collen, D.; Daemen, M.; Carmeliet, P. Disruption of the plasminogen gene in mice abolishes wound healing after myocardial infarction. *Am. J. Pathol.* **2000**, *156*, 1865–1873. [CrossRef]

28. Duffy, M.J.; Duggan, C. Re: Urokinase and urokinase receptor: Association with in vitro invasiveness of human bladder cancer cell lines. *J. Natl. Cancer Inst.* **1997**, *89*, 1628–1630. [CrossRef]

29. Duffy, M.J.; Duggan, C.; Maguire, T.; Mulcahy, K.; Elvin, P.; McDermott, E.; Fennelly, J.J.; O'Higgins, N. Urokinase plasminogen activator as a predictor of aggressive disease in breast cancer. *Enzym. Protein* **1996**, *49*, 85–93. [CrossRef]

30. Duffy, M.J.; Duggan, C.; Mulcahy, H.E.; McDermott, E.W.; O'Higgins, N.J. Urokinase plasminogen activator: A prognostic marker in breast cancer including patients with axillary node-negative disease. *Clin. Chem.* **1998**, *44*, 1177–1183. Available online: https://www.ncbi.nlm.nih.gov/pubmed/9625040 (accessed on 1 May 2022). [CrossRef]

31. Duffy, M.J.; O'Grady, P.; Devaney, D.; O'Siorain, L.; Fennelly, J.J.; Lijnen, H.J. Urokinase-plasminogen activator, a marker for aggressive breast carcinomas. Preliminary report. *Cancer* **1988**, *62*, 531–533. [CrossRef]

32. Duffy, M.J.; Reilly, D.; O'Sullivan, C.; O'Higgins, N.; Fennelly, J.J.; Andreasen, P. Urokinase-plasminogen activator, a new and independent prognostic marker in breast cancer. *Cancer Res.* **1990**, *50*, 6827–6829. Available online: https://www.ncbi.nlm.nih.gov/pubmed/2119883 (accessed on 1 May 2022). [PubMed]

33. Skelly, M.M.; Troy, A.; Duffy, M.J.; Mulcahy, H.E.; Duggan, C.; Connell, T.G.; O'Donoghue, D.P.; Sheahan, K. Urokinase-type plasminogen activator in colorectal cancer: Relationship with clinicopathological features and patient outcome. *Clin. Cancer Res.* **1997**, *3*, 1837–1840. Available online: https://www.ncbi.nlm.nih.gov/pubmed/9815571 (accessed on 1 May 2022). [PubMed]

34. Look, M.; van Putten, W.; Duffy, M.; Harbeck, N.; Christensen, I.J.; Thomssen, C.; Kates, R.; Spyratos, F.; Ferno, M.; Eppenberger-Castori, S.; et al. Pooled analysis of prognostic impact of uPA and PAI-1 in breast cancer patients. *Thromb. Haemost.* **2003**, *90*, 538–548. [CrossRef] [PubMed]

35. Harbeck, N.; Alt, U.; Berger, U.; Kruger, A.; Thomssen, C.; Janicke, F.; Hofler, H.; Kates, R.E.; Schmitt, M. Prognostic impact of proteolytic factors (urokinase-type plasminogen activator, plasminogen activator inhibitor 1, and cathepsins B, D, and L) in primary breast cancer reflects effects of adjuvant systemic therapy. *Clin. Cancer Res.* **2001**, *7*, 2757–2764. Available online: https://www.ncbi.nlm.nih.gov/pubmed/11555589 (accessed on 1 May 2022). [PubMed]

36. Janicke, F.; Prechtl, A.; Thomssen, C.; Harbeck, N.; Meisner, C.; Untch, M.; Sweep, C.G.; Selbmann, H.K.; Graeff, H.; Schmitt, M.; et al. Randomized adjuvant chemotherapy trial in high-risk, lymph node-negative breast cancer patients identified by urokinase-type plasminogen activator and plasminogen activator inhibitor type 1. *J. Natl. Cancer Inst.* **2001**, *93*, 913–920. [CrossRef] [PubMed]

37. Cantero, D.; Friess, H.; Deflorin, J.; Zimmermann, A.; Brundler, M.A.; Riesle, E.; Korc, M.; Buchler, M.W. Enhanced expression of urokinase plasminogen activator and its receptor in pancreatic carcinoma. *Br. J. Cancer* **1997**, *75*, 388–395. [CrossRef]

38. Gutova, M.; Najbauer, J.; Gevorgyan, A.; Metz, M.Z.; Weng, Y.; Shih, C.C.; Aboody, K.S. Identification of uPAR-positive chemoresistant cells in small cell lung cancer. *PLoS ONE* **2007**, *2*, e243. [CrossRef]

39. Soff, G.A.; Sanderowitz, J.; Gately, S.; Verrusio, E.; Weiss, I.; Brem, S.; Kwaan, H.C. Expression of plasminogen activator inhibitor type 1 by human prostate carcinoma cells inhibits primary tumor growth, tumor-associated angiogenesis, and metastasis to lung and liver in an athymic mouse model. *J. Clin. Investig.* **1995**, *96*, 2593–2600. [CrossRef]

40. Kwaan, H.C. The Central Role of Fibrinolytic Response in Trauma-Induced Coagulopathy: A Hematologist's Perspective. *Semin. Thromb. Hemost.* **2020**, *46*, 116–124. [CrossRef]

41. Moore, H.B.; Cohen, M.J.; Moore, E.E. Comment on "The S100A10 Pathway Mediates an Occult Hyperfibrinolytic Subtype in Trauma Patients". *Ann. Surg.* **2020**, *271*, e110–e111. [CrossRef] [PubMed]

42. Moore, H.B.; Moore, E.E. TEG Lysis Shutdown Represents Coagulopathy in Bleeding Trauma Patients: Analysis of the PROPPR Cohort. *Shock* **2019**, *52*, 639–640. [CrossRef] [PubMed]

43. Moore, H.B.; Moore, E.E.; Neal, M.D.; Sheppard, F.R.; Kornblith, L.Z.; Draxler, D.F.; Walsh, M.; Medcalf, R.L.; Cohen, M.J.; Cotton, B.A.; et al. Fibrinolysis Shutdown in Trauma: Historical Review and Clinical Implications. *Anesth. Analg.* **2019**, *129*, 762–773. [CrossRef] [PubMed]

44. Moore, H.B.; Neeves, K.B. Tranexamic acid for trauma: Repackaged and redelivered. *J. Thromb. Haemost.* **2019**, *17*, 1626–1628. [CrossRef] [PubMed]

45. Moore, H.B.; Walsh, M.; Kwaan, H.C.; Medcalf, R.L. The Complexity of Trauma-Induced Coagulopathy. *Semin. Thromb. Hemost.* **2020**, *46*, 114–115. [CrossRef]

46. Walsh, M.; Fries, D.; Moore, E.; Moore, H.; Thomas, S.; Kwaan, H.C.; Marsee, M.K.; Grisoli, A.; McCauley, R.; Vande Lune, S.; et al. Whole Blood for Civilian Urban Trauma Resuscitation: Historical, Present, and Future Considerations. *Semin. Thromb. Hemost.* **2020**, *46*, 221–234. [CrossRef]

47. Meizoso, J.P.; Karcutskie, C.A.; Ray, J.J.; Namias, N.; Schulman, C.I.; Proctor, K.G. Persistent Fibrinolysis Shutdown Is Associated with Increased Mortality in Severely Injured Trauma Patients. *J. Am. Coll. Surg.* **2017**, *224*, 575–582. [CrossRef]

48. Roberts, D.J.; Kalkwarf, K.J.; Moore, H.B.; Cohen, M.J.; Fox, E.E.; Wade, C.E.; Cotton, B.A. Time course and outcomes associated with transient versus persistent fibrinolytic phenotypes after injury: A nested, prospective, multicenter cohort study. *J. Trauma Acute Care Surg.* **2019**, *86*, 206–213. [CrossRef]

49. Sandrini, L.; Ieraci, A.; Amadio, P.; Zara, M.; Barbieri, S.S. Impact of Acute and Chronic Stress on Thrombosis in Healthy Individuals and Cardiovascular Disease Patients. *Int. J. Mol. Sci.* **2020**, *21*, 7818. [CrossRef]

50. Hoirisch-Clapauch, S. Mechanisms affecting brain remodeling in depression: Do all roads lead to impaired fibrinolysis? *Mol. Psychiatry* **2022**, *27*, 525–533. [CrossRef]

51. Malemud, C.J. Matrix metalloproteinases (MMPs) in health and disease: An overview. *Front. Biosci.* **2006**, *11*, 1696–1701. [CrossRef] [PubMed]

52. Strickland, S.; Beers, W.H. Studies on the role of plasminogen activator in ovulation. In vitro response of granulosa cells to gonadotropins, cyclic nucleotides, and prostaglandins. *J. Biol. Chem.* **1976**, *251*, 5694–5702. Available online: https://www.ncbi.nlm.nih.gov/pubmed/965386 (accessed on 1 May 2022). [CrossRef]

53. Ohlsson, R.; Pfeifer-Ohlsson, S. [Hereditary memory: Genomic imprinting and its importance for embryonal development and carcinogenesis]. *Lakartidningen* **1991**, *88*, 1087–1090. Available online: https://www.ncbi.nlm.nih.gov/pubmed/2016942 (accessed on 1 May 2022). [PubMed]

54. Seeds, N.W.; Basham, M.E.; Haffke, S.P. Neuronal migration is retarded in mice lacking the tissue plasminogen activator gene. *Proc. Natl. Acad. Sci. USA* **1999**, *96*, 14118–14123. [CrossRef]

55. Melchor, J.P.; Strickland, S. Tissue plasminogen activator in central nervous system physiology and pathology. *Thromb. Haemost.* **2005**, *93*, 655–660. [CrossRef]

56. Bennur, S.; Shankaranarayana Rao, B.S.; Pawlak, R.; Strickland, S.; McEwen, B.S.; Chattarji, S. Stress-induced spine loss in the medial amygdala is mediated by tissue-plasminogen activator. *Neuroscience* **2007**, *144*, 8–16. [CrossRef]

57. Su, E.J.; Fredriksson, L.; Geyer, M.; Folestad, E.; Cale, J.; Andrae, J.; Gao, Y.; Pietras, K.; Mann, K.; Yepes, M.; et al. Activation of PDGF-CC by tissue plasminogen activator impairs blood-brain barrier integrity during ischemic stroke. *Nat. Med.* **2008**, *14*, 731–737. [CrossRef]

58. Boyd, B.J.; Galle, A.; Daglas, M.; Rosenfeld, J.V.; Medcalf, R. Traumatic brain injury opens blood-brain barrier to stealth liposomes via an enhanced permeability and retention (EPR)-like effect. *J. Drug Target.* **2015**, *23*, 847–853. [CrossRef]

59. Lees, K.R.; Emberson, J.; Blackwell, L.; Bluhmki, E.; Davis, S.M.; Donnan, G.A.; Grotta, J.C.; Kaste, M.; von Kummer, R.; Lansberg, M.G.; et al. Effects of Alteplase for Acute Stroke on the Distribution of Functional Outcomes: A Pooled Analysis of 9 Trials. *Stroke* **2016**, *47*, 2373–2379. [CrossRef]

60. Whiteley, W.N.; Emberson, J.; Lees, K.R.; Blackwell, L.; Albers, G.; Bluhmki, E.; Brott, T.; Cohen, G.; Davis, S.; Donnan, G.; et al. Risk of intracerebral haemorrhage with alteplase after acute ischaemic stroke: A secondary analysis of an individual patient data meta-analysis. *Lancet Neurol.* **2016**, *15*, 925–933. [CrossRef]

61. Kris-Etherton, P.M.; Stewart, P.W.; Ginsberg, H.N.; Tracy, R.P.; Lefevre, M.; Elmer, P.J.; Berglund, L.; Ershow, A.G.; Pearson, T.A.; Ramakrishnan, R.; et al. The Type and Amount of Dietary Fat Affect Plasma Factor VIIc, Fibrinogen, and PAI-1 in Healthy Individuals and Individuals at High Cardiovascular Disease Risk: 2 Randomized Controlled Trials. *J. Nutr.* **2020**, *150*, 2089–2100. [CrossRef] [PubMed]

62. Sillen, M.; Declerck, P.J. Targeting PAI-1 in Cardiovascular Disease: Structural Insights Into PAI-1 Functionality and Inhibition. *Front. Cardiovasc. Med.* **2020**, *7*, 622473. [CrossRef] [PubMed]

63. Iacoviello, L.; Agnoli, C.; De Curtis, A.; di Castelnuovo, A.; Giurdanella, M.C.; Krogh, V.; Mattiello, A.; Matullo, G.; Sacerdote, C.; Tumino, R.; et al. Type 1 plasminogen activator inhibitor as a common risk factor for cancer and ischaemic vascular disease: The EPICOR study. *BMJ Open* **2013**, *3*, e003725. [CrossRef] [PubMed]

64. Alessi, M.C.; Juhan-Vague, I. PAI-1 and the metabolic syndrome: Links, causes, and consequences. *Arter. Thromb. Vasc. Biol.* **2006**, *26*, 2200–2207. [CrossRef]
65. Visse, R.; Nagase, H. Matrix metalloproteinases and tissue inhibitors of metalloproteinases: Structure, function, and biochemistry. *Circ. Res.* **2003**, *92*, 827–839. [CrossRef]
66. Ghosh, A.K.; Quaggin, S.E.; Vaughan, D.E. Molecular basis of organ fibrosis: Potential therapeutic approaches. *Exp. Biol. Med.* **2013**, *238*, 461–481. [CrossRef]
67. Ghosh, A.K.; Vaughan, D.E. PAI-1 in tissue fibrosis. *J. Cell. Physiol.* **2012**, *227*, 493–507. [CrossRef]
68. Khan, S.S.; Shah, S.J.; Strande, J.L.; Baldridge, A.S.; Flevaris, P.; Puckelwartz, M.J.; McNally, E.M.; Rasmussen-Torvik, L.J.; Lee, D.C.; Carr, J.C.; et al. Identification of Cardiac Fibrosis in Young Adults With a Homozygous Frameshift Variant in SERPINE1. *JAMA Cardiol.* **2021**, *6*, 841–846. [CrossRef]
69. Kwaan, H.C.; Lindholm, P.F. The Central Role of Fibrinolytic Response in COVID-19-A Hematologist's Perspective. *Int. J. Mol. Sci.* **2021**, *22*, 1283. [CrossRef]
70. Kwaan, H.C.; Mazar, A.P. More on the Source of D-Dimer in COVID-19. *Thromb. Haemost.* **2022**, *122*, 158–159. [CrossRef]
71. Walsh, M.M.; Khan, R.; Kwaan, H.C.; Neal, M.D. Fibrinolysis Shutdown in COVID-19-Associated Coagulopathy: A Crosstalk among Immunity, Coagulation, and Specialists in Medicine and Surgery. *J. Am. Coll. Surg.* **2021**, *232*, 1003–1006. [CrossRef] [PubMed]
72. Vaughan, D.E.; Lazos, S.A.; Tong, K. Angiotensin II regulates the expression of plasminogen activator inhibitor-1 in cultured endothelial cells. A potential link between the renin-angiotensin system and thrombosis. *J. Clin. Investig.* **1995**, *95*, 995–1001. [CrossRef] [PubMed]
73. Kuba, K.; Imai, Y.; Rao, S.; Gao, H.; Guo, F.; Guan, B.; Huan, Y.; Yang, P.; Zhang, Y.; Deng, W.; et al. A crucial role of angiotensin converting enzyme 2 (ACE2) in SARS coronavirus-induced lung injury. *Nat. Med.* **2005**, *11*, 875–879. [CrossRef] [PubMed]
74. Idell, S.; Kueppers, F.; Lippmann, M.; Rosen, H.; Niederman, M.; Fein, A. Angiotensin converting enzyme in bronchoalveolar lavage in ARDS. *Chest* **1987**, *91*, 52–56. [CrossRef]
75. Bhandary, Y.P.; Shetty, S.K.; Marudamuthu, A.S.; Ji, H.L.; Neuenschwander, P.F.; Boggaram, V.; Morris, G.F.; Fu, J.; Idell, S.; Shetty, S. Regulation of lung injury and fibrosis by p53-mediated changes in urokinase and plasminogen activator inhibitor-1. *Am. J. Pathol.* **2013**, *183*, 131–143. [CrossRef]

International Journal of
Molecular Sciences

MDPI

Article

Reduced Expression of Urokinase Plasminogen Activator in Brown Adipose Tissue of Obese Mouse Models

Chung-Ze Wu [1,2], Li-Chien Chang [3], Chao-Wen Cheng [4], Te-Chao Fang [5,6,7], Yuh-Feng Lin [4,8], Dee Pei [9,10] and Jin-Shuen Chen [11,12,*]

[1] Division of Endocrinology and Metabolism, Department of Internal Medicine, School of Medicine, College of Medicine, Taipei Medical University, Taipei 11031, Taiwan; chungze@yahoo.com.tw
[2] Division of Endocrinology and Metabolism, Department of Internal Medicine, Shuang Ho Hospital, Taipei Medical University, New Taipei City 23561, Taiwan
[3] School of Pharmacy, National Defense Medical Center, Taipei 11490, Taiwan; lichien@ndmctsgh.edu.tw
[4] Graduate Institute of Clinical Medicine, College of Medicine, Taipei Medical University, Taipei 11031, Taiwan; ccheng@tmu.edu.tw (C.-W.C.); linyf@shh.org.tw (Y.-F.L.)
[5] TMU Research Center of Urology and Kidney, Taipei Medical University, Taipei 11031, Taiwan; fangtc@tmu.edu.tw
[6] Division of Nephrology, Department of Internal Medicine, School of Medicine, College of Medicine, Taipei Medical University, Taipei 11031, Taiwan
[7] Division of Nephrology, Department of Internal Medicine, Taipei Medical University Hospital, Taipei 11031, Taiwan
[8] Deputy Superintendent, Shuang Ho Hospital, Taipei Medical University, New Taipei City 23561, Taiwan
[9] School of Medicine, College of Medicine, Fu Jen Catholic University, New Taipei City 24205, Taiwan; peidee@gmail.com
[10] Division of Endocrinology and Metabolism, Department of Internal Medicine, Fu Jen Catholic University Hospital, New Taipei City 24352, Taiwan
[11] Department of Medical Education and Research, Kaohsiung Veterans General Hospital, No 386, Dazhong 1st Rd., Zuoying Dist., Kaohsiung City 81362, Taiwan
[12] Division of Nephrology, Department of Internal Medicine, Tri-Service General Hospital, National Defense Medical Center, Taipei 11490, Taiwan
* Correspondence: dgschen@vghks.gov.tw; Tel.: +886-7-342-2121; Fax: +886-7-342-2288

Citation: Wu, C.-Z.; Chang, L.-C.; Cheng, C.-W.; Fang, T.-C.; Lin, Y.-F.; Pei, D.; Chen, J.-S. Reduced Expression of Urokinase Plasminogen Activator in Brown Adipose Tissue of Obese Mouse Models. *Int. J. Mol. Sci.* **2021**, 22, 3407. https://doi.org/10.3390/ijms22073407

Academic Editor: Hau C. Kwaan

Received: 28 February 2021
Accepted: 23 March 2021
Published: 26 March 2021

Publisher's Note: MDPI stays neutral with regard to jurisdictional claims in published maps and institutional affiliations.

Abstract: In recent decades, the obesity epidemic has resulted in morbidity and mortality rates increasing globally. In this study, using obese mouse models, we investigated the relationship among urokinase plasminogen activator (uPA), metabolic disorders, glomerular filtration rate, and adipose tissues. Two groups, each comprised of C57BL/6J and BALB/c male mice, were fed a chow diet (CD) and a high fat diet (HFD), respectively. Within the two HFD groups, half of each group were euthanized at 8 weeks (W8) or 16 weeks (W16). Blood, urine and adipose tissues were collected and harvested for evaluation of the effects of obesity. In both mouse models, triglyceride with insulin resistance and body weight increased with duration when fed a HFD in comparison to those in the groups on a CD. In both C57BL/6J and BALB/c HFD mice, levels of serum uPA initially increased significantly in the W8 group, and then the increment decreased in the W16 group. The glomerular filtration rate declined in both HFD groups. The expression of uPA significantly decreased in brown adipose tissue (BAT), but not in white adipose tissue, when compared with that in the CD group. The results suggest a decline in the expression of uPA in BAT in obese m models as the serum uPA increases. There is possibly an association with BAT fibrosis and dysfunction, which may need further study.

Keywords: urokinase plasminogen activator; brown adipose tissue; obesity

1. Introduction

In recent decades, along with high-fat diets (HFD) and sedentary lifestyles, the prevalence of obesity has increased, resulting in a worldwide epidemic of type 2 diabetes mellitus, hypertension and cardiovascular events [1]. Obesity also contributes to non-alcoholic steatohepatitis progressing to liver cirrhosis and hepatoma [2], and leads to glomerular hyperfiltration, reduction of nephron mass, and glomerulopathy in the kidney [3]. Although the causality of obesity is complex, the excessive accumulation of fat mass is believed to be a crucial factor. Generally, there are two different types of adipose tissues—white adipose tissue (WAT) and brown adipose tissue (BAT) [4]. WAT is distributed in subcutaneous and peri-visceral regions and is mainly responsible for energy storage. BAT, containing abundant mitochondria in adipocytes, contributes to non-shivering thermogenesis, which can burn up to 20% of daily energy intake per 50 g of BAT. Recently, several studies have focused on BAT and its anti-obesity effect [5].

In the pathology of obesity-related complications, in the islets of patients with type 2 diabetes mellitus, amyloid deposition is a critical feature associated with loss of β cell mass [6]. Non-alcoholic steatohepatitis, in its aggressive necro-inflammatory form, may accumulate fibrosis, resulting in cirrhosis and end-stage liver disease [2]. Although most patients with obesity-related glomerulopathy have stable or slowly progressive proteinuria, up to one-third develop progressive renal failure and end-stage renal disease [7]. Recently, dysfunction of adipose tissue has been widely noted to be associated with hyperglycemia, dyslipidemia, and macrophage infiltration in peri-visceral fat [8–10]. Meanwhile, hypoxia and fibrosis of adipose tissue, abundant collagen, and extracellular matrix deposits contributing to inflammation and the infiltration of macrophages are regarded as important pathogenic mechanisms [11,12]. However, the exact pathophysiology of major organs with regard to adipose tissue fibrosis remains unclear.

Urokinase plasminogen activator (uPA), well known as a fibrinolytic protein, binds to its receptor (uPAR) and activates plasminogen, converting it to plasmin in the fibrinolytic process of thrombosis in the extracellular matrix. Apart from its functions in the fibrinolytic cascade, uPA is a pluripotent protease participating in activating the innate immune response, which regulates immune cell migration, recruitment, and lymphocytes proliferation [13,14]. Kawao et al. found uPA to play an important role in the activation of macrophage phagocytosis during liver repair [15]. In our previous study, we found uPAR to be related to various forms of kidney disease and its soluble form to be associated with the glomerular filtration rate and the amount of proteinuria present [16]. To date, the role of uPA in major organs during the development of obesity and adipose tissue dysfunction remains unknown.

It is essential to clarify the role of uPA in the causation of obesity-related complications. In the present study, we investigated uPA expression in WAT, BAT and major organs during the progression of obesity.

2. Results

2.1. The Different Presentations of Obesity and Metabolic Profiles in C57BL/6J and BALB/c Mice on HFD

In Table 1, we summarize the general characteristics of C57BL/6J and BALB/c mice in both groups. In the HFD group of C57BL/6J mice, body weight (BW) and total cholesterol (TC) increased over 8 weeks (W8). As time progressed, by 16 weeks (W16), levels of triglyceride (TG), and homeostasis model assessment–insulin resistance (HOMA-IR) were significantly elevated in the HFD group of C57BL/6J mice. The fractional excretion of sodium (FENa) was slightly decreased in the HFD group of C57BL/6J mice. Accordingly, the C57BL/6J HFD mice were prone to obesity, insulin resistance, glomerular hypofiltration, and dyslipidemia on a HFD. On the other hand, initially in BALB/c mice, metabolic profiles of the chow diet (CD) group and the HFD group did not differ significantly. However, by W16, BW, blood glucose (BG), TG, and HOMA-IR significantly increased, but FENa decreased in the HFD group of BALB/c mice. In addition, the presentation of renal function in BALB/c mice were assessed. The blood urea nitrogen (BUN) levels showed no significant difference in CD and HFD BALB/c mice (58.7 ± 14.1 ng/mL vs. 69.6 ± 7.7 ng/mL; $p = 0.166$). The proteinuria (urine protein/urine creatinine ratio) significantly increased in HFD BALB/c mice at W8 (1.98 ± 0.96; 3.93 ± 2.59; $p = 0.038$). However, the significant increase in proteinuria was not found by W16. BALB/c HFD mice were prone to obesity, insulin resistance, hyperglycemia, glomerular hypofiltraion and dyslipidemia. In our findings, the dysmetabolic phenotypes of obese mice induced by a HFD differed according to the strain.

Table 1. Relationship between glomerular hyperfiltration and metabolic status in C57BL/6J and BALB/c mice fed a chow diet (CD) or high fat diet (HFD) at 8 weeks (W8) and 16 weeks (W16). ($n = 5$, in each group).

| | C57BL/6J | | | | BALB/c | | | |
| | W8 | | W16 | | W8 | | W16 | |
	CD	HFD	CD	HFD	CD	HFD	CD	HFD
Body weight (g)	25.80 ± 0.94	29.46 ± 1.32 ***	27.04 ± 2.14	39.1 ± 1.86 ***	29.26 ± 2.05	30.22 ± 1.23	30.04 ± 3.30	33.95 ± 2.73 *
BG (mmol/L)	6.97 ± 1.22	9.24 ± 3.21	7.43 ± 0.90	9.50 ± 3.96	5.08 ± 1.00	6.36 ± 2.23	5.40 ± 1.25	6.96 ± 1.73 *
TC (mmol/L)	2.60 ± 0.03	3.12 ± 0.35 *	2.59 ± 0.01	3.68 ± 0.58 *	4.94 ± 0.44	4.92 ± 0.81	3.42 ± 0.06	4.24 ± 1.36
TG (mmol/L)	1.11 ± 0.26	1.19 ± 0.09	1.07 ± 0.19	1.34 ± 0.13 *	1.06 ± 0.09	1.07 ± 0.14	0.85 ± 0.11	0.71 ± 0.08 *
HOMA-IR	2.58 ± 1.00	5.42 ± 2.88	2.96 ± 1.03	8.05 ± 1.61 ***	1.80 ± 1.06	5.16 ± 3.75	1.76 ± 1.11	9.21 ± 5.92 *
HOMA-ß	26.42 ± 3.09	23.27 ± 6.06	41.30 ± 17.45	34.71 ± 18.95	258.4 ± 151.1	618.4 ± 502.5	366.8 ± 458.9	122.4 ± 64.8
FENa	1.00 ± 0.76	0.69 ± 0.48	1.08 ± 0.70	0.33 ± 0.24 **	0.81 ± 0.78	0.84 ± 0.83	1.14 ± 0.54	0.53 ± 0.37 **
uPA (μg/mL)	0.87 ± 0.35	2.56 ± 0.65 *	1.24 ± 0.55	1.92 ± 0.19	1.88 ± 0.48	3.34 ± 0.84 *	1.31 ± 0.37	2.52 ± 0.95
suPAR (ng/mL)	3.26 ± 0.19	3.43 ± 0.36	2.56 ± 0.74	2.27 ± 0.53	1.67 ± 0.60	2.22 ± 0.66	1.64 ± 0.56	3.09 ± 1.23
PAI-1 (ng/mL)	4.95 ± 0.77	4.06 ± 0.45	3.86 ± 0.58	7.76 ± 1.01	28.80 ± 2.03	36.22 ± 4.39 *	22.54 ± 2.12	56.16 ± 4.60 ***
Adiponectin	7.92 ± 0.43	9.06 ± 0.68	8.86 ± 0.45	6.98 ± 0.45 *	9.12 ± 0.38	6.96 ± 0.13 **	8.87 ± 0.45	8.31 ± 0.19

Data shown as mean ± SD; BG: blood glucose, TC: Total cholesterol, TG: Triglyceride, HOMA-IR: homeostatic model assessment–insulin resistance, HOMA-ß: homeostatic model assessment-ß, FENa: functional excretion of sodium; * $p < 0.05$, ** $p < 0.01$, *** $p < 0.001$, compared with the CD group.

2.2. Circulating uPA, Soluble uPAR (suPAR) and Plasminogen Activator Inhibitor-1 (PAI-1) Levels and Adiponectin in C57BL/6J and BALB/c Obese Mice

Table 1 also shows the serum uPA, suPAR, PAI-1 and adiponectin levels in C57BL/6J and BALB/c mice on a CD or a HFD. The uPA levels in the HFD group of both strains of mice were significantly increased at W8. However, at W16, the uPA levels of the CD and HFD groups in both strains of mice did not differ significantly. On the other hand, the suPAR levels were similar between the CD and HFD groups in both strains of mice at both W8 and W16. The PAI-1 levels in BALB/c mice were obviously higher than those in C57BL/6J mice. However, the PAI-1 levels significantly increased in BALB/c mice with HFD at both W8 and W16. Similar changes were not found in C57BL/6J mice. The adiponectin levels showed a significant decrease in the HFD W16 group in C57BL/6J mice and in the HFD W8 group in BALB/c mice in comparison to related CD groups.

2.3. uPA Expression on Subcutaneous White Adipose Tissue (sWAT), Visceral White Adipose Tissue (vWAT), BAT, Liver, Kidney, and Pancreas in C57BL/6J and BALB/c Obese Mice

Figures 1 and 2 show the expression of uPA in various adipose tissues, the liver, the kidney, and the islets in C57BL/6J and BALB/c obese mice. In gross appearance, the size of fat droplets in sWAT, vWAT, BAT and the liver in the HFD group were larger than those in the CD groups in both strains of mice. In addition, the histology of the kidney showed no obvious change (Supplementary Materials Figure S1). The collagen IV expression of the liver mildly increased in HFD groups at W16 in both strains of mice (Figure S1). According to the findings of the immunohistochemistry (IHC) stain and Western blot (WB) for uPA, uPA expression in the liver, kidney, islet, sWAT and vWAT did not differ significantly between CD and HFD groups in either strain of mice. The uPA expression in the islet is mainly on β cells (Figure S2). In the kidney, the majority of uPA expression is on the tubular region. The uPA expression on the renal glomerular region is weak. However, the uPA expression in the kidney showed no prominent difference between CD and HFD groups in both strains of mice. Interestingly, we found a significant decline in uPA expression in BAT by IHC stain and WB early at W8, which persisted at W16 in the HFD groups in both strains of mice.

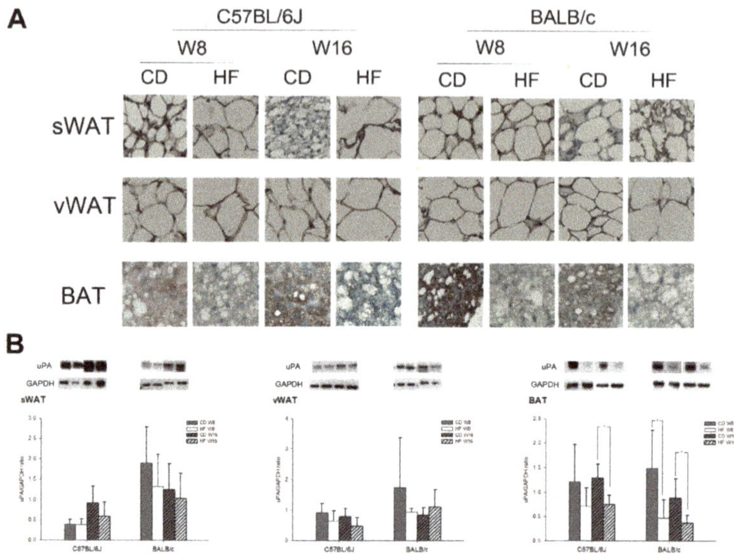

Figure 1. The (**A**) immunohistochemical stain and (**B**) Western blot for uPA expression on subcutaneous white adipose tissue (sWAT), visceral white adipose tissue (vWAT) and brown adipose tissue (BAT) in C57BL/6J and BALB/c mice ($n = 5$, in each group). The uPA expression in sWAT and vWAT of CD and HFD groups did not differ significantly, but there was a significant decline in the BAT in the HFD group. All figures of IHC stain: 400×.

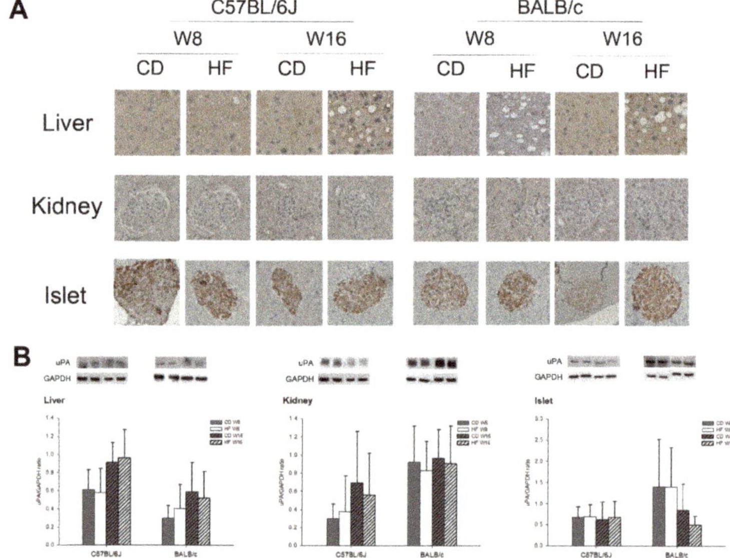

Figure 2. The (**A**) immunohistochemical stain and (**B**) Western blot for uPA expression on liver, kidney and islet tissue in C57BL/6J and BALB/c mice ($n = 5$, in each group). In the CD and HFD groups in both strains of mice, the uPA expression in the liver, kidney, and islet did not differ significantly. All figures of IHC stain: 400×.

3. Discussion

In the present study, in different strains of HFD mice, we found different presentations of dysmetabolic profiles. Serum uPA levels initially increased and then declined with time in both C57BL/6J and BALB/c mice fed a HFD. In both C57BL/6J and BALB/c mice, the expression of uPA in BAT significantly declined after feeding a HFD. However, with a CD and a HFD, uPA expression was similar in the liver, kidney and WAT in both strains of mice. In HFD mice, uPAR expression in BAT and WAT showed no significant change.

C57BL/6J and BALB/c mice are commonly used for studies of mouse strains on immunoregulation in various disease models. As C57BL/6J mice preferentially develop the Th1 immune response and BALB/c mice, Th2-type cytokine polarization, they are regarded as prototypic Th1- and Th2-type mouse strains, respectively [17,18]. In addition to their distinct T-cell responses, macrophages from these two mouse strains exert different reactions in response to various stimuli [19]. Recent evidence indicates that the balance between the M1/M2 macrophages and the Th1/Th2 lymphocytes is of critical importance for the outcome of many diseases, including obesity-related metabolic disorders [20]. Jovicic et al. explored liver steatosis and immune cells in C57BL/6J and BALB/c mice fed a HFD for 24 weeks and found different immune–metabolic profiles between the two strains of mice [21]. C57BL/6J mice fed with a HFD were prone to obesity, hyperglycemia, increasing visceral adipose tissue, liver inflammation, and fibrosis. BALB/c mice fed with a HFD were susceptible to liver steatosis. Our findings do not contradict those of the empirical studies discussed above. BW significantly increased and FENa decreased in C57BL/6J on a HFD, which may imply that HFD-induced obesity, dyslipidemia and glomerular hypofiltration easily develop in Th1-prone mice. Contrastingly, in BALB/c mice fed a HFD, the increment in BG, TG and HOMA-IR significantly increased, and FENa decreased, which may imply that HFD-induced hyperglycemia, hypertriglycemia, insulin resistance and glomerular hypofiltration may be easily found on Th2-prone mice. In addition, adiponectin levels significantly decreased at W8 in HFD BALB/c mice, but not until W16 in HFD C57BL/6J mice. Adiponectin, one of the anti-inflammatory adipokines, decreases in inflammatory status. According to our results, it is suggested that HFD-related pro-inflammation may be found in the early phase of obesity in Th2-prone mice and in the late phase of obesity in Th1-prone mice.

The uPA is secreted from various cells and contributes to the degradation of the extracellular matrix and cellular remodeling and repair. Circulating uPA is predominantly excreted from hemopoietic cells and is responsible for the cascade activation of the fibrinolytic process and immune modulation [22]. The relationship between the uPA/uPAR system and atherosclerosis has been researched extensively. Some studies indicate high circulating uPA levels as possibly being involved in the migration of foamy cells or the stability of atheroma [23–25]. Our results indicate a significant increase in uPA in the early phase of obesity in both strains of mice. It was speculated that an increase in uPA may be a passive response to the accumulation of the extracellular matrix around endothelial cells and the modulation of fibrinogenesis and fibrinolysis. However, in our study, the increment in circulating uPA levels decreased in the late phase of obesity. It was presumed that some subsequent obesity-related inflammatory cytokines, such as PAI-1, might have inhibited the circulating uPA. Moreover, Zhou et al. investigated the change in circulating uPA in patients with chronic hepatitis B. The uPA levels increased in the acute phase but decreased in the late phase of hepatitis [26]. The findings are similar to the results of our HFD mice. They also highlighted the association between inflammation and uPA change. However, the real mechanism of uPA level change needs further study. Although some studies showed PAI-1 to be associated with adipocyte differentiation and regulating recruitment of inflammatory cells within adipose tissue [27], we found that serum PAI-1 levels of CD and HFD mice did not differ significantly in C57BL/6J mice. However, the PAI-1 levels significantly increased in HFD BLAB/c mice. In our previous studies of clinical investigation, PAI-1 was found to be positively related to the BMI percentile in boys and to body fat in girls [28]. The majority of studies explored PAI-1 expression in different tissues with an obese model. Morange et al. found that serum PAI-1 increased in obese mice fed a HFD for 17 weeks [29] which was similar to our BALB/c mice. Presumably, because the time frame over which we induced obesity in our C57BL/6J HFD mice was relatively short, the change in PAI-1 might not have had time to manifest in our C57BL/6J obese mice. In addition, PAI-1 is an inhibitor of plasminogen activator. PAI-1 levels non-significantly increased at W16 in HFD C57BL/6J mice and significantly increased at W16 in HFD BALB/c mice, which may also explain the decreasing uPA levels in both strains of mice at W16.

During the course of obesity development, WAT mass and the cellular size of adipocytes expand rapidly in the body. The rapid growth impedes the prompt delivery of sufficient oxygen to WAT from circulation, resulting in WAT dysfunction [30]. Consequently, several cytokines released from adipocytes induce systemic pro-inflammation and WAT fibrosis [9,12]. On the other hand, enlarged intracellular fat droplets and the reduced number of mitochondria in brown adipocytes have been noted during the process of obesity [31]. Obesity-related molecules involved in pro-inflammation and extracellular matrix turnover deteriorate BAT function [32]. Fibrotic BAT may exacerbate the development of obesity. We found decreasing expression of uPA in the BAT of obese mice in both strains. It was assumed that the decline in uPA expression in the BAT may contribute to adipocyte fibrosis and dysfunction, in turn leading to obesity-related complications. Spencer et al. also found increased collagen V expression in adipose tissue in obese subjects and presumed the extracellular matrix to be associated with insulin resistance [33]. However, an insufficiency of uPA in BAT would impair the necessary degradation of the extracellular matrix, resulting in the compromised migration and remodeling of brown adipocytes. Although we did not detect histologic fibrosis of BAT in our mice, other studies had similar findings. Alcalá et al. investigated mice fed with a HFD for 20 weeks, during which they developed obesity and mild hyperglycemia [34]. The pathology of BAT in HFD mice showed cellular hypertrophy and no obvious fibrotic change. However, they found inflammation, oxidative stress, and some anti-oxidative enzyme activity reflectively increased in BAT in HFD mice. In addition, Trayhurn et al. found an increase in some markers related to fibrosis during deoxygenation following BAT expansion in obesity [30]. Consequently, uPA may have responded before the formation of histologic fibrosis in the BAT. However, further research into the cause of uPA decline in BAT on a HFD is necessary.

There are some limitations to our study. First, several factors may have influenced the activation of BAT, including cold temperatures and adrenergic stimulation [35]. Our mice were caged at room temperature during the whole period of the experiment so that we did not observe changes in uPA expression in the BAT in cold temperatures. Second, the real mechanism of decreasing uPA expression on BAT, not WAT, on a HFD remains unknown. We presume that lipotoxicity in brown adipocytes may be one possible factor. Further study of the cellular model is needed to explore the exact pathway. Third, our numbers of mice were only five per group. The statistic power of serum biochemistries in each group of mice may not be enough. However, the trends of metabolic parameters between CD and HFD groups in different strands of mice may be informative. Fourth, we did not have the data of baseline biochemistries in each group of mice for assessing the change in each variable. However, there were two points of time in each group for evaluating the difference during the treating interval. In addition, we did not explore the change in various organs after obesity. The hypertrophy of an organ after obesity may be associated with uPA expression. However, our research is the first to explore the relationships between uPA changes and adipose tissue in obesity. We used strains of mice with different immune-prone characteristics to arrive at our circulating and histologic findings. Whether changes in uPA in the BAT induced BAT dysfunction in obesity or the two phenomena develop simultaneously is a worthy subject for advanced studies in the future.

4. Materials and Methods

4.1. Induced Obese Mouse Model

The male wild-type C57BL/6J and BALB/c mice, obtained from the Laboratory Animal Center (National Taiwan University College of Medicine, Taipei, Taiwan), Taiwan, were housed in laboratory cages and, from the age of 5 weeks, fed a normal CD in the CD group and a HFD (40% fat) in the HFD group, respectively. We measured BW and BG weekly. These mice were euthanized at W8 or W16 after feeding with a CD or HFD. Blood samples were collected before euthanasia. The sWAT from inguinal WAT, and vWAT from the WAT of the epididymis were harvested. The BAT from the posterior neck region, liver, kidney, and pancreatic tissues were harvested from the mice after euthanasia. The Institutional Animal Care and Use Committee at National Defense Medical Center, Taipei, Taiwan approved the experimental animal protocol (Approval No: IACUC-13-199).

4.2. Measurement of uPA, suPAR, PAI-1 and Metabolic Biochemistry

After collecting blood, we separated plasma and serum by centrifugation and stored them at $-80\ ^\circ$C prior to analysis. The uPA was measured by using the Mouse uPA Total Antigen Assay enzyme-linked immunosorbent assay (ELISA) kit (Molecular Innovations, Novi, MI, USA), with the intra- and inter-assay coefficients of variation being 6.18% and 7.47%, respectively. The soluble uPAR were measured by using the Mouse uPAR DuoSet ELISA kit (DY531, R&D system, Minneapolis, MN, USA). The mean coefficient of variation in these assays was 5%. Murine PAI-1 was measured by using the PAI-1 Total Mouse ELISA kit (ab157529, Abcam, Cambridge, MA, USA) with the intra- and inter-assay coefficients of variation being 7.9% and 12.9%, respectively. Total cholesterol, TG, serum and urine sodium, and creatinine were analyzed by spectrophotometry (Fuji Dri-Chem 3000, Fuji Film, Kanagawa, Japan). The FENa was applied to evaluate renal filtration and calculated by the ratio of urine and plasma sodium and creatinine. Serum insulin was measured using the Mouse Insulin ELISA Kit (Mercodia AB, Uppsala, Sweden) with the intra- and inter-assay coefficients of variation being 3.4% and 3.6%, respectively. All samples were assayed in duplicate. HOMA-IR was calculated to assess insulin resistance [36].

4.3. IHC Stain and WB of sWAT, vWAT, BAT, Liver, Kidney, and Islet

The sWAT, vWAT, BAT, liver, kidney and pancreatic tissues were fixed in 10% formaldehyde fixative solution and embedded tPBShem in paraffin. The sections of formalin-fixed tissue were immersed in xylene for 5 min three times and incubated with phosphate-buffered saline (PBS) and 1% bovine serum albumin (BSA) at room temperature (RT) for 30 min for blocking. After removing paraffin and rehydrating, the slices were incubated with 1:400 dilution of the primary antibody (anti-uPA antibody (ab28230, Abcam, Cambridge, MA, USA)) in PBS at 4 $^\circ$C overnight. Subsequently, the slices were incubated with 1:50 dilution of the secondary antibody (biotinylated anti-rabbit antibody (Vector Laboratories, BA-1300, CA, USA)) for 40 min and washed with Tris-buffered saline containing 0.05% Tween 20 (TBST; pH 7.4). We then treated the sections with VECTASTAIN ABC (Vector Laboratories, CA, USA) working solution for 30 min. The peroxidase activity was visualized with 3,3′-diaminobenzidine (DAB) using a DAB substrate kit for peroxidase (BD Pharmingen™). The slices were observed with an optical photomicroscope.

For WB, equal amounts of protein (30 μg) from each tissue of various whole organs were separated after homogenization by 8% SDS-PAGE gel, which was electro-blotted onto a nitrocellulose membrane and incubated for 1 h in blocking buffer (TBST, 2% bovine serum albumin). It was then washed three times in TBST and incubated with 1:2000 dilutions of anti-uPA antibody (ab28230, Abcam, Cambridge, MA, USA) or 1:10000 dilutions of anti-glyceraldehyde-3-phosphate dehydrogenase (GAPDH) antibody (ab181602, Abcam, Cambridge, MA, USA), respectively, in TBST at 4 $^\circ$C overnight. The membranes were washed blots and incubated in horseradish peroxidase-conjugated goat-anti-rabbit-IgG-HRP antibody (Cat#3053-S-Ex, EPITOMICS, CA, USA) for 1 h at room temperature. After washing the membranes, we detected and incubated the membrane-bound antibody with a Western blot detection system and captured it on X-ray film.

4.4. Statistical Analysis

The PASW statistics version 18.0 package for Windows (IBM SPSS Statistics) was used for data analysis. The continuous variables were expressed as mean $\pm$ SD. A nonparametric Mann–Whitney U test was used for comparison of the two groups. All statistical data are expressed as two-sided, and p values < 0.05 considered to be statistically significant.

5. Conclusions

In summary, HFD-induced obesity presents different dysmetabolic profiles in different immune responses. Serum uPA increased in the early phase of obesity in our model. Decreasing uPA expression in the BAT may contribute to BAT dysfunction in obesity.

Supplementary Materials: The following are available online at https://www.mdpi.com/1422-0067/22/7/3407/s1.

Author Contributions: All authors contributed significantly. C.-Z.W. gathered all results and wrote the manuscript. L.-C.C. and C.-W.C. conducted the animal model and edited the manuscript. T.-C.F., Y.-F.L. and D.P. supervised the whole process of study and analyzed the results. J.-S.C. developed the experimental concept and designed the process of the whole study. All authors have read and agreed to the published version of the manuscript.

Funding: This work was supported by National Defense Medical Center, Tri-Service General Hospital, Taiwan under Grant TSGH-C106-093, TSGH-C107-097, MAB-106-091 and Taipei Medical University, Taiwan under Grant TMU105-AE1-B28.

Acknowledgments: All authors acknowledge the help of Mary Goodwin, English Department, National Taiwan Normal University, in manuscript editing.

Conflicts of Interest: All authors declare no conflict of interest.

References

1. Huang, K.C. Obesity and its related diseases in Taiwan. *Obes. Rev.* **2008**, *9* (Suppl. S1), 32–34. [CrossRef]
2. Fazel, Y.; Koenig, A.B.; Sayiner, M.; Goodman, Z.D.; Younossi, Z.M. Epidemiology and natural history of non-alcoholic fatty liver disease. *Metabolism* **2016**, *65*, 1017–1025. [CrossRef]
3. Tsuboi, N.; Utsunomiya, Y.; Hosoya, T. Obesity-related glomerulopathy and the nephron complement. *Nephrol. Dial. Transplant.* **2013**, *28* (Suppl. S4), 108–113. [CrossRef]
4. Giralt, M.; Villarroya, F. White, brown, beige/brite: Different adipose cells for different functions? *Endocrinology* **2013**, *154*, 2992–3000. [CrossRef] [PubMed]
5. Zafrir, B. Brown adipose tissue: Research milestones of a potential player in human energy balance and obesity. *Horm. Metab. Res.* **2013**, *45*, 774–785. [CrossRef] [PubMed]
6. Hull, R.L.; Westermark, G.T.; Westermark, P.; Kahn, S.E. Islet amyloid: A critical entity in the pathogenesis of type 2 diabetes. *J. Clin. Endocrinol. Metab.* **2004**, *89*, 3629–3643. [CrossRef]
7. D'Agati, V.D.; Chagnac, A.; de Vries, A.P.; Levi, M.; Porrini, E.; Herman-Edelstein, M.; Praga, M. Obesity-related glomerulopathy: Clinical and pathologic characteristics and pathogenesis. *Nat. Rev. Nephrol.* **2016**, *12*, 453–471. [CrossRef]
8. Hajer, G.R.; van Haeften, T.W.; Visseren, F.L. Adipose tissue dysfunction in obesity, diabetes, and vascular diseases. *Eur. Heart J.* **2008**, *29*, 2959–2971. [CrossRef]
9. Goossens, G.H. The role of adipose tissue dysfunction in the pathogenesis of obesity-related insulin resistance. *Physiol. Behav.* **2008**, *94*, 206–218. [CrossRef] [PubMed]
10. Blüher, M. Adipose tissue dysfunction contributes to obesity related metabolic diseases. *Best Pract. Res. Clin. Endocrinol. Metab.* **2013**, *27*, 163–177. [CrossRef] [PubMed]
11. Sun, K.; Tordjman, J.; Clément, K.; Scherer, P.E. Fibrosis and adipose tissue dysfunction. *Cell Metab.* **2013**, *18*, 470–477. [CrossRef]
12. Buechler, C.; Krautbauer, S.; Eisinger, K. Adipose tissue fibrosis. *World J. Diabetes* **2015**, *6*, 548–553. [CrossRef] [PubMed]
13. Mondino, A.; Blasi, F. uPA and uPAR in fibrinolysis, immunity and pathology. *Trends Immunol.* **2004**, *25*, 450–455. [CrossRef]
14. Gyetko, M.R.; Libre, E.A.; Fuller, J.A.; Chen, G.H.; Toews, G. Urokinase is required for T lymphocyte proliferation and activation in vitro. *J. Lab. Clin. Med.* **1999**, *133*, 274–288. [CrossRef]
15. Kawao, N.; Nagai, N.; Tamura, Y.; Horiuchi, Y.; Okumoto, K.; Okada, K.; Suzuki, Y.; Umemura, K.; Yano, M.; Ueshima, S.; et al. Urokinase-type plasminogen activator and plasminogen mediate activation of macrophage phagocytosis during liver repair in vivo. *Thromb. Haemost.* **2012**, *107*, 749–759. [CrossRef] [PubMed]

16. Wu, C.Z.; Chang, L.C.; Lin, Y.F.; Hung, Y.J.; Pei, D.; Chu, N.F.; Chen, J.S. Urokinase plasminogen activator receptor and its soluble form in common biopsy-proven kidney diseases and in staging of diabetic nephropathy. *Clin. Biochem.* **2015**, *48*, 1324–1329. [CrossRef]

17. Stewart, D.; Fulton, W.B.; Wilson, C.; Monitto, C.L.; Paidas, C.N.; Reeves, R.H.; De Maio, A. Genetic contribution to the septic response in a mouse model. *Shock* **2002**, *18*, 342–347. [CrossRef] [PubMed]

18. Mills, C.D.; Kincaid, K.; Alt, J.M.; Heilman, M.J.; Hill, A.M. M-1/M-2 macrophages and the Th1/Th2 paradigm. *J. Immunol.* **2000**, *164*, 6166–6173. [CrossRef] [PubMed]

19. Watanabe, H.; Numata, K.; Ito, T.; Takagi, K.; Matsukawa, A. Innate immune response in Th1- and Th2-dominant mouse strains. *Shock* **2004**, *22*, 460–466. [CrossRef]

20. Lumeng, C.N.; Bodzin, J.L.; Saltiel, A.R. Obesity induces a phenotypic switch in adipose tissue macrophage polarization. *J. Clin. Investig.* **2007**, *117*, 175–184. [CrossRef]

21. Jovicic, N.; Jeftic, I.; Jovanovic, I.; Radosavljevic, G.; Arsenijevic, N.; Lukic, M.L.; Pejnovic, N. Differential Immunometabolic Phenotype in Th1 and Th2 Dominant Mouse Strains in Response to High-Fat Feeding. *PLoS ONE* **2015**, *10*, e0134089. [CrossRef]

22. Mazar, A.P.; Henkin, J.; Goldfarb, R.H. The urokinase plasminogen activator system in cancer: Implications for tumor angiogenesis and metastasis. *Angiogenesis* **1999**, *3*, 15–32. [CrossRef] [PubMed]

23. Pawlak, K.; Pawlak, D.; Myśliwiec, M. Urokinase-type plasminogen activator and metalloproteinase-2 are independently related to the carotid atherosclerosis in haemodialysis patients. *Thromb. Res.* **2008**, *121*, 543–548. [CrossRef]

24. Gyöngyösi, M.; Glogar, D.; Weidinger, F.; Domanovits, H.; Laggner, A.; Wojta, J.; Zorn, G.; Iordanova, N.; Huber, K. Association between plasmin activation system and intravascular ultrasound signs of plaque instability in patients with unstable angina and non-st-segment elevation myocardial infarction. *Am. Heart J.* **2004**, *147*, 158–164. [CrossRef] [PubMed]

25. Dellas, C.; Schremmer, C.; Hasenfuss, G.; Konstantinides, S.V.; Schäfer, K. Lack of urokinase plasminogen activator promotes progression and instability of atherosclerotic lesions in apolipoprotein E-knockout mice. *Thromb. Haemost.* **2007**, *98*, 220–227. [CrossRef]

26. Zhou, H.; Wu, X.; Lu, X.; Chen, G.; Ye, X.; Huang, J. Evaluation of plasma urokinase-type plasminogen activator and urokinase-type plasminogen-activator receptor in patients with acute and chronic hepatitis B. *Thromb. Res.* **2009**, *123*, 537–542. [CrossRef]

27. Juhan-Vague, I.; Alessi, M.C.; Mavri, A.; Morange, P.E. Plasminogen activator inhibitor-1, inflammation, obesity, insulin resistance and vascular risk. *J. Thromb. Haemost.* **2003**, *1*, 1575–1579. [CrossRef]

28. Chen, J.S.; Wu, C.Z.; Chu, N.F.; Chang, L.C.; Pei, D.; Lin, Y.F. Association among fibrinolytic proteins, metabolic syndrome components, insulin secretion and resistance in schoolchildren. *Int. J. Endocrinol.* **2015**, *2015*, 170987. [CrossRef]

29. Morange, P.E.; Lijnen, H.R.; Alessi, M.C.; Kopp, F.; Collen, D.; Juhan-Vague, I. Influence of PAI-1 on adipose tissue growth and metabolic parameters in a murine model of diet-induced obesity. *Arterioscler. Thromb. Vasc. Biol.* **2000**, *20*, 1150–1154. [CrossRef] [PubMed]

30. Trayhurn, P.; Alomar, S.Y. Oxygen deprivation and the cellular response to hypoxia in adipocytes—Perspectives on white and brown adipose tissues in obesity. *Front. Endocrinol.* **2015**, *6*, 19. [CrossRef] [PubMed]

31. Pellegrinelli, V.; Carobbio, S.; Vidal-Puig, A. Adipose tissue plasticity: How fat depots respond differently to pathophysiological cues. *Diabetologia* **2016**, *59*, 1075–1088. [CrossRef] [PubMed]

32. Lee, P.; Greenfield, J.R.; Ho, K.K.; Fulham, M.J. A critical appraisal of the prevalence and metabolic significance of brown adipose tissue in adult humans. *Am. J. Physiol. Endocrinol. Metab.* **2010**, *299*, E601–E606. [CrossRef] [PubMed]

33. Spencer, M.; Unal, R.; Zhu, B.; Rasouli, N.; McGehee, R.E., Jr.; Peterson, C.A.; Kern, P.A. Adipose tissue extracellular matrix and vascular abnormalities in obesity and insulin resistance. *J. Clin. Endocrinol. Metab.* **2011**, *96*, E1990–E1998. [CrossRef] [PubMed]

34. Alcalá, M.; Calderon-Dominguez, M.; Bustos, E.; Ramos, P.; Casals, N.; Serra, D.; Viana, M.; Herrero, L. Increased inflammation, oxidative stress and mitochondrial respiration in brown adipose tissue from obese mice. *Sci. Rep.* **2017**, *7*, 16082. [CrossRef]

35. Fenzl, A.; Kiefer, F.W. Brown adipose tissue and thermogenesis. *Horm. Mol. Biol. Clin. Investig.* **2014**, *19*, 25–37. [CrossRef]

36. Matthews, D.R.; Hosker, J.P.; Rudenski, A.S.; Naylor, B.A.; Treacher, D.F.; Turner, R.C. Homeostasis model assessment: Insulin resistance and beta-cell function from fasting plasma glucose and insulin concentrations in man. *Diabetologia* **1985**, *28*, 412–419. [CrossRef] [PubMed]

International Journal of
Molecular Sciences

Review

Fibrinolysis in Platelet Thrombi

Rahim Kanji [1,2], Ying X. Gue [3], Vassilios Memtsas [2] and Diana A. Gorog [1,2,4,*]

1   Faculty of Medicine, National Heart and Lung Institute, Imperial College, London SW3 6LY, UK; rahim.kanji04@imperial.ac.uk
2   Cardiology Department, East and North Hertfordshire NHS Trust, Stevenage, Hertfordshire SG1 4AB, UK; vassilios.memtsas@nhs.net
3   Liverpool Centre for Cardiovascular Science, University of Liverpool and Liverpool Heart & Chest Hospital, Liverpool L69 7TX, UK; y.gue@nhs.net
4   School of Life and Medical Sciences, Postgraduate Medical School, University of Hertfordshire, Hatfield AL10 9AB, UK
*   Correspondence: d.gorog@imperial.ac.uk; Tel.: +44-207-0348841

Abstract: The extent and duration of occlusive thrombus formation following an arterial atherothrombotic plaque disruption may be determined by the effectiveness of endogenous fibrinolysis. The determinants of endogenous fibrinolysis are the subject of much research, and it is now broadly accepted that clot composition as well as the environment in which the thrombus was formed play a significant role. Thrombi with a high platelet content demonstrate significant resistance to fibrinolysis, and this may be attributable to an augmented ability for thrombin generation and the release of fibrinolysis inhibitors, resulting in a fibrin-dense, stable thrombus. Additional platelet activators may augment thrombin generation further, and in the case of coronary stenosis, high shear has been shown to strengthen the attachment of the thrombus to the vessel wall. Neutrophil extracellular traps contribute to fibrinolysis resistance. Additionally, platelet-mediated clot retraction, release of Factor XIII and resultant crosslinking with fibrinolysis inhibitors impart structural stability to the thrombus against dislodgment by flow. Further work is needed in this rapidly evolving field, and efforts to mimic the pathophysiological environment in vitro are essential to further elucidate the mechanism of fibrinolysis resistance and in providing models to assess the effects of pharmacotherapy.

Keywords: fibrinolysis; thrombin; shear; clot retraction; Factor XIII; clot stability; NETs; platelets

Citation: Kanji, R.; Gue, Y.X.; Memtsas, V.; Gorog, D.A. Fibrinolysis in Platelet Thrombi. *Int. J. Mol. Sci.* **2021**, 22, 5135. https://doi.org/10.3390/ijms22105135

Academic Editor: Hau C. Kwaan

Received: 2 April 2021
Accepted: 9 May 2021
Published: 12 May 2021

Publisher's Note: MDPI stays neutral with regard to jurisdictional claims in published maps and institutional affiliations.

1. Introduction

Atherothrombotic events are a considerable cause of morbidity and mortality. Much focus and treatment thus far has surrounded the inhibition of platelets due to their crucial role in arterial thrombus formation. However, despite antiplatelet therapy, some patients remain at risk of recurrent thrombotic events. Optimising risk assessment is essential to help identify these patients, with the ultimate aim to reduce events. Emerging data suggest that assessment of endogenous fibrinolysis may help, through identifying patients with impaired endogenous fibrinolysis who are at markedly increased risk of ischaemic events [1,2]. Understanding the determinants of endogenous fibrinolysis is, therefore, paramount, and may highlight new treatment targets (Figure 1).

Interestingly, platelets again seem to be significant modulators of endogenous fibrinolysis, and in this review, we will discuss the mechanisms for mediating resistance to fibrinolysis and the evidence supporting the role of platelets. Areas of discussion will include the role of thrombin, thrombin generation, high shear-induced platelet activation and clot stabilisation, clot retraction, Factor XIII and neutrophil extracellular traps (NETs).

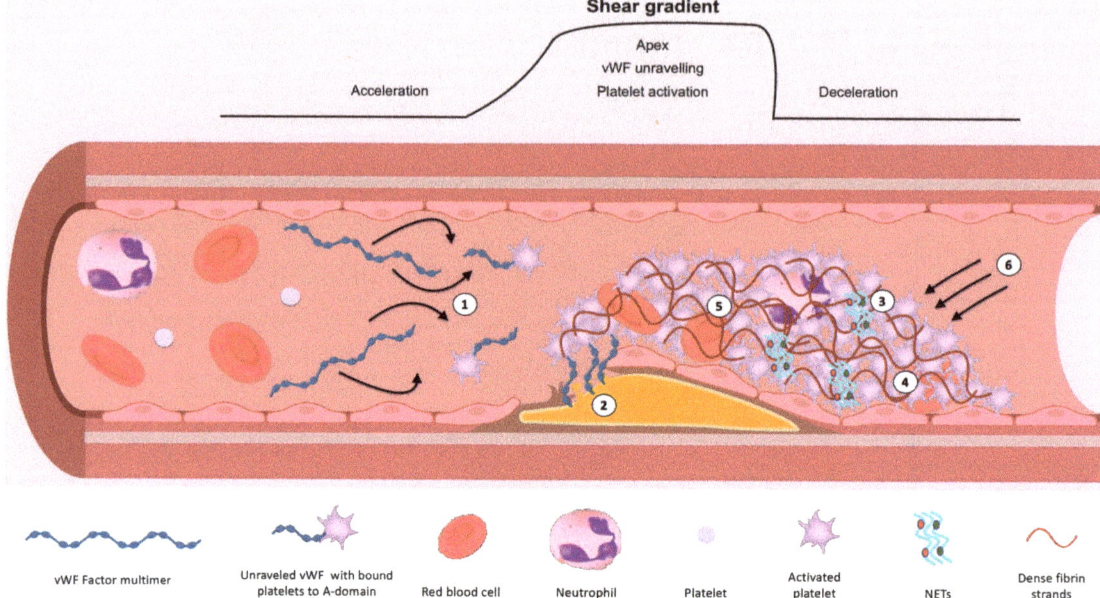

Figure 1. Illustration demonstrating the determinants of endogenous fibrinolysis. (1) High shear/shear gradient results in platelet activation and unravelling of vWF, exposing A-domains for platelet and extracellular matrix binding. This further enhances the affinity of GPIIb/IIIa for fibrinogen. (2) High shear flow contributes to the strength of attachment to the vessel wall. (3) NETs augment coagulation and inhibit fibrinolysis. (4) Platelet-rich clots augment thrombin generation and release of inhibitors including TAFI and PAI-1. This results in densely packed, thin fibrin strands, which are resistant to fibrinolysis. (5) Factor XIIIa-mediated crosslinking of α2-antiplasmin, TAFI and PAI-1 with fibrin inhibits fibrinolysis. (6) Myosin-mediated clot retraction results in increased fibrin density and reduced clot permeability.

2. Role of Thrombin in Endogenous Fibrinolysis

Thrombin is a serine protease that confers significant resistance to endogenous fibrinolysis. It does so through a number of means, which ultimately lead to the formation of a platelet-rich, stable thrombus. Being a potent platelet activator, it maximises the recruitment and aggregation of platelets, which further enhances thrombin release, generating a feed-forward loop. The resultant high thrombin concentration facilitates the formation of a stable clot by cleaving fibrinogen to form insoluble fibrin, which binds platelets together. In addition, it directly inhibits endogenous fibrinolysis through the activation of thrombin activatable fibrinolysis inhibitor (TAFI), which binds and lyses carboxy-terminal lysine residues on fibrin. This prevents the binding and activation of plasminogen to plasmin, which cleaves fibrin into fibrin degradation products [3]. It additionally indirectly inhibits endogenous fibrinolysis through the release of plasminogen activator inhibitor-1 (PAI-1) from platelets, which is a potent inhibitor of tissue-plasminogen activator (t-PA) [4]. The release is activated through intracellular signalling initiated by the g-protein-coupled protease-activated receptor (PAR) [5]. There are two functionally active thrombin receptors found on human platelets, namely PAR-1 and PAR-4 [6,7]. PAR-1 contains a hirudin-like domain possessing a high affinity for thrombin [8], unlike PAR-4, and therefore results in activation at lower concentrations [7]. The activation of these receptors results in the release of the contents of the α-granules of platelets [4], which is responsible for >90% of the circulating PAI-1 detectable during acute arterial thrombosis [9,10].

In addition to this, a number of studies have assessed the relationship between thrombin concentration and fibrin fibre thickness and density [11]. In the presence of a high thrombin concentration, arterial thrombi exhibit thin, densely packed fibrin strands, whilst the converse is seen at lower thrombin concentrations [11–15]. Such structural changes directly impact the resistance of the thrombus endogenous fibrinolysis, as the tiny pores and thin strands impair plasminogen entry and binding with the thrombus, as recently demonstrated in a study of ST-segment elevation myocardial infarction (STEMI) patients using the Global Thrombosis Test and electron microscopy [16]. In this study, impaired endogenous fibrinolysis, assessed in whole blood using a point-of-care technique, showed a correlation with certain structural characteristics of thrombi on electron microscopy, namely reduced fibrin fibre thickness in both the core and periphery of the thrombus and a more densely packed fibrin meshwork, compared to patients with effective endogenous fibrinolysis.

In addition, since the platelet surface plays a central role in the promotion and regulation of thrombin generation, as demonstrated by insignificant generation of thrombin in platelet-poor plasma and a positive correlation observed between platelet number and the extent of thrombin generation [17], platelets are arguably one of the major determinants of endogenous fibrinolysis. This partially explains the difference in resistance to fibrinolysis between arterial (platelet-rich) and venous (erythrocyte-rich) thrombi [18]. Furthermore, the incorporation of red blood cells leads to areas of reduced fibrin fibre density [19].

Understanding the roles of thrombin and platelets is, therefore, key to understanding the determinants of endogenous fibrinolysis.

3. Thrombin Generation

After early studies demonstrated the key role of platelets in thrombin generation, subsequent efforts have focused on elucidating the mechanism. Reports of increased exposure of phosphatidylserine from 2% to 12% in an almost on/off phenomenon upon platelet activation [20], and the association seen between the amount of phosphatidylserine on the platelet surface and the extent of thrombin generation [21] indicate that phosphatidylserine is a determinant of thrombin generation. However, the augmentation in thrombin generation is not exclusively related to phosphatidylserine, as it does not replicate the full procoagulant potential of platelets [22]; furthermore, phosphatidylserine has been found on endothelial cells, which are not prothrombotic [23].

More complex platelet interactions and morphological changes are additionally involved, including platelet degranulation with the release of α-granules, platelet ballooning and protein binding. Platelet degranulation in response to thrombin results in the release of coagulation factors, including Factor V, which are activated on the platelet surface [24]. This procoagulant surface is further enhanced as a result of platelet ballooning [25], facilitating greater coagulation factor binding, activation, and assembly of complexes, all amplifying thrombin generation [26]. For those coagulation factors which possess a y-glutamyl carboxyl acid (GIa) domain, (prothrombin, Factor VII, Factor IX and Factor X), PS shows a high affinity [27], whilst other factors such as FVIII are brought close to the platelet surface for thrombin-driven activation through von Willebrand Factor (vWF) binding with the glycoprotein (GP) Ib-IX-V complex [28].

Complexes formed on the platelet surface ultimately increase thrombin generation. This includes the tenase complex, consisting of Factor VIIIa and Factor IXa, which activates Factor X, and the prothrombinase complex made up of Factor Va and Factor Xa, which activates prothrombin [29,30]. Furthermore, in phosphatidylserine-exposed platelets, the above complexes are co-localised, which amplifies their activity 1000-fold [31].

The binding of ligands to platelet glycoproteins further enhances thrombin generation, which is exemplified by GPIIb/IIIa. Upon platelet activation, the affinity of this glycoprotein for fibrinogen increases [32], eventually resulting in a stable thrombus; however, upon binding, outside-in signalling results in further platelet PS exposure and thrombin generation. This, too, is significant, as blocking this receptor with a monoclonal antibody (abciximab) reduces thrombin generation by 40–70% [33–36].

Clearly, many complex interactions take place; however, the above description may still be an oversimplification. It is now generally accepted that platelet activation and coagulation are not separate processes, and both interplay to generate a stable thrombus by maximising thrombin generation.

4. Synergistic Effects of Shear Stress on Platelet Activation

Atherosclerotic plaque rupture, and the resultant exposure of prothrombotic material, including collagen and tissue factor, was previously thought to be the main mechanism behind platelet adhesion, activation and eventual thrombus formation in atherothrombotic events, including myocardial infarction (MI) and stroke. The mechanism behind this includes the binding of GPVI and integrin $\alpha_2\beta_1$ on the platelet surface to collagen [37,38], resulting in intracellular signalling and platelet activation. As discussed above, this leads to a morphological change in platelets and degranulation, including the release of adenosine diphosphate [39]. This is a potent platelet activator and, through paracrine effects, leads to significant platelet recruitment. It is unsurprising, therefore, that antagonising the $P2Y_{12}$ receptor is the standard of care for patients with ischaemic stroke and MI (in addition to aspirin).

Plaque erosion is another cause of plaque disruption, which has been shown to account for one-third of acute coronary syndromes [40]. The pathophysiological process differs significantly from that of plaque rupture, and as a result, thrombus content and fibrinolysis potential may differ. Desquamation secondary to degradation of the basement membrane by matrix metalloproteinases [40], and/or apoptosis of the endothelial cells by potent oxidant species produced by myeloperoxidase [41] or shear stress [42], have been proposed. Histological studies have confirmed differences in clot composition, with those formed secondary to plaque erosion being more platelet-rich [43]. This could be attributable to a burst of tissue factor expression by endothelial cells in response to potent reactive oxygen species produced by myeloperoxidase [41], and also by the migration and recruitment of neutrophils secondary to the release of chemokines and endothelial cell injury, with resultant neutrophil extracellular trap (NET) formation [44]. NETs have the ability to acquire tissue factor, platelets and fibrin, facilitating the formation of a platelet-rich thrombus [45,46].

Another important mechanism of arterial thrombosis is shear-induced platelet activation. At low shear levels, vWF circulates as a large multimer [47]. In this conformation, its A-domains, which are required for binding to the platelet and extracellular matrix, are concealed. However, high shear leads to unravelling/uncoiling of vWF [48]; in fact, in the presence of a shear gradient, such as that seen at sites of coronary stenosis, this unravelling occurs with greater efficiency as the proximal and distal ends of the multimer experience differing pulling forces, resulting in unravelling at lower shear [49]. Furthermore, in the presence of an atherosclerotic plaque rupture and growing thrombus, shear forces and gradients will increase further, leading to further unravelling, thus initiating a cycle.

This unravelling and elongation exposes the A1 domain, which binds to the platelet GPIbα receptor, leading to platelet adhesion and subsequent aggregation [50,51]. This occurs in addition to the aggregation driven by the binding of fibrinogen to GPIIb/IIIa, with the two processes therefore working in synergy to form a stable thrombus. In fact, recent evidence suggests that the binding of GPIbα with vWF increases the affinity of GPIIb/IIIa for fibrinogen [52]; thus, it enhances platelet aggregation.

The role of platelet activation under high shear flow conditions, causing platelet aggregation and being a determinant of fibrinolysis, is supported by recent data from the RISK-PPCI study, in which endogenous fibrinolysis was assessed using the Global Thrombosis Test in patients presenting with STEMI [1]. In these patients, time to form an occlusive thrombus under high shear conditions in vitro correlated inversely with the effectiveness of endogenous fibrinolysis, implying that shear-activated platelets contribute to impaired endogenous fibrinolysis.

This is very relevant, as current pharmacotherapy for patients with arterial thrombosis is directed mainly at antagonising the $P2Y_{12}$ receptor and inhibiting cyclo-oxygenase, which have no effect on this shear-driven, vWF-dependent pathway of platelet aggregation and activation. Furthermore, inhibition of the P2Y$_{12}$ receptor has not been shown to affect endogenous fibrinolysis [53].

Additionally, this shear-driven mechanism for aggregation is not dependent upon plaque rupture, and so patients with significant stenoses are at risk of both thrombus formation and impaired endogenous fibrinolysis. Furthermore, this risk may be dependent on the degree of stenosis and plaque burden. With increasing luminal narrowing, the shear gradient increases, resulting in increasing platelet activation at the apex of the lesion. Studies have demonstrated platelet activation [54], including microparticle formation in response to shear [55], and increased activation secondary to platelet hammering (exposure to repeated hyper-shear) [56]. Additionally, increased phosphatidylserine externalisation and procoagulant activity, including thrombin generation, have been observed at high shear rates [57], which appear to be dependent upon the binding of vWF and the GPIbα platelet receptor [58,59]. This receptor has a mechanosensitive domain, which unfolds when bound to the A1 domain of vWF, leading to intracellular signalling [60] and intermediary activation of other integrins, increasing their affinity for ligand and facilitation of outside-in signalling [60]. Furthermore, since GPIbα can also bind soluble vWF [61], and since platelets remain sensitised after exposure to high shear [62], hyper-aggregation can be seen downstream from the site of maximal luminal stenosis [63], where deceleration and low shear favour thrombus formation.

Studies aiming to block this high shear-driven platelet aggregation using monoclonal antibodies to the A1 domain of vWF have shown reduced thrombin generation, adhesion and aggregation [64]. Furthermore, they have shown reduced bleeding times when compared with abciximab [65], highlighting specificity for shear-driven activation and therefore allowing aggregation of platelets at low shear with fibrinogen and GPIIb/IIIa. However, whether this formally affects endogenous fibrinolysis is unclear.

5. Clot Retraction

Clot retraction is a physiological mechanism to aid healing during haemostasis. Through the expulsion of serum, which has been depleted of clotting factors, the volume of the clot reduces, leading to the coming-together of wound edges [66]. It therefore has favourable effects in physiology and is platelet-mediated. After the binding of fibrinogen to the GPIIb/IIIa receptor, phosphorylation of the receptor leads to outside-in signalling, resulting in myosin binding [67] and co-localisation of the ANK domain containing Bcl-3 with the cytoskeleton, which is tyrosine kinase dependent [68].

Thus, when contractile forces are generated by myosin [69], this leads to clot retraction through its connection with the GPIIb/IIIa receptor and fibrin(ogen), which is crosslinked with other fibrin strands and platelets [70]. Therefore, the greater the platelet number and fibrin crosslinking are, the greater the force and effects of clot retraction are. This has the overall effect of increasing fibrin density and reducing clot permeability [71], which, as mentioned above, affects fibrinolysis. Thin, densely packed fibrin strands are resistant to fibrinolysis, and reduced permeability and pore size impairs the entry of plasminogen and t-PA [72].

This effect on fibrinolysis has been shown both in vitro and in vivo. In a mouse model of thrombus generation through mesenteric vein injury and thrombin injection, clot retraction was seen to occur over a period of 3 h, which was inhibited by blebbistatin, a potent myosin IIa inhibitor [69]. Furthermore, when recombinant t-PA was infused over the thrombi, the lysis of unretracted thrombi was far greater than that of retracted thrombi; in fact, a relationship was seen between the degree of lysis and clot retraction. However, what was unexpected was the potential role of early limited endogenous fibrinolysis in clot retraction. When t-PA was infused early following clot formation, fibrinolysis was seen with a reduction in both thrombus volume and fibrin. However, after 30 min, thrombus volume reduced, with no effect on fibrin. Furthermore, pre-treatment of mice with tranexamic acid, an inhibitor of fibrinolysis, led to impairment of early clot retraction. This effect was confirmed in vitro, where low concentrations of t-PA, in fact, facilitated clot retraction; however, at higher doses, lysis was observed.

These findings have been replicated somewhat using human blood in vitro [73]. Retracted clots were found to be resistant to external fibrinolysis; however, this was not the case for endogenous fibrinolysis. For retracted clots that had been bathed in t-PA prior to their formation with thrombin, the rate of endogenous fibrinolysis was higher when compared with clots that had been prepared in a similar manner but with the addition of inhibitors of retraction.

Clearly a link exists between platelet-mediated clot retraction and endogenous fibrinolysis.

6. Factor XIIIa

Factor XIII is a coagulation factor that has many functions, primarily directed at promoting clot stability. Its timely release from activated platelets with fibrinogen, prothrombin, Factor V and Factor VIII (Table 1), all required during the later stages of clot formation, ensures its abundance when required to stabilise the formed thrombus. It can also be found in plasma complexed with fibrinogen, where it is activated by thrombin.

Table 1. Site of synthesis and main source of major pro- and anti-fibrinolytic proteins discussed within this review.

Protein	Origin	Major Source
Factor XIII	Cellular (megakaryocytes, platelets, monocytes, osteoblasts) Plasma	Platelets
Factor V	Liver, platelets and plasma	Liver
Factor X	Liver and plasma	Liver
Factor VIII	Liver, endothelial cells and plasma	Liver
Prothrombin	Liver and plasma	Liver
Plasminogen activator inhibitor (PAI)	Endothelial cells, liver, adipose tissue, plasma and platelets	Platelets
Tissue plasminogen activator (t-PA)	Endothelial cells, mesothelial cells, megakaryocytes and plasma	Endothelial cells
Fibrinogen	Liver, plasma (predominantly), platelets, lymph and interstitial fluid	Liver
α2-antiplasmin	Liver and plasma	Liver

One of its many stabilising effects includes the crosslinking of fibrin [74], which is particularly important in environments of high shear [75]. Furthermore, through promoting coupling with protofibrils, deformation at low shear is prevented through stiffening [76]. Coupling also has the effect of reducing the size of the pores and impairing the entry and diffusion of fibrinolytic enzymes, including t-PA, into the clot [77].

Factor XIIIa also facilitates the crosslinking of α2-antiplasmin [78], TAFI and PAI-1 with fibrin. Crosslinked α2-antiplasmin prevents adsorption of plasminogen with fibrin, preventing its activation and lysis of fibrin. Furthermore, in plasma, it binds and inhibits plasmin. Therefore, it has a central role in inhibiting endogenous fibrinolysis.

However, in vivo studies representing its effects on endogenous fibrinolysis are limited. Reed et al. undertook a study involving anaesthetised ferrets with pulmonary embolism and found significantly enhanced endogenous fibrinolysis activity in ferrets treated with a Factor XIIIa inhibitor [79]. Furthermore, total Factor XIIIa inhibition resulted in greater endogenous fibrinolytic activity compared with only α2-antiplasmin-inhibited crosslinking, suggesting that Factor XIIIa-mediated fibrin crosslinking also plays a major role in endogenous fibrinolysis.

In vitro human studies are greater in number and confirm the role of Factor XIII in fibrinolysis. In a study by Jansen et al., t-PA was added to fresh human whole blood prior to clot formation with the addition of thrombin [80]. Fibrinolysis was greater in the samples that had antibody-inhibited Factor XIIIa activity. These findings were reproduced in a study using plasma clots and a Chandler loop [81]. Interestingly, they both also concluded that the majority of the inhibitory effect of Factor XIIIa on fibrinolysis was mediated through α2-antiplasmin.

7. Activated Neutrophils and NETs

Neutrophil extracellular traps (NETs) are web-like structures composed of DNA and histones [82]. They are released by activated neutrophils, in addition to elastases, and have a significant effect on coagulation. Histones specifically activate platelets [83], inhibit activated protein C-mediated inhibition of coagulation [84] and support thrombin activation [85]. DNA can activate Factor XII and initiate coagulation [86], whilst elastases can break down inhibitors of coagulation [87].

There is now evolving evidence that further highlights the effect of NETs on fibrinolysis. One group reported on the effects of histone–DNA complexes, which resulted in the formation of thrombi with reduced permeability, in both fibrin [88] and plasma clots [89], and prolongation of t-PA-mediated fibrinolysis. This effect was reproduced when activated neutrophils themselves were added to plasma (with confirmation of NET formation using electron microscopy) and reversed with the addition of DNAse, implicating a contributory role for DNA (in NETs) in inhibiting fibrinolysis. Further evidence suggests that elastases bound to DNA in NETs are responsible for plasminogen degradation, and this may be one mechanism behind fibrinolysis resistance [90].

Further human data are limited to ex vivo and in vitro studies. A histological analysis of clots retrieved from 108 patients undergoing endovascular therapy for acute ischaemic stroke confirmed the presence of NETs, which correlated with procedure time [91]. When the retrieved clots were then treated with t-PA, the administration of DNAse hastened lysis time. In 126 patients treated in hospital for pulmonary embolism, raised lactate levels were associated with a 29% higher neutrophil count, 45% higher plasma citrullinated histone H3 level, reduced plasma fibrin clot permeability and longer clot lysis time [92]. Furthermore, lactate positively correlated with plasma citrullinated histone H3 concentration, plasma clot lysis time and PAI-1 level.

Thus, NETs confer resistance to endogenous fibrinolysis. The mechanism may be multifactorial and includes the protection of thrombin from degradation (and resultant dense fibrin clot formation) and the promotion of plasminogen breakdown by bound elastases.

8. Clot Stability

Clot stability refers to the ability of a thrombus to resist fibrinolysis and dislodgement from the vessel wall by flowing blood. The former has been discussed extensively within this review, but the mechanisms determining the strength of attachment to the vessel wall have not been addressed. The latter appears to be mediated through shear stress. With increasing wall shear, an increasing number of platelets are recruited to the growing thrombus [93]. Under high wall shear conditions, the formed thrombus has a thicker shell and a more densely packed core [94]. This may be facilitated by the shell preventing washout of platelet activators, thus promoting paracrine activity [95], with the resultant thrombus being resistant to fibrinolysis. Furthermore, the strength of attachment to the vessel wall is increased by high shear flow, and this may be secondary to the high affinity state of the A1 domain of vWF for GPIbα under these conditions [50]. However, a point is reached where the risk of dislodgement is greater; furthermore, Shi et al. suggest that wall shear may also have a contributory role, demonstrating a parabolic relationship between wall shear and thrombus area [94].

Clearly, increasing wall shear stress and shear flow play a role in clot stability; however, their effects are not linear. A point is reached where the bond with the vessel wall is overcome, leading to thrombus dislodgment. This may have an effect on endogenous fibrinolysis potential, as microemboli may have exposed areas for the entry of fibrinolytic enzymes, resulting in more rapid fibrinolysis than the original mother thrombus.

ADP signalling also appears to be involved in clot stability. Administration of $P2Y_{12}$ inhibitors to whole blood has been shown to destabilise thrombus formation under high shear in vitro, resulting in microbleeds [53]. The effect was more profound with more potent inhibitors such as cangrelor, and this also enhanced endogenous fibrinolysis.

Furthermore, in an in vitro study, the administration of ticagrelor after the initiation of clot formation through the exposure of ADP and collagen led to dispersion, confirmed by aggregometry [96]. This has also been shown in vivo in a murine model, where early arterial thrombotic occlusion was partially reversed with the administration of ticagrelor.

Factors that potentiate clot stability may, therefore, confer resistance to endogenous fibrinolysis.

9. Conclusions

In conclusion, there are many mechanisms involved in controlling endogenous fibrinolysis. There is a significant contributory and complex role of cellular components, particularly platelets and NETs, in determining resistance to endogenous fibrinolysis. Additionally, high shear flow conditions further impact platelet activation and thrombus stability. Further work is needed in this rapidly evolving field, and efforts to mimic the pathophysiological environment in vitro are essential to further elucidate the mechanism of fibrinolysis resistance and in providing models to assess the effects of pharmacotherapy.

Author Contributions: Design, literature search and drafting of the manuscript, R.K., Y.X.G. and V.M.; Concept, design, critical revision for important intellectual content and final approval of the manuscript submitted D.A.G. All authors have read and agreed to the published version of the manuscript.

Funding: This research received no external funding.

Institutional Review Board Statement: Not applicable.

Informed Consent Statement: Not applicable.

Data Availability Statement: Not applicable.

Conflicts of Interest: The authors declare no conflict of interest.

Abbreviations

GP	Glycoprotein
GTT	Global Thrombosis Test
MI	Myocardial infarction
NETs	Neutrophil extracellular traps
PAI-1	Plasminogen activator inhibitor-1
PAR	Protease-activated receptor
PS	Phosphatidylserine
STEMI	ST-segment elevation myocardial infarction
TAFI	Thrombin activatable fibrinolysis inhibitor
t-PA	Tissue-plasminogen activator
vWF	von Willebrand Factor

References

1. Farag, M.; Spinthakis, N.; Gue, Y.X.; Srinivasan, M.; Sullivan, K.; Wellsted, D.; Gorog, D.A. Impaired endogenous fibrinolysis in ST-segment elevation myocardial infarction patients undergoing primary percutaneous coronary intervention is a predictor of recurrent cardiovascular events: The RISK PPCI study. *Eur. Heart J.* **2019**, *40*, 295–305. [CrossRef]
2. Sumaya, W.; Wallentin, L.; James, S.K.; Siegbahn, A.; Gabrysch, K.; Bertilsson, M.; Himmelmann, A.; Ajjan, R.A.; Storey, R.F. Fibrin clot properties independently predict adverse clinical outcome following acute coronary syndrome: A PLATO substudy. *Eur. Heart J.* **2018**, *39*, 1078–1085. [CrossRef]
3. Bouma, B.N.; Mosnier, L.O. Thrombin Activatable Fibrinolysis Inhibitor (TAFI) at the Interface between Coagulation and Fibrinolysis. *Pathophysiol. Haemost. Thromb.* **2003**, *33*, 375–381. [CrossRef] [PubMed]
4. Huebner, B.R.; Moore, E.E.; Moore, H.B.; Stettler, G.R.; Nunns, G.R.; Lawson, P.; Sauaia, A.; Kelher, M.; Banerjee, A.; Silliman, C.C. Thrombin Provokes Degranulation of Platelet α-Granules Leading to the Release of Active Plasminogen Activator Inhibitor-1 (PAI-1). *Shock* **2018**, *50*, 671–676. [CrossRef] [PubMed]

5. Nylander, M.; Osman, A.; Ramström, S.; Åklint, E.; Larsson, A.; Lindahl, T.L. The role of thrombin receptors PAR1 and PAR4 for PAI-1 storage, synthesis and secretion by human platelets. *Thromb. Res.* **2012**, *129*, e51–e58. [CrossRef] [PubMed]

6. Kahn, M.L.; Nakanishi-Matsui, M.; Shapiro, M.J.; Ishihara, H.; Coughlin, S.R. Protease-activated receptors 1 and 4 mediate activation of human platelets by thrombin. *J. Clin. Investig.* **1999**, *103*, 879–887. [CrossRef] [PubMed]

7. Xu, W.F.; Andersen, H.; Whitmore, T.E.; Presnell, S.R.; Yee, D.P.; Ching, A.; Gilbert, T.; Davie, E.W.; Foster, D.C. Cloning and characterization of human protease-activated receptor 4. *Proc. Natl. Acad. Sci. USA* **1998**, *95*, 6642–6646. [CrossRef] [PubMed]

8. Vu, T.; Wheaton, V.; Hung, D.; Charo, I.; Coughlin, S. Domains specifying thrombin-receptor interaction. *Nature* **1991**, *353*, 674–677. [CrossRef]

9. Abu Assab, T.; Raveh-brawer, D.; Abramowitz, J.; Naamad, M.; Rowe, J.M.; Ganzel, C. The Predictive Value of Thromboelastogram in the Evaluation of Patients with Suspected Acute Venous Thromboembolism. *Blood* **2018**, *132*, 5052. [CrossRef]

10. Zhu, Y.; Carmeliet, P.; Fay, W.P. Plasminogen Activator Inhibitor-1 Is a Major Determinant of Arterial Thrombolysis Resistance. *Circulation* **1999**, *99*, 3050–3055. [CrossRef]

11. Wolberg, A.S. Thrombin generation and fibrin clot structure. *Blood Rev.* **2007**, *21*, 131–142. [CrossRef]

12. Ryan, E.A.; Mockros, L.F.; Weisel, J.W.; Lorand, L. Structural origins of fibrin clot rheology. *Biophys. J.* **1999**, *77*, 2813–2826. [CrossRef]

13. Blombäck, B.; Carlsson, K.; Fatah, K.; Hessel, B.; Procyk, R. Fibrin in human plasma: Gel architectures governed by rate and nature of fibrinogen activation. *Thromb. Res.* **1994**, *75*, 521–538. [CrossRef]

14. Wolberg, A.S.; Monroe, D.M.; Roberts, H.R.; Hoffman, M. Elevated prothrombin results in clots with an altered fiber structure: A possible mechanism of the increased thrombotic risk. *Blood* **2003**, *101*, 3008–3013. [CrossRef]

15. Blombäck, B.; Carlsson, K.; Hessel, B.; Liljeborg, A.; Procyk, R.; Aslund, N. Native fibrin gel networks observed by 3D microscopy, permeation and turbidity. *Biochim. Biophys. Acta* **1989**, *997*, 96–110. [CrossRef]

16. Spinthakis, N.; Gue, Y.; Farag, M.; Ren, G.; Srinivasan, M.; Baydoun, A.; Gorog, D.A. Impaired endogenous fibrinolysis at high shear using a point-of-care test in STEMI is associated with alterations in clot architecture. *J. Thromb. Thrombolysis* **2019**, *47*, 392–395. [CrossRef]

17. Buckwalter, J.; Blythe, W.; Brinkhous, K. Effect of blood platelets on prothrombin utilization of dog and human plasmas. *Am. J. Physiol.* **1949**, *159*, 316–321. [CrossRef]

18. Kanji, R.; Kubica, J.; Navarese, E.P.; Gorog, D.A. Endogenous fibrinolysis—Relevance to clinical thrombosis risk assessment. *Eur. J. Clin. Investig.* **2021**, *51*, e13471. [CrossRef]

19. Gersh, K.C.; Nagaswami, C.; Weisel, J.W. Fibrin network structure and clot mechanical properties are altered by incorporation of erythrocytes. *Thromb. Haemost.* **2009**, *102*, 1169–1175. [CrossRef]

20. Solum, N.O. Procoagulant Expression in Platelets and Defects Leading to Clinical Disorders. *Arterioscler. Thromb. Vasc. Biol.* **1999**, *19*, 2841–2846. [CrossRef]

21. Lindhout, T.; Govers-Riemslag, J.; van de Waart, P.; Hemker, H.; Rosing, J. Factor Va-factor Xa interaction: Effects of phospholipid vesicles of varying composition. *Biochemisrty* **1982**, *21*, 5494–5502. [CrossRef] [PubMed]

22. Walsh, P.N.; Lipscomb, M.S. Comparison of the coagulant activities of platelets and phospholipids. *Br. J. Haematol.* **1976**, *33*, 9–18. [CrossRef] [PubMed]

23. Ravanat, C.; Archipoff, G.; Beretz, A.; Freund, G.; Cazenave, J.P.; Freyssinet, J.M. Use of annexin-V to demonstrate the role of phosphatidylserine exposure in the maintenance of haemostatic balance by endothelial cells. *Biochem. J.* **1992**, *282 Pt 1*, 7–13. [CrossRef]

24. Alberios, L.; Safa, O.; Clemetson, K.; Esmon, C.; Dale, G. Surface expression and functional characterization of alpha-granule factor V in human platelets: Effects of ionophore A23187, thrombin, collagen, and convulxin. *Blood* **2000**, *95*, 1694–1702. [CrossRef]

25. Mattheij, N.J.A.; Braun, A.; van Kruchten, R.; Castoldi, E.; Pircher, J.; Baaten, C.C.F.M.J.; Wülling, M.; Kuijpers, M.J.E.; Köhler, R.; Poole, A.W.; et al. Survival protein anoctamin-6 controls multiple platelet responses including phospholipid scrambling, swelling, and protein cleavage. *FASEB J.* **2016**, *30*, 727–737. [CrossRef]

26. Agbani, E.O.; van den Bosch, M.T.J.; Brown, E.; Williams, C.M.; Mattheij, N.J.A.; Cosemans, J.M.E.M.; Collins, P.W.; Heemskerk, J.W.M.; Hers, I.; Poole, A.W. Coordinated Membrane Ballooning and Procoagulant Spreading in Human Platelets. *Circulation* **2015**, *132*, 1414–1424. [CrossRef]

27. Monroe, D.M.; Hoffman, M.; Roberts, H.R. Platelets and Thrombin Generation. *Arterioscler. Thromb. Vasc. Biol.* **2002**, *22*, 1381–1389. [CrossRef]

28. Pieters, J.; Lindhout, T.; Hemker, H. In situ-generated thrombin is the only enzyme that effectively activates factor VIII and factor V in thromboplastin-activated plasma. *Blood* **1989**, *74*, 1021–1024. [CrossRef]

29. Berny, M.A.; Munnix, I.C.A.; Auger, J.M.; Schols, S.E.M.; Cosemans, J.M.E.M.; Panizzi, P.; Bock, P.E.; Watson, S.P.; McCarty, O.J.T.; Heemskerk, J.W.M. Spatial Distribution of Factor Xa, Thrombin, and Fibrin(ogen) on Thrombi at Venous Shear. *PLoS ONE* **2010**, *5*, e10415. [CrossRef]

30. Swieringa, F.; Kuijpers, M.J.E.; Lamers, M.M.E.; van der Meijden, P.E.J.; Heemskerk, J.W.M. Rate-limiting roles of the tenase complex of factors VIII and IX in platelet procoagulant activity and formation of platelet-fibrin thrombi under flow. *Haematologica* **2015**, *100*, 748–756. [CrossRef]

31. Heemskerk, J.; Bevers, E.; Lindhout, T. Platelet activation and blood coagulation. *Thromb. Haemost.* **2002**, *88*, 186–193.

32. Marguerie, G.; Plow, E.; Edgington, T. Human platelets possess an inducible and saturable receptor specific for fibrinogen. *J. Biol. Chem.* **1979**, *254*, 5357–5363. [CrossRef]
33. Kumar, R.; Beguin, S.; Hemker, H. The effect of fibrin clots and clot-bound thrombin on the development of platelet procoagulant activity. *Thromb. Haemost.* **1995**, *74*, 962–968. [CrossRef]
34. Butenas, S.; Cawthern, K.M.; van't Veer, C.; DiLorenzo, M.E.; Lock, J.B.; Mann, K.G. Antiplatelet agents in tissue factor–induced blood coagulation. *Blood* **2001**, *97*, 2314–2322. [CrossRef]
35. Li, Y.; Spencer, F.A.; Ball, S.; Becker, R.C. Inhibition of platelet-dependent prothrombinase activity and thrombin generation by glycoprotein IIb/IIIa receptor-directed antagonists: Potential contributing mechanism of benefit in acute coronary syndromes. *J. Thromb. Thrombolysis* **2000**, *10*, 69–76. [CrossRef]
36. Reverter, J.C.; Béguin, S.; Kessels, H.; Kumar, R.; Hemker, H.C.; Coller, B.S. Inhibition of platelet-mediated, tissue factor-induced thrombin generation by the mouse/human chimeric 7E3 antibody. Potential implications for the effect of c7E3 Fab treatment on acute thrombosis and "clinical restenosis". *J. Clin. Investig.* **1996**, *98*, 863–874. [CrossRef]
37. Staatz, W.; Walsh, J.; Pexton, T.; Santoro, S. The alpha 2 beta 1 integrin cell surface collagen receptor binds to the alpha 1 (I)-CB3 peptide of collagen. *J. Biol. Chem.* **1990**, *265*, 4778–4781. [CrossRef]
38. Jung, S.M.; Moroi, M. Platelet glycoprotein VI. *Adv. Exp. Med. Biol.* **2008**, *640*, 53–63. [CrossRef]
39. Blair, P.; Flaumenhaft, R. Platelet alpha-granules: Basic biology and clinical correlates. *Blood Rev.* **2009**, *23*, 177–189. [CrossRef]
40. Partida, R.A.; Libby, P.; Crea, F.; Jang, I.-K. Plaque erosion: A new in vivo diagnosis and a potential major shift in the management of patients with acute coronary syndromes. *Eur. Heart J.* **2018**, *39*, 2070–2076. [CrossRef]
41. Sugiyama, S.; Kugiyama, K.; Aikawa, M.; Nakamura, S.; Ogawa, H.; Libby, P. Hypochlorous acid, a macrophage product, induces endothelial apoptosis and tissue factor expression: Involvement of myeloperoxidase-mediated oxidant in plaque erosion and thrombogenesis. *Arterioscler. Thromb. Vasc. Biol.* **2004**, *24*, 1309–1314. [CrossRef] [PubMed]
42. Tricot, O.; Mallat, Z.; Heymes, C.; Belmin, J.; Lesèche, G.; Tedgui, A. Relation between endothelial cell apoptosis and blood flow direction in human atherosclerotic plaques. *Circulation* **2000**, *101*, 2450–2453. [CrossRef] [PubMed]
43. Hu, S.; Yonetsu, T.; Jia, H.; Karanasos, A.; Aguirre, A.D.; Tian, J.; Abtahian, F.; Vergallo, R.; Soeda, T.; Lee, H.; et al. Residual thrombus pattern in patients with ST-segment elevation myocardial infarction caused by plaque erosion versus plaque rupture after successful fibrinolysis: An optical coherence tomography study. *J. Am. Coll. Cardiol.* **2014**, *63*, 1336–1338. [CrossRef] [PubMed]
44. Libby, P.; Pasterkamp, G.; Crea, F.; Jang, I.-K. Reassessing the Mechanisms of Acute Coronary Syndromes. *Circ. Res.* **2019**, *124*, 150–160. [CrossRef]
45. Martinod, K.; Wagner, D.D. Thrombosis: Tangled up in NETs. *Blood* **2014**, *123*, 2768–2776. [CrossRef]
46. Fuchs, T.A.; Brill, A.; Duerschmied, D.; Schatzberg, D.; Monestier, M.; Myers, D.D.; Wrobleski, S.K.; Wakefield, T.W.; Hartwig, J.H.; Wagner, D.D. Extracellular DNA traps promote thrombosis. *Proc. Natl. Acad. Sci. USA* **2010**, *107*, 15880–15885. [CrossRef]
47. Ulrichts, H.; Vanhoorelbeke, K.; Girma, J.P.; Lenting, P.J.; Vauterin, S.; Deckmyn, H. The von Willebrand factor self-association is modulated by a multiple domain interaction. *J. Thromb. Haemost.* **2005**, *3*, 552–561. [CrossRef]
48. Alexander-Katz, A.; Schneider, M.F.; Schneider, S.W.; Wixforth, A.; Netz, R.R. Shear-Flow-Induced Unfolding of Polymeric Globules. *Phys. Rev. Lett.* **2006**, *97*, 138101. [CrossRef]
49. Sing, C.E.; Alexander-Katz, A. Elongational flow induces the unfolding of von Willebrand factor at physiological flow rates. *Biophys. J.* **2010**, *98*, L35–L37. [CrossRef]
50. Fu, H.; Jiang, Y.; Yang, D.; Scheiflinger, F.; Wong, W.P.; Springer, T.A. Flow-induced elongation of von Willebrand factor precedes tension-dependent activation. *Nat. Commun.* **2017**, *8*, 324. [CrossRef]
51. Savage, B.; Saldívar, E.; Ruggeri, Z.M. Initiation of platelet adhesion by arrest onto fibrinogen or translocation on von Willebrand factor. *Cell* **1996**, *84*, 289–297. [CrossRef]
52. Chen, Y.; Ju, L.A.; Zhou, F.; Liao, J.; Xue, L.; Su, Q.P.; Jin, D.; Yuan, Y.; Lu, H.; Jackson, S.P.; et al. An integrin αIIbβ3 intermediate affinity state mediates biomechanical platelet aggregation. *Nat. Mater.* **2019**, *18*, 760–769. [CrossRef]
53. Spinthakis, N.; Farag, M.; Gue, Y.X.; Srinivasan, M.; Wellsted, D.M.; Gorog, D.A. Effect of P2Y12 inhibitors on thrombus stability and endogenous fibrinolysis. *Thromb. Res.* **2019**, *173*, 102–108. [CrossRef]
54. Holme, P.A.; Ørvim, U.; Hamers, M.J.A.G.; Solum, N.O.; Brosstad, F.R.; Barstad, R.M.; Sakariassen, K.S. Shear-Induced Platelet Activation and Platelet Microparticle Formation at Blood Flow Conditions as in Arteries With a Severe Stenosis. *Arterioscler. Thromb. Vasc. Biol.* **1997**, *17*, 646–653. [CrossRef]
55. Sakariassen, K.S.; Holme, P.A.; Ørvim, U.; Marius Barstad, R.; Solum, N.O.; Brosstad, F.R. Shear-induced platelet activation and platelet microparticle formation in native human blood. *Thromb. Res.* **1998**, *92*, S33–S41. [CrossRef]
56. Sheriff, J.; Tran, P.L.; Hutchinson, M.; DeCook, T.; Slepian, M.J.; Bluestein, D.; Jesty, J. The platelet hammer: In vitro platelet activation under repetitive hypershear. In Proceedings of the 2015 37th Annual International Conference of the IEEE Engineering in Medicine and Biology Society (EMBC), Milano, Italy, 25–29 August 2015; Volume 2015, pp. 262–265.
57. Roka-Moiia, Y.; Walk, R.; Palomares, D.E.; Ammann, K.R.; Dimasi, A.; Italiano, J.E.; Sheriff, J.; Bluestein, D.; Slepian, M.J. Platelet Activation via Shear Stress Exposure Induces a Differing Pattern of Biomarkers of Activation versus Biochemical Agonists. *Thromb. Haemost.* **2020**, *120*, 776–792. [CrossRef]

58. Ikeda, Y.; Handa, M.; Kawano, K.; Kamata, T.; Murata, M.; Araki, Y.; Anbo, H.; Kawai, Y.; Watanabe, K.; Itagaki, I. The role of von Willebrand factor and fibrinogen in platelet aggregation under varying shear stress. *J. Clin. Investig.* **1991**, *87*, 1234–1240. [CrossRef]

59. Kulkarni, S.; Dopheide, S.M.; Yap, C.L.; Ravanat, C.; Freund, M.; Mangin, P.; Heel, K.A.; Street, A.; Harper, I.S.; Lanza, F.; et al. A revised model of platelet aggregation. *J. Clin. Investig.* **2000**, *105*, 783–791. [CrossRef]

60. Zhang, W.; Deng, W.; Zhou, L.; Xu, Y.; Yang, W.; Liang, X.; Wang, Y.; Kulman, J.D.; Zhang, X.F.; Li, R. Identification of a juxtamembrane mechanosensitive domain in the platelet mechanosensor glycoprotein Ib-IX complex. *Blood* **2015**, *125*, 562–569. [CrossRef]

61. Ruggeri, Z.M.; Orje, J.N.; Habermann, R.; Federici, A.B.; Reininger, A.J. Activation-independent platelet adhesion and aggregation under elevated shear stress. *Blood* **2006**, *108*, 1903–1910. [CrossRef]

62. Sheriff, J.; Bluestein, D.; Girdhar, G.; Jesty, J. High-Shear Stress Sensitizes Platelets to Subsequent Low-Shear Conditions. *Ann. Biomed. Eng.* **2010**, *38*, 1442–1450. [CrossRef]

63. Nesbitt, W.S.; Westein, E.; Tovar-Lopez, F.J.; Tolouei, E.; Mitchell, A.; Fu, J.; Carberry, J.; Fouras, A.; Jackson, S.P. A shear gradient-dependent platelet aggregation mechanism drives thrombus formation. *Nat. Med.* **2009**, *15*, 665–673. [CrossRef]

64. Rana, A.; Westein, E.; Niego, B.; Hagemeyer, C.E. Shear-Dependent Platelet Aggregation: Mechanisms and Therapeutic Opportunities. *Front. Cardiovasc. Med.* **2019**, *6*, 141. [CrossRef]

65. Kageyama, S.; Yamamoto, H.; Nakazawa, H.; Matsushita, J.; Kouyama, T.; Gonsho, A.; Ikeda, Y.; Yoshimoto, R. Pharmacokinetics and pharmacodynamics of AJW200, a humanized monoclonal antibody to von Willebrand factor, in monkeys. *Arterioscler. Thromb. Vasc. Biol.* **2002**, *22*, 187–192. [CrossRef]

66. Fox, J. The platelet cytoskeleton. *Thromb. Haemost.* **1993**, *70*, 884–893. [CrossRef]

67. Shattil, S.J.; Kashiwagi, H.; Pampori, N. Integrin Signaling: The Platelet Paradigm. *Blood* **1998**, *91*, 2645–2657. [CrossRef]

68. Weyrich, A.S.; Denis, M.M.; Schwertz, H.; Tolley, N.D.; Foulks, J.; Spencer, E.; Kraiss, L.W.; Albertine, K.H.; McIntyre, T.M.; Zimmerman, G.A. mTOR-dependent synthesis of Bcl-3 controls the retraction of fibrin clots by activated human platelets. *Blood* **2007**, *109*, 1975–1983. [CrossRef]

69. Samson, A.L.; Alwis, I.; Maclean, J.A.A.; Priyananda, P.; Hawkett, B.; Schoenwaelder, S.M.; Jackson, S.P. Endogenous fibrinolysis facilitates clot retraction in vivo. *Blood* **2017**, *130*, 2453–2462. [CrossRef]

70. Lam, W.A.; Chaudhuri, O.; Crow, A.; Webster, K.D.; Li, T.-D.; Kita, A.; Huang, J.; Fletcher, D.A. Mechanics and contraction dynamics of single platelets and implications for clot stiffening. *Nat. Mater.* **2011**, *10*, 61–66. [CrossRef]

71. Weisel, J.W. Biophysics. Enigmas of blood clot elasticity. *Science* **2008**, *320*, 456–457. [CrossRef] [PubMed]

72. Kunitada, S.; FitzGerald, G.; Fitzgerald, D. Inhibition of clot lysis and decreased binding of tissue-type plasminogen activator as a consequence of clot retraction. *Blood* **1992**, *79*, 1420–1427. [CrossRef] [PubMed]

73. Tutwiler, V.; Peshkova, A.D.; Le Minh, G.; Zaitsev, S.; Litvinov, R.I.; Cines, D.B.; Weisel, J.W. Blood clot contraction differentially modulates internal and external fibrinolysis. *J. Thromb. Haemost.* **2019**, *17*, 361–370. [CrossRef] [PubMed]

74. Komáromi, I.; Bagoly, Z.; Muszbek, L. Factor XIII: Novel structural and functional aspects. *J. Thromb. Haemost.* **2011**, *9*, 9–20. [CrossRef] [PubMed]

75. Durda, M.A.; Wolberg, A.S.; Kerlin, B.A. State of the art in factor XIII laboratory assessment. *Transfus. Apher. Sci.* **2018**, *57*, 700–704. [CrossRef]

76. Kurniawan, N.A.; Grimbergen, J.; Koopman, J.; Koenderink, G.H. Factor XIII stiffens fibrin clots by causing fiber compaction. *J. Thromb. Haemost.* **2014**, *12*, 1687–1696. [CrossRef]

77. Byrnes, J.R.; Wolberg, A.S. Newly-Recognized Roles of Factor XIII in Thrombosis. *Semin. Thromb. Hemost.* **2016**, *42*, 445–454. [CrossRef]

78. Van Giezen, J.; Minkema, K.; Bouma, B.; Jansen, J. Cross-linking of alpha 2-antiplasmin to fibrin is a key factor in regulating blood clot lysis: Species differences. *Blood Coagul. Fibrinolysis* **1993**, *4*, 869–875. [CrossRef]

79. Reed, G.L.; Houng, A.K. The contribution of activated factor XIII to fibrinolytic resistance in experimental pulmonary embolism. *Circulation* **1999**, *99*, 299–304. [CrossRef]

80. Jansen, J.W.C.M.; Haverkate, F.; Koopman, J.; Nieuwenhuis, H.K.; Kluft, C.; Boschman, T.A.C. Influence of Factor XIIIa Activity on Human Whole Blood Clot Lysis In Vitro. *Thromb. Haemost.* **1987**, *57*, 171–175. [CrossRef]

81. Fraser, S.R.; Booth, N.A.; Mutch, N.J. The antifibrinolytic function of factor XIII is exclusively expressed through α2-antiplasmin cross-linking. *Blood* **2011**, *117*, 6371–6374. [CrossRef]

82. Brinkmann, V.; Reichard, U.; Goosmann, C.; Fauler, B.; Uhlemann, Y.; Weiss, D.S.; Weinrauch, Y.; Zychlinsky, A. Neutrophil extracellular traps kill bacteria. *Science* **2004**, *303*, 1532–1535. [CrossRef]

83. Fuchs, T.A.; Bhandari, A.A.; Wagner, D.D. Histones induce rapid and profound thrombocytopenia in mice. *Blood* **2011**, *118*, 3708–3714. [CrossRef]

84. Ammollo, C.T.; Semeraro, F.; Xu, J.; Esmon, N.L.; Esmon, C.T. Extracellular histones increase plasma thrombin generation by impairing thrombomodulin-dependent protein C activation. *J. Thromb. Haemost.* **2011**, *9*, 1795–1803. [CrossRef]

85. Barranco-Medina, S.; Pozzi, N.; Vogt, A.D.; Di Cera, E. Histone H4 promotes prothrombin autoactivation. *J. Biol. Chem.* **2013**, *288*, 35749–35757. [CrossRef]

86. Von Brühl, M.-L.; Stark, K.; Steinhart, A.; Chandraratne, S.; Konrad, I.; Lorenz, M.; Khandoga, A.; Tirniceriu, A.; Coletti, R.; Köllnberger, M.; et al. Monocytes, neutrophils, and platelets cooperate to initiate and propagate venous thrombosis in mice in vivo. *J. Exp. Med.* **2012**, *209*, 819–835. [CrossRef]

87. Brinkmann, V.; Zychlinsky, A. Neutrophil extracellular traps: Is immunity the second function of chromatin? *J. Cell Biol.* **2012**, *198*, 773–783. [CrossRef]

88. Longstaff, C.; Varjú, I.; Sótonyi, P.; Szabó, L.; Krumrey, M.; Hoell, A.; Bóta, A.; Varga, Z.; Komorowicz, E.; Kolev, K. Mechanical stability and fibrinolytic resistance of clots containing fibrin, DNA, and histones. *J. Biol. Chem.* **2013**, *288*, 6946–6956. [CrossRef]

89. Varju, I.; Longstaff, C.; Szabo, L.; Farkas, A.Z.; Varga-Szabo, V.J. DNA, histones and neutrophil extracellular traps exert anti-fibrinolytic effects in a plasma environment. *Coagul. Fibrinolysis* **2015**, *113*, 1289–1298. [CrossRef]

90. Barbosa da Cruz, D.; Helms, J.; Aquino, L.R.; Stiel, L.; Cougourdan, L.; Broussard, C.; Chafey, P.; Riès-Kautt, M.; Meziani, F.; Toti, F.; et al. DNA-bound elastase of neutrophil extracellular traps degrades plasminogen, reduces plasmin formation, and decreases fibrinolysis: Proof of concept in septic shock plasma. *FASEB J.* **2019**, *33*, 14270–14280. [CrossRef]

91. Ducroux, C.; Di Meglio, L.; Loyau, S.; Delbosc, S.; Boisseau, W.; Deschildre, C. Thrombus Neutrophil Extracellular Traps Content Impair tPA-Induced Thrombolysis in Acute Ischemic Stroke. *Stroke* **2018**, *49*, 754–757. [CrossRef]

92. Ząbczyk, M.; Natorska, J.; Janion-Sadowska, A.; Malinowski, K.P.; Janion, M.; Undas, A. Elevated Lactate Levels in Acute Pulmonary Embolism Are Associated with Prothrombotic Fibrin Clot Properties: Contribution of NETs Formation. *J. Clin. Med.* **2020**, *9*, 953. [CrossRef] [PubMed]

93. Maxwell, M.J.; Westein, E.; Nesbitt, W.S.; Giuliano, S.; Dopheide, S.M.; Jackson, S.P. Identification of a 2-stage platelet aggregation process mediating shear-dependent thrombus formation. *Blood* **2007**, *109*, 566–576. [CrossRef] [PubMed]

94. Shi, X.; Yang, J.; Huang, J.; Long, Z.; Ruan, Z.; Xiao, B.; Xi, X. Effects of different shear rates on the attachment and detachment of platelet thrombi. *Mol. Med. Rep.* **2016**, *13*, 2447–2456. [CrossRef] [PubMed]

95. Stalker, T.J.; Traxler, E.A.; Wu, J.; Wannemacher, K.M.; Cermignano, S.L.; Voronov, R.; Diamond, S.L.; Brass, L.F. Hierarchical organization in the hemostatic response and its relationship to the platelet-signaling network. *Blood* **2013**, *121*, 1875–1885. [CrossRef]

96. Speich, H.E.; Bhal, V.; Houser, K.H.; Caughran, A.T.; Lands, L.T.; Houng, A.K.; Bäckstrom, J.; Enerbäck, M.; Reed, G.L.; Jennings, L.K. Signaling via P2Y12 may be critical for early stabilization of platelet aggregates. *J. Cardiovasc. Pharmacol.* **2014**, *63*, 520–527. [CrossRef]

International Journal of
Molecular Sciences

MDPI

Review

Plasminogen Activators in Neurovascular and Neurodegenerative Disorders

Manuel Yepes [1,2,3,*], Yena Woo [2] and Cynthia Martin-Jimenez [2]

[1] Department of Neurology, Emory University, Atlanta, GA 30329, USA
[2] Neuropharmacology and Neurological Diseases Section, Yerkes National Primate Research Center, Atlanta, GA 30329, USA; yena.woo@emory.edu (Y.W.); cynthia.alexandra.martin-jimenez@emory.edu (C.M.-J.)
[3] Department of Neurology, Veterans Affairs Medical Center, Decatur, GA 30329, USA
[*] Correspondence: myepes@emory.edu

Abstract: The neurovascular unit (NVU) is a dynamic structure assembled by endothelial cells surrounded by a basement membrane, pericytes, astrocytes, microglia and neurons. A carefully coordinated interplay between these cellular and non-cellular components is required to maintain normal neuronal function, and in line with these observations, a growing body of evidence has linked NVU dysfunction to neurodegeneration. Plasminogen activators catalyze the conversion of the zymogen plasminogen into the two-chain protease plasmin, which in turn triggers a plethora of physiological events including wound healing, angiogenesis, cell migration and inflammation. The last four decades of research have revealed that the two mammalian plasminogen activators, tissue-type plasminogen activator (tPA) and urokinase-type plasminogen activator (uPA), are pivotal regulators of NVU function during physiological and pathological conditions. Here, we will review the most relevant data on their expression and function in the NVU and their role in neurovascular and neurodegenerative disorders.

Keywords: tissue-type plasminogen activator (tPA); urokinase-type plasminogen activator (uPA); neurodegeneration

Citation: Yepes, M.; Woo, Y.; Martin-Jimenez, C. Plasminogen Activators in Neurovascular and Neurodegenerative Disorders. *Int. J. Mol. Sci.* **2021**, *22*, 4380. https://doi.org/10.3390/ijms22094380

Academic Editor: Hau C. Kwaan

Received: 18 March 2021
Accepted: 20 April 2021
Published: 22 April 2021

Publisher's Note: MDPI stays neutral with regard to jurisdictional claims in published maps and institutional affiliations.

1. Introduction

The plasminogen activation system is assembled by a cascade of proteases and their inhibitors that catalyze the conversion of the zymogen plasminogen into the two-chain protease plasmin (Figure 1). Plasminogen is a 90 kDa single-chain multidomain glycoprotein produced mainly in the liver [1] and assembled by 791 amino acids distributed in seven different structural domains: an N-terminal pre-activation peptide, 5 kringle domains and a C-terminal trypsin-like serine protease domain that harbors the catalytic triad His603, Asp646 and Ser741 [2]. Plasminogen binds to a plethora of highly heterogenous receptors on the cell surface, and this interaction not only triggers the generation of plasmin, but also activates cell signaling pathways that orchestrate a wide variety of functions including wound healing, angiogenesis, cell migration and inflammation [3]. Cleavage of plasminogen at the Arg561–Val562 bond by one of the two main plasminogen activators [tissue-type plasminogen activator (tPA) and urokinase-type plasminogen activator (uPA)] generates a two-chain plasmin molecule assembled by an N-terminal heavy-chain and a disulfide-linked C-terminal light chain containing the proteolytically active site. Importantly, the conversion of plasminogen into plasmin is enhanced when plasminogen is bound to fibrin or to the cell surface [4]. The generation of plasmin is tightly controlled at different steps of the plasminogen activating system. Accordingly, while plasminogen activator inhibitor-1 (PAI-1) and plasminogen activator inhibitor-2 (PAI-2) antagonize the proteolytic effect of tPA and uPA [5], α_2-antiplasmin inhibits plasmin. In the intravascular space, plasmin acts not only as an effector of the fibrinolytic system by degrading fibrin, but also as an immune regulator [6]. Likewise, on the cell surface plasmin triggers the degradation of multiple components of the extracellular matrix (ECM) and basement membrane, including collagen, vitronectin, laminin, fibronectin and proteoglycans.

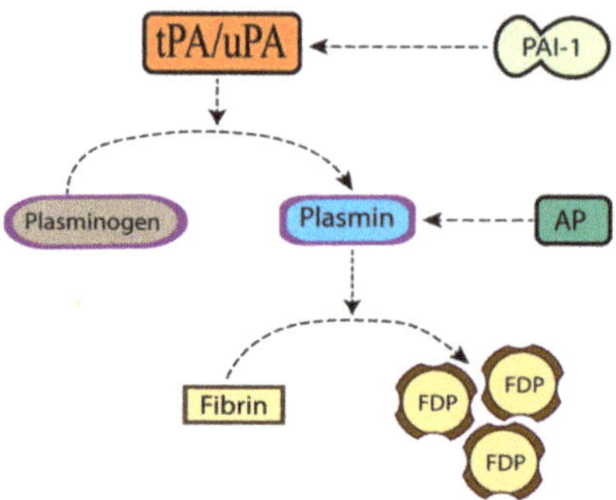

Figure 1. The plasminogen activating system. tPA: tissue-type plasminogen activator. uPA: urokinase-type plasminogen activator. AP: antiplasmin. PAI-1: plasminogen activator inhibitor-1. FDP: fibrin degradation products.

2. The Neurovascular Unit

The concept of the neurovascular unit (NVU) describes a dynamic interaction in the central nervous system between endothelial cells ensheathed by a basement membrane, and surrounding pericytes, astrocytes, microglia and neurons (Figure 2). The nature of the interplay between these cellular and non-cellular components has led to the proposal that the NVU is a single functioning unit responsible for the maintenance of cerebral hemostasis [7].

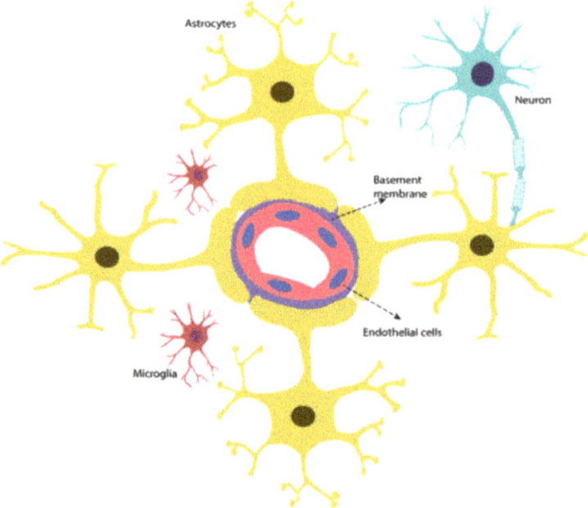

Figure 2. The neurovascular unit. Schematic representation of the cellular and non-cellular components of the neurovascular unit.

3. Plasminogen Activators in the Neurovascular Unit under Physiological Conditions

3.1. Tissue-Type Plasminogen Activator

Tissue-type plasminogen activator (tPA) is a 70-kDa serine proteinase secreted as a single-chain form (that upon its cleavage by plasmin at Arg275-Ile276 generates an active two-chain form held together by disulfide bonds). The tPA molecule is assembled by four domains: an amino-terminal region (fibronectin-like or finger domain), an EGF-like domain, two kringles and one serine protease region that harbors the active site residues His322,

Asp371 and Ser478 [8]. In the neurovascular unit (NVU), tPA is found in endothelial cells, perivascular astrocytes, microglia, pericytes and neurons [9,10].

3.2. Tissue-Type Plasminogen Activator in the Neurovascular Unit

3.2.1. Cerebral Endothelial Cells

In endothelial cells tPA is stored in Weibel–Palade bodies, the specialized endothelial storage granules for von Willebrand factor [11]. The expression of tPA is increased at the transcriptional level in endothelial cells by a variety of stimuli, including vascular endothelial growth factor (VEGF), fluid shear stress, thrombin and histamine [12–14]. In turn, its release from a preformed pool is triggered by physical activity, β-adrenergic drugs, cholinergic agents, disseminated intravascular coagulation and hypoxia [15,16]. Studies with a primary monoclonal antibody that detected free and PAI-1-complexed human tPA revealed that in the non-human primate brain, tPA is found in a reduced number of endothelial cells of the microvasculature, most of them pre-capillary arterioles and post-capillary veins [17]. However, despite the relevance of these data, it is important to consider that since this report was published almost 3 decades ago no further effort has been made to characterize the expression of tPA with newer antibodies in endothelial cells of the brain. Finally, although tPA has been detected in in vitro lines of human microvascular endothelial cells [18], no in vivo studies have been published describing the expression of tPA in blood vessels of the human brain.

3.2.2. Pericytes

Very few studies have assessed the expression and function of tPA in pericytes. However, the few that have been published to this date indicate that zinc prompts the release of tPA from these cells [19], and that pericytes negatively regulate fibrinolysis-triggered endothelial cell-derived tPA [20].

3.2.3. Astrocytes

tPA is abundantly found in astrocytes, and several stimuli including hypoxia [21] and mechanical injury [22] trigger its release, both in vivo and in vitro. In line with these observations, tPA activates the NF-κB pathway in astrocytes [23], and its release into the basement membrane increases the permeability of the NVU [21]. The mechanism whereby tPA activates the NF-κB pathway is not completely understood. However, work with cerebral cortical astrocytes and rat kidney interstitial fibroblasts (NRK-49F) [24] revealed that the interaction between tPA and LRP1 triggers the phosphorylation of IKKα, which then allows p65/p50 to translocate to the nucleus [25]. Further work with kidney cells has shown that another mechanism whereby tPA activates the NF-κB is by triggering annexin 2-mediated aggregation of the integrin CD11B, which in turn prompts the phosphorylation of IKβ with the resultant nuclear translocation of p65/p50 [26] (Figure 3). In addition to its proinflammatory effect, an increase in the expression and activity of astrocytic tPA induced by multipotent mesenchymal stromal cells has been associated with neurite growth [27]. Furthermore, it has been proposed that astrocytes recycle tPA released in the synaptic cleft in response to glutamatergic signals [28], and that tPA released from astrocytes modulates the microglial response to endotoxins [29]. Together, these data indicate that the release of astrocytic tPA plays a pivotal role in the NVU as a regulator of the permeability of the astrocyte–basement membrane–endothelial cell interface, synaptic transmission, neuroinflammation and microglial function.

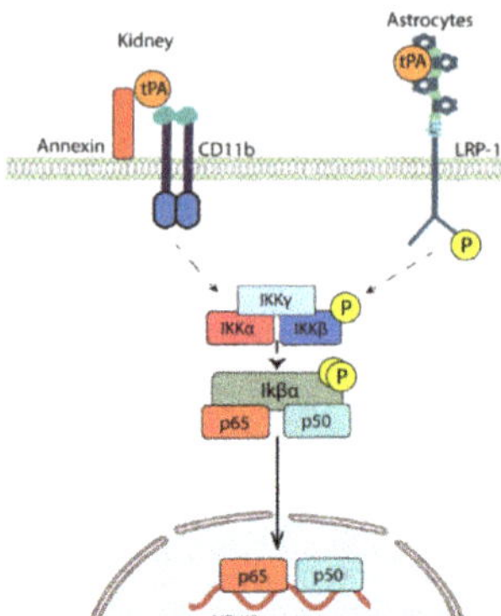

Figure 3. Mechanisms of tPA-induced NF-κB activation. Representative diagram of the proposed mechanisms whereby tPA activates the NF-κB pathway in the kidney and cerebral cortical astrocytes. In both cases, IKBα phosphorylation is followed by the nuclear translocation of p65/p50.

3.2.4. tPA in Microglia

Inasmuch as a functional link between tPA and microglial activation has been experimentally demonstrated, it is not clear if microglia release tPA. However, it has been proposed that injured neurons release tPA, and that this tPA triggers the release of microglial tPA, which in turn causes neurodegeneration [30]. Independently of these considerations, experimental evidence indicates that tPA mediates endotoxin- and kainic acid-induced microglial activation via an annexin II-mediated mechanism [31] that does not require plasmin generation [32] and triggers neuronal apoptosis [33].

3.2.5. Neuronal tPA

Neurons are a major reservoir of tPA in the central nervous system (CNS), and the release of neuronal tPA in the developing and mature brain plays a central role in the regulation of synaptic function and the response of the CNS to a variety of injuries. Indeed, the release of tPA by neuronal growth cones in the developing brain [34] induces neuronal migration [35] and neurite outgrowth and remodeling [36]. Noticeably, a similar sequence of events in the mature brain promotes neuronal recovery following an ischemic injury [37,38]. In contrast with the developing CNS, in situ zymography studies have revealed that only well-defined areas of the mature brain exhibit tPA-catalyzed proteolytic activity, namely the hippocampus, hypothalamus, thalamus, amygdala, cerebellum and meningeal blood vessels [39]. Furthermore, the interaction of tPA with *N*-methyl-D-aspartate (NMDA) receptors in these structures [40] regulates glutamatergic neurotransmission [41] and promotes the development of synaptic plasticity, as demonstrated in in vitro and in vivo models of long-term potentiation [42], learning [43,44], stress-induced anxiety [45] and visual cortex plasticity [37].

3.3. Urokinase-Type Plasminogen Activator

Urokinase-type plasminogen activator (uPA) is a 53 kDa serine proteinase secreted as a single-chain uPA (scuPA) with 411 amino acids assembled into three domains: an N-terminal domain homologous to the epidermal growth factor (responsible for its binding to uPAR), a kringle domain that interacts with plasminogen activator inhibitor-1 (PAI-1) and a C-terminal catalytic domain that harbors the active site with the His204, Asp255 and Ser356 amino acids triad [46]. The binding of scuPA to its receptor (uPAR) triggers its cleavage at K158-I159, thus prompting its conversion into an active two-chain form (tcuPA), with an A chain with the epidermal growth factor and kringle domains, and a B chain with the proteolytic domain [47]. In turn, tcuPA catalyzes the conversion of plasminogen into plasmin on the cell surface [48].

The receptor for uPA (uPAR) has 270 amino acids assembled into three cysteine-rich Cd59-like sequence domains (D1, D2 and D3) connected by short linker regions and bound to the surface of the plasma membrane by a glycosyl phosphatidylinositol (GPI) tail. Regulation of uPAR is accomplished by either an inactivating uPA-induced cleavage of the receptor between D1 and D2, or by endocytosis of a PAI-1–uPA–uPAR low-density lipoprotein receptor-related protein-1 (LRP1) complex assembled on the cell surface, which then recycles free uPAR back to the membrane to bind to more uPA [49].

3.4. Urokinase-Type Plasminogen Activator in the Neurovascular Unit (NVU)
3.4.1. Cerebral Endothelial Cells

A substantial body of experimental evidence indicates that uPA and uPAR are found in endothelial cells [50], and that the release of this uPA and the increase in the expression of uPAR in endothelial cells triggered by a variety of stimuli [51,52] induces cell migration, angiogenesis [53] and capillary branching. However, it is important to take into account that most of these studies have been performed with cell lines, and that very few studies have assessed the in vivo expression of uPA in cerebral endothelial cells. With that in mind, it has been reported that *Cryptococcus neoformans* increases the expression of uPA in cerebral endothelial cells [54], and that microvascular endothelial cells upregulate uPA following an ischemic injury to the spinal cord in vivo [55].

3.4.2. Astrocytes and Microglia

The abundance of uPA and uPAR is increased in glial cell tumors, particularly glioblastoma multiforme, where they have been reported to play a role in tumor growth [56]. In contrast, the role of astrocytic uPA and uPAR under physiological conditions is less well understood. However, recent studies indicate that the release of uPA under physiological conditions triggers astrocytic activation [57] and induces the formation of peripheral astrocytic processes [58]. The expression and role of microglial uPA and uPAR under non-pathological conditions is largely unknown, although in vitro studies have shown that endotoxins, kainic acid and neurogeneration increase their abundance in microglia [59].

3.4.3. Neurons

Despite the fact that uPA and uPAR are abundantly found in developing neurons [60–62], their expression changes dramatically over a few days. Hence, while day in vitro (DIV) 3 neurons harbor uPAR in their cell body and neurites, at DIV 7 the expression of this receptor shifts to the axon shaft and growth cones, and at DIV 16 is restricted to the distal segment of some axons and very few growth cones [60]. Significantly, uPA/uPAR binding during the early stages of development induces neuritogenesis and neuronal migration via a plethora of mechanisms that do not always require plasmin generation [63,64]. More specifically, by promoting activation of integrins and the focal kinase adhesion (FAK) pathway, uPAR regulates the reorganization of the cytoskeleton [63], thus triggering axonal growth, neuronal migration [65] and dendritic branching [66]. In line with these observations, uPAR seems to be pivotal for the formation of neuronal circuits that underlie the development of language and cognition, and dysregulation of the uPA/uPAR system has been linked to epilepsy and cognitive and developmental disorders [67].

In contrast, the expression and role of uPA/uPAR in the mature brain have been less studied. However, recent studies with human, murine and non-human primate brains indicate that uPA is abundantly found in synapses of the second and fifth cortical layers of the cerebral cortex, and that uPA/uPAR binding modulates excitatory neurotransmission by regulating the synaptic expression of neuronal cadherin (NCAD) [68]. Additionally, these studies showed that uPA induces the expression of NCAD in the synapse, and that the resultant generation of NCAD-dimers leads to the formation of synaptic contacts in neurons maintained under physiological conditions [68].

4. Plasminogen Activators in the Neurovascular Unit (NVU) under Ischemic Conditions

Ischemic stroke is a leading cause of mortality and disability in the world [69]. Significantly, plasminogen activators are pivotal for the protection and repair of the NVU that has suffered an ischemic injury. Indeed, while acute cerebral ischemia causes the rapid release of tPA from each cellular compartment of the NVU [70], the abundance and activity of uPA increase only during the recovery stages from the ischemic insult [61]. These data have led to the proposal that while the early release of tPA restores the patency of the occluded blood vessel and has a neuroprotective effect, the delayed release of uPA promotes the repair of the damaged NVU. Below we will review data on the role of tPA and uPA in each component of the NVU under hypoxic/ischemic conditions.

4.1. Endothelial Cells

4.1.1. Tissue-Type Plasminogen Activator

The effect of hypoxia on the release of tPA from human cerebrovascular endothelial cells has been poorly characterized. However, studies with human saphenous and umbilical veins [71,72] have shown that hypoxia decreases the abundance and activity of tPA in endothelial cells, and that this effect is accompanied by an increase in the expression and activity of PAI-1. Furthermore, in vitro studies with rat brain microvascular endothelial cells indicate that tPA has a harmful effect on endothelial cells exposed to oxygen and glucose deprivation conditions [4,73]. In contrast with these in vitro studies, in vivo observations have revealed an increase in the concentrations of tPA and PAI-1 in the intravascular space of patients suffering an acute ischemic stroke [74], suggesting that endothelial cells release tPA into the intravascular space as an attempt to restore the patency of the occluded blood vessel. This is supported by the observation of complete or nearly complete recovery of neurological function in acute ischemic stroke patients intravascularly treated with recombinant tPA within 3–4.5 h of the onset of symptoms [75,76].

A growing body of experimental evidence indicates a link between plasminogen activation and the immune system. Indeed, while some studies have revealed that plasminogen activators play a role in bradykinin-mediated endothelial cell activation [77], others show that an interaction between plasmin and factor XII increases the permeability of the neurovascular unit in neurodegenerative diseases [78,79]. Importantly, in apparent discrepancy with a proinflammatory role of plasmin, in vivo studies with an animal model of cerebral ischemia suggest that tPA attenuates the activation of the immune response in the neurovascular unit that has suffered an ischemic injury [80].

4.1.2. Urokinase-Type Plasminogen Activator

The role of uPA in hypoxic/ischemic cerebral endothelial cells is even less well studied. Indeed, although studies with human umbilical endothelial cells (HUVEC) [81] and human microvascular endothelial cells [82] have revealed that a hypoxia-induced, hypoxia-inducible factor (HIF)-mediated increase in uPAR expression in endothelial cells triggers angiogenesis and cell migration [83], the effect of hypoxia on uPA and uPAR expression and function in cerebral endothelial cells has not been characterized. Independently of these considerations, clinical studies indicate that the intravascular administration of recombinant uPA effectively restores the patency of the occluded blood vessel and improves neurological outcome in acute ischemic stroke patients [84,85].

4.2. Astrocytes

4.2.1. Tissue-Type Plasminogen Activator

As discussed above, tPA is abundantly found in astrocytes [21], and the release of astrocytic tPA has a direct effect on the permeability of the NVU. Indeed, the interaction between tPA, released from perivascular astrocytes in response to the ischemic injury, and the low-density lipoprotein receptor-related protein-1 (LRP-1) found in astrocytic end-feet processes, activates an NF-κ-regulated inflammatory response [86] and triggers the detachment of perivascular astrocytes from the basement membrane, which in turn increases the permeability of the blood–brain barrier, thus causing ischemic cerebral edema [21]. In line with these observations, the intracerebroventricular administration of recombinant tPA induces an LRP-1-mediated increase in the permeability of the NVU [87]. The translational relevance of these observations is underscored by neuroradiological studies showing that the intravenous administration of recombinant tPA to acute ischemic stroke patients also increases the permeability of the blood–brain barrier [88] which is in line with a reported 10-fold increase in the risk of hemorrhagic complications in recombinant tPA (rtPA)-treated stroke patients [75]. Interestingly, besides its effect on the permeability of the NVU, experimental evidence indicates that multipotent mesenchymal stromal cell-induced tPA activity in astrocytes promotes neurorepair after stroke by facilitating neurite outgrowth in the ischemic area [27,89].

4.2.2. Urokinase-Type Plasminogen Activator

The roles of astrocytic uPA and uPAR in the ischemic brain have only recently been studied. Accordingly, it was reported that the binding of uPA released from neurons to uPAR recruited to the astrocytic plasma membrane in response to the ischemic injury, induces astrocytic activation [57]. Ezrin is a protein that regulates the reorganization of the actin cytoskeleton [90] and the formation of microvilli, filopodia and lamellipodia [91]. In the cytosol, ezrin exists in an inactive conformation. However, following its recruitment to the plasma membrane, ezrin is activated by phosphorylation at a conserved Thr567 residue [92]. Ezrin is abundantly found in astrocytic filopodia [93], and its activation is required for the formation of peripheral astrocytic processes [94]. Significantly, uPA induces the expression of ezrin in astrocytes, thus triggering the formation of peripheral astrocytic processes that, upon embracing the pre- and post-synaptic compartments, preserve the integrity of the tripartite synapse that has suffered an ischemic insult [58].

4.3. Microglia

4.3.1. Tissue-Type Plasminogen Activator

Microglial activation is a key step in a sequence of events that trigger not only cell death but also neurorepair in the ischemic brain [95]. Remarkably, tPA is pivotal for microglial activation [31], and in support of these observations, genetic deficiency of tPA attenuates cerebral ischemia-induced microglial activation [32]. Interestingly, the N-terminal fibronectin type III finger domain of tPA also mediates endotoxin-induced microglial activation, most likely by its interaction with annexin II on the cell membrane [96]. Further work has revealed that LRP-1 mediates the effect of tPA on microglial activation in the ischemic brain [97], and that the resultant downstream activation of latent platelet-derived growth factor-CC (PDGF-CC) increases the permeability of the NVU [98]. Additionally, it was reported that by modulating the release of cytokines, interferon-β attenuated the effect of tPA-induced microglial activation on the permeability of the NVU [99].

4.3.2. Urokinase-Type Plasminogen Activator

It has been recognized that cultured human microglia express uPAR [100], and that the abundance of this receptor in microglia is greatly increased by treatment with endotoxins. More importantly, experimental studies have shown that uPAR is able to induce microglial activation by a mechanism that always requires uPA [101], but that in some cell lines is mediated by MMP-9 [102]. Strikingly, despite the importance of these observations, the role of uPA/uPAR in cerebral ischemia-induced microglial activation is still poorly understood.

4.4. Neurons

4.4.1. Tissue-Type Plasminogen Activator

Despite the fact that a large number of studies agree on the fact that hypoxia and ischemia trigger the release of neuronal tPA [70,103–105], there is significant disagreement on the effect that this tPA has on cell survival. Indeed, results from early studies showing that mice genetically deficient in tPA (tPA$^{-/-}$) have a significant decrease in the volume of the ischemic lesion following transient occlusion of the middle cerebral artery (tMCAo) [104,106] seeded the idea that tPA has a neurotoxic effect in the ischemic brain. Strikingly, this idea lingered for a long time despite subsequent publications by other groups describing an increase in the volume of the ischemic lesion in tPA$^{-/-}$ mice [106], and either a beneficial [107] or even a lack of effect [108] of rtPA treatment on the volume of the ischemic lesion following tMCAo.

This discrepancy was dramatically brought to the forefront of the scientific discussion by the publication of a National Institute of Neurological Disorders and Stroke (NINDS)-led clinical study showing that treatment with recombinant tPA within 3 h of the onset of symptoms was associated with complete or nearly complete recovery in neurological function in a significant number of acute ischemic stroke patients [75,109], and by the subsequent incorporation of rtPA in the protocols used for the treatment of these patients [110]. Notably, although treatment with rtPA also increases the risk of intracerebral hemorrhage [109] and augments the permeability of the NVU [89], to this date no clinical study has shown a neurotoxic effect caused by rtPA treatment. The translational impact of this disagreement between basic and clinical researchers has been heightened by the observation that, following its intravenous administration, rtPA crosses through the blood–brain barrier and permeates the ischemic tissue [111]. In other words, if findings published by basic researchers are true, then clinicians are treating acute ischemic stroke patients with a neurotoxic agent. For obvious reasons this discrepancy needs to be resolved as it has called into question the clinical translatability of basic science research.

Early studies proposed that tPA mediated excitotoxin-induced neuronal death, which is a pivotal mechanism of cell death in the ischemic brain. Indeed, it was found that genetic deficiency of tPA attenuated kainic acid-induced hippocampal cell death [112] via plasmin-induced proteolysis of laminin in the extracellular matrix [113], and that tPA$^{-/-}$ mice were resistant to KA-induced seizures [112]. This study was followed by work from a different group of researchers that measured the volume of the ischemic lesion in rodents injected with NMDA into the striatum and then intravascularly treated with 10 mg/Kg/IV of rtPA [114]. These investigators found that rtPA treatment enhanced the harmful excitotoxic effect of NMDA, which was interpreted as another demonstration of a neurotoxic effect of tPA. In contrast, a different group of investigators using a similar experimental paradigm but a different dose of rtPA (0.9 mg/Kg/IV, the same dose used to treat acute ischemic stroke patients), found an opposite effect: a decrease in the volume of the necrotic lesion in rtPA-treated animals [115]. Furthermore, they also found that the damage induced by the intrastrial injection of NMDA was significantly attenuated in mice overexpressing tPA only in neurons. Additionally, it was soon clear that the intracerebral injection of an excitotoxin (kainic acid) caused a transient increase in the activity of tPA in cells of the hippocampal CA1 layer that survived the excitotoxic injury [116], and this was followed by a report indicating that tPA protected hippocampal cells from the harmful effects of the excitotoxic injury [117].

This led a different group of investigators to quantify neuronal survival in cerebral cortical neurons incubated with NMDA in the presence of 0–500 nM of either proteolytically active tPA or a mutant of tPA with an alanine for serine substitution at the active site Ser481 that rendered it unable to catalyze the conversion of plasminogen into plasmin (proteolytically inactive tPA) [115]. These experiments revealed that tPA caused a modest increase in NMDA-induced neuronal death only at doses greater than 100 nM, which are not found in in vivo systems, even after the intravenous administration of rtPA. Furthermore, it was discovered that at concentrations found in the ischemic brain, tPA attenuated NMDA-induced neuronal death by a mechanism that did not entail plasmin generation but required the co-receptor function of a member of the low-density lipoprotein receptor (LDLR) family, most likely LRP1. In an attempt to explain these discrepancies, it was proposed that selective activation of NMDA receptors by single-chain but not two-chain tPA is responsible for the neurotoxic effect of tPA [118], and therefore that treatment with two-chain tPA is more efficient than single-chain tPA to reduce the volume of the ischemic lesion and promote functional recovery after the experimental induction of an ischemic stroke [119]. Together, these results show that a causal link between tPA and cerebral ischemia- and excitotoxin-induced neuronal death was difficult to establish, as it seemed to depend on the chemical structure and dose of rtPA as well as the specific experimental paradigm used in each report.

The resultant renewed interest of the scientific community to understand the role of neuronal tPA in the ischemic brain led a group of investigators using an in vitro model of oxygen and glucose deprivation (OGD) to discover that treatment with 5 nM of rtPA prevented cell death in cerebral cortical neurons exposed to 55 min of OGD conditions, and that this effect was mediated by LRP1 and open synaptic NMDA receptors [104]. Remarkably, the detection of a maximal neuroprotective effect within the first three hours after OGD bears a notable resemblance with the maximal neurological recovery observed in acute ischemic stroke patients treated with rtPA within three hours of the onset of symptoms [75]. The obvious lack of a clot in this in vitro system indicated that a mechanism other than thrombolysis mediates tPA's neuroprotective effect, and this possibility was confirmed by the finding that treatment with recombinant tPA after tMCAo also decreased the volume of the ischemic lesion in animals genetically deficient in plasminogen ($Plg^{-/-}$). These data indicate that tPA has a neuroprotective effect in the ischemic brain that is not mediated by the generation of plasmin and instead requires the co-receptor function of the NMDAR and a member of the LDLR family.

4.4.2. Urokinase-Type Plasminogen Activator

The role of uPA in the ischemic NVU is less well understood. Indeed, early studies with an animal model of permanent cerebral ischemia induced by occluding a distal branch of the middle cerebral artery with a surgical suture showed a decrease in the volume of the ischemic lesion in mice genetically deficient in uPAR [105] but not uPA [120]. Interestingly, using a similar animal model of cerebral ischemia, a different group of investigators detected a large increase in uPA-catalyzed proteolysis 72 h after the onset of the ischemic injury [121]. This was followed by the observation that the concentrations of uPA in the culture medium of cerebral cortical neurons remained unchanged during 60 min of exposure to OGD conditions [61].

However, in an unexpected turn of events, it was found that these neurons released large amounts of uPA after they were returned to normoxic conditions. Importantly, this uPA did not seem to have an effect on cell death, as there was no difference in neuronal survival between Wt and $uPA^{-/-}$ neurons exposed to OGD conditions [61]. The in vivo significance of these observations was supported by the finding that although cerebral ischemia did not have an effect on the abundance of uPA during the acute phase of the ischemic injury, the expression of uPA in the ischemic tissue increased during the recovery period.

The finding that the delayed release of uPA following a hypoxic/ischemic injury did not have an effect on neuronal survival or the volume of the ischemic lesion, led researchers to investigate if uPA plays a role in neurorepair. Noticeably, this possibility was supported by the observation that compared to wild-type (Wt) littermate controls, $uPA^{-/-}$ and $uPAR^{-/-}$ mice had a protracted recovery in neurological function following tMCAo, and that treatment with ruPA or the release of endogenous uPA prompted functional recovery in Wt and $uPA^{-/-}$, but not in $uPAR^{-/-}$ mice [60,61].

Further studies showed that the release of uPA promoted the recovery of axonal boutons and post-synaptic terminals disassembled by the ischemic injury. More specifically, it was found that by regulating the expression and activity of GAP-43, neuronal uPA promoted the regeneration of axons damaged by the ischemic injury [122]. Furthermore, by its ability to regulate the expression of ezrin, uPA was able to reorganize the cytoskeleton of the post-synaptic density, prompting the recovery of dendritic spines that disappeared in the earlier stages of the ischemic insult [61]. In line with these observations, in vivo studies indicated that intravenous treatment with recombinant uPA 24 h after the onset of the ischemic injury increased the number of synaptic contacts in the area that surrounds the necrotic core [57].

In summary, the data available to this date indicate that the expression of uPA and uPAR increases in the recovery stages of an ischemic stroke, and suggest that uPA binding to uPAR plays a central role in the process of neurorepair following an acute ischemic injury. These observations are supported by reports from other groups indicating that uPAR modulates peripheral nerve regeneration after a crushed nerve [67], and that genetic deficiency of uPA aggravates the motor deficit and increases neuronal death in an animal model of traumatic brain injury [121].

5. Plasminogen Activators in Neurodegenerative Disorders

The concept of neurodegenerative disorders encompasses several clinical entities including Alzheimer's disease (AD), Parkinson's disease (PD) and amyotrophic lateral sclerosis (ALS), all characterized by the progressive decline of neuronal function. Remarkably, a rapidly growing knowledge of the pathophysiology of these disorders has led to two important conclusions. First, that they are not caused by isolated neuronal pathology, but instead that a dysfunctional NVU is a contributory factor in many of them [122,123]; and second, that a dysfunctional plasminogen activating system plays a still poorly understood role in their pathophysiology. Together, the data reviewed below underscore the relevance of the interaction between the plasminogen activating system and neurodegeneration, and how research on this interaction may unveil potential targets for the development of strategies for their prevention and treatment.

5.1. Plasminogen Activators in Alzheimer's Disease

AD affects approximately 46.8 million people worldwide, and this number is expected to reach 131.5 million by 2050 [124]. It is a dual proteinopathy, that accounts for almost 60–80% of all dementias, and is characterized by the extracellular deposition of Aβ 1–40 and 1–42 fibrils in neuritic plaques and intracellular aggregates of hyperphosphorylated tau in neurofibrillary tangles (NFT). Importantly, a substantial number of studies have found that even in the early stages of this disease the NVU is dysfunctional. Accordingly, a long-time accepted neurocentric theory of the genesis of AD has slowly been integrated into a more holistic model that includes all the cellular and non-cellular components of the NVU.

5.2. Endothelial Cells

There is ample evidence implicating endothelial cell dysfunction in the pathophysiology of AD. Indeed, virtually all AD patients exhibit endothelial cell degeneration and abnormal thickening of the perivascular basement membrane in zones with Aβ deposition [125]. These morphological changes underlie the reduction in cerebral blood flow, and impaired cerebrovascular reactivity and neurovascular coupling observed even in early stages of the disease [126,127]. Importantly, the few studies published to this date on plasminogen activators and endothelial cells in AD indicate that although Aβ does not have an effect on the release of endothelial tPA [19], deficiency of this plasminogen activator, likely caused by increased PAI-1, underlies the impairment in neurovascular coupling observed in mice expressing the Swedish mutation of the amyloid precursor protein (APP; tg2576) [128]. It has also been postulated that plasminogen derived from the intravascular space causes an inflammatory response and Aβ deposition. More specifically, it has been reported that depletion of plasminogen in the intravascular space attenuates microglial activation and improves AD pathology in mice transgenic for human APP/Presinilin 1 with five early-onset familial AD mutations [78]. In contrast with these studies, the role of uPA in endothelial cell dysfunction in AD has been addressed by fewer investigators. However, it has been reported that Aβ induces the expression of uPA in cultured human cerebrovascular smooth muscle cells [129], and that LRP1 in endothelial cells regulates the efflux of Aβ into the intravascular space [130]. The translational relevance of these observations, performed in the murine brain, should be understood in the light of studies indicating that human brain endothelial cells do not express LRP1 [131].

5.3. Astrocytes

Several studies have reported an association between early astrocytic activation [132] and poor prognosis in advanced phases of this disease [133]. Interestingly, this process seems to affect only a sub-population of astrocytes with an increased abundance of aquaporin-4 in their end-feet processes [134]. This is of special interest, because the interaction between astrocytic end-feet processes and endothelial cells modulates the permeability of the NVU [135]. In line with these observations, in vitro and in vivo studies have revealed an increase in the permeability of the NVU in different animal models of AD [129] and in the brain of AD patients [136]. Finally, astrocytes are tightly associated with Aβ catabolism, and these cells display an abnormal response upon exposure to Aβ [137]. Strikingly, no studies have directly addressed the specific role of astrocytic tPA and uPA in the pathogenesis of Alzheimer's disease.

5.4. Microglia

Microglial activation in the brain of AD patients [138] has been linked to the triggering of a neuroinflammatory response that promotes Aβ-containing plaque formation and neuronal degeneration [139]. However, more recent studies have revealed that microglial activation in AD is a heterogenous process, and in line with these observations, specific and well-differentiated subpopulations of microglia also seem to have a protective effect in the brain of AD patients [140]. Furthermore, it has been reported that the plasminogen activating system modulates the activation of microglia in AD [141] and that treatment with rtPA triggers the activation of the above-mentioned neuroprotective microglial phenotype [142]. In contrast, the role of uPA in microglial activation in AD has been less well studied. However, it has been reported that Aβ-treated human microglia upregulate uPA and uPAR [143] and that uPAR is a marker of microglial activation in the brain of AD patients [144].

5.5. Neurons

The amyloid hypothesis of AD is a neurocentric model in which Aβ deposition leads to progressive tau hyperphosphorylation, synaptic dysfunction and neuronal loss. However, despite its importance and long-time acceptance, a growing body of experimental evidence suggests that not only neurons, but all cellular and non-cellular elements of the NVU, play key roles in the pathogenesis of this disease [145]. Independently of these considerations, knowledge gathered over the last 3 decades has resulted in a better understanding of the biochemical process that leads to the production and accumulation of Aβ. More specifically, the proteolytic processing of APP by α-secretase on the cell membrane generates soluble APPα, which has been implicated in neuronal plasticity and synaptogenesis [146]. However, those APP molecules that are not processed by α-secretase are endocytosed and cleaved by β-secretase 1 (BACE1) and γ-secretase to generate Aβ 1–40 and 1–42 peptides [147–149], which have a harmful effect on cell survival and synaptic structure and function [150,151]. Inasmuch as our understanding of this process has grown, it has led to the discovery of different therapeutic strategies to minimize Aβ deposition that have been successfully tested in animal models of AD [152], but unsuccessful [153] or only partially effective in AD patients [154].

The last three decades of research on the role of the plasminogen activating system in the pathophysiology of AD have focused almost exclusively on the ability of tPA and uPA to cleave Aβ deposits. However, recent studies have also discovered a role for uPA in the pathogenesis of synaptic dysfunction in AD that does not require the conversion of plasminogen into plasmin.

5.5.1. Plasminogen Activators and the Formation of Aβ Deposits

A role for plasminogen activators in the pathogenesis of AD was suggested by in vitro studies showing that plasmin triggers α-secretase-induced cleavage of Aβ in lipid rafts [155] and cleaves insoluble Aβ fibrils [156,157]. This was followed by work revealing a reduction in the expression and activity of plasmin in AD brains [158], most likely due to a decrease in tPA activity [157]. These observations were contested by a different group of investigators that detected normal concentrations of plasminogen and plasmin in the brains of AD patients [159], and postulated that the reported decrease in plasmin was actually due to the disruption of lipid rafts by abnormal cholesterol metabolism in the neuronal membrane [160].

Most of the studies on the role of plasminogen activators on the formation of Aβ deposits have been performed with tPA. Hence, it has been found that insoluble Aβ activates tPA [161] and increases the expression of tPA mRNA in cerebral cortical neurons, purportedly as an attempt to trigger plasmin-induced cleavage of extracellular insoluble Aβ-containing plaques [158]. In discrepancy with these studies, in vivo studies with mouse models of AD have revealed that chronic elevation of Aβ actually decreases tPA activity by enhancing the inhibitory effect of PAI-1 on tPA, and that the intracerebral injection of Aβ causes neuronal degeneration in animals genetically deficient in either tPA or plasminogen [162].

More specifically, the expression of PAI-1 is increased in the cerebrospinal fluid [163] and the brains of AD patients [164]. The clinical relevance of these observations is underscored by experimental work indicating that the genetic deletion of PAI-1 in the brain of a murine model of AD reduces the deposition of Aβ by increasing tPA-induced plasmin-mediated cleavage of Aβ-containing plaques [164]. Together, these data have led to the proposal of a model in which increased PAI-1 activity in the brain of AD patients abrogates tPA-induced plasmin-triggered cleavage of Aβ deposits. In seeming contradiction with a protective role of tPA in AD, other studies have shown that this plasminogen activator actually mediates the neurotoxic effect of Aβ via ERK $\frac{1}{2}$ activation [165]. The role of uPA on the formation of Aβ deposits has been less well studied. Nevertheless, it has been reported that Aβ increases the abundance of uPA mRNA [156], and that as described for tPA, uPA also induces plasmin-mediated cleavage of insoluble Aβ-containing extracellular plaques [166].

5.5.2. Role of Plasminogen Activators in Synaptic Dysfunction in AD

The idea that the extracellular accumulation of insoluble Aβ peptides is the cause of the cognitive decline observed in AD patients [167] has been challenged by neuropathological and clinical studies indicating that the development of cognitive deficits in these individuals correlates with abnormalities in synaptic structure and function more than with the number of tangles and insoluble Aβ-containing plaques [168,169]. This concept is of significant translational importance, because synaptic dysfunction is an early event in the pathogenesis of AD that is amenable to therapeutic interventions to prevent its development.

Our knowledge of the synaptic role of Aβ has increased significantly during the last 30 years. Hence, we know that the production of Aβ increases during neuronal activity [150], and that while at low concentrations, soluble Aβ induces presynaptic facilitation, at high concentrations triggers post-synaptic depression [170] by decreasing the abundance of glutamatergic receptors in the postsynaptic density [161,171] and enhances the excitotoxic effect of glutamate by blocking its reuptake from the synaptic cleft [172]. Additionally, high concentrations of Aβ impair long-term potentiation (LTP) and enhance long-term depression (LTD) by blocking α-amino-3-hydroxy-5-methyl-4-isoxazolepropionic acid (AMPA) and NMDA receptor function [173], and augment the excitotoxic effect of glutamate by blocking its reuptake from the synaptic cleft [172]. Importantly, increased soluble Aβ has been linked to synaptic depression and the disruption of neuronal network activity in the brains of AD patients [171]. Furthermore, it has been reported that uPA antagonizes the harmful effect of Aβ on synaptic structure and function by a mechanism independent of its ability to trigger proteolytic cleavage of Aβ-containing plaques [68].

Recent studies have shown that cleavage of Aβ-containing plaques is not the only role of uPA in AD brains. Hence, it has been found that the expression of uPA, but not of its receptor (uPAR), is decreased in the synapses of AD patients and 5XFAD mice (express human APP with the Swedish (KM670/671NL), Florida (I716V) and London (V717I) mutations together with a mutant presenilin 1 (M146L, L286V) under the control of the murine Thy-1 promoter), by the ability of Aβ to halt the transcription of uPA mRNA in neurons but not in astrocytes [68]. The translational importance of these findings is supported by observations indicating that treatment with recombinant uPA abrogates the harmful effect of soluble Aβ on synaptic structure and function, via its ability to induce the expression of neuronal cadherin (NCAD). Remarkably, in contrast with the reported effect of tPA and uPA on the proteolytic cleavage of Aβ-containing plaques, the effect of uPA on the synapses of AD patients and animal models of AD does not require the generation of plasmin.

5.5.3. Plasminogen Activators, Physical Activity and Alzheimer's Disease

Physical activity has a direct effect on the expression and activity of components of the plasminogen activating system. More specifically, 6 months of intensive physical activity increase the intravascular concentration of tPA and uPA, and this effect is accompanied by a substantial decrease in the levels of PAI-1 [15,174]. These observations are of significant importance when interpreted in the context of studies showing that physical activity decreases the risk of AD [175] and improves the attention span, memory and executive functioning of healthy individuals [176]. Thus, it is tempting to postulate that plasminogen activators mediate the protective effect of exercise on cognitive function and the risk of developing AD. However, although scientifically plausible, to this date there are no data to support this hypothesis.

5.6. Plasminogen Activators in Parkinson's Disease

Parkinsonism is a clinical syndrome characterized by bradykinesia, resting tremor, rigidity and postural and gait impairment. Most cases of parkinsonism are caused by Parkinson's disease (PD), which is a neurodegenerative disease that affects 3% of the population older than 65 years of age [177]. The neuropathological hallmarks of this disease are the loss of dopamine-containing neurons in the substantia nigra and the presence of α-synuclein-containing intracellular inclusions. The extracellular levels of α-synuclein depend not only on its release from neurons, but also on its removal by proteolytic degradation. It is unclear if α-synuclein induces the expression or activity of tPA or uPA. However, plasmin cleaves and degrades α-synuclein, and α-synuclein upregulates PAI-1 [178]. It has thus been proposed that an excess of PAI-1 in the brains of PD patients prevents plasmin-induced clearance of α-synuclein aggregates [179]. The translational relevance of these findings is supported by the fact that increased levels of PAI-1 have been linked to a worse clinical prognosis in PD patients [180]. Despite the importance of these observations, to this date it is unclear if tPA- or uPA-catalyzed plasmin generation plays a role in the pathogenesis of this disease.

5.7. Plasminogen Activators in Amyotrophic Lateral Sclerosis (ALS)

Amyotrophic lateral sclerosis (ALS) is a motor neuron disease characterized by a progressive decline in motor function caused by weakness and spasticity without sensory loss. Knowledge on the role of plasminogen activators in the pathogenesis of ALS is still in its earlier stages. However, it has been reported that plasminogen from ALS patients, and recombinant tPA and plasmin, induce motoneuron degeneration in BALB/c mice [181]. Likewise, experimental work with G93 mice (with a SOD1 mutation linked to familial ALS) and samples from ALS patients revealed that the abundance of uPAR increased in the ventral horn of the spinal cord of ALS patients and G93 mice, along with enhanced uPA-dependent plasminogen activation in advanced stages of this disease. Remarkably, treatment with an inhibitor of uPA prolonged survival in these animals [182].

Author Contributions: Writing, editing and funding acquisition, M.Y.; writing and investigation, Y.W.; writing—original draft preparation, C.M.-J. All authors have read and agreed to the published version of the manuscript.

Funding: This research was funded in part by National Institutes of Health Grant NS-NS091201 (to M.Y.) and VA MERIT Award IO1BX003441 (to M.Y.).

Data Availability Statement: Not Applicable.

Conflicts of Interest: The authors declare no conflict of interest.

References

1. Raum, D.; Marcus, D.; Alper, C.A.; Levey, R.; Taylor, P.D.; Starzl, T.E. Synthesis of Human Plasminogen by the Liver. *Science* **1980**, *208*, 1036–1037. [CrossRef] [PubMed]
2. Petersen, T.E.; Martzen, M.R.; Ichinose, A.; Davie, E.W. Characterization of the Gene for Human Plasminogen, A Key Proenzyme in the Fibrinolytic System. *J. Biol. Chem.* **1990**, *265*, 6104–6111. [CrossRef]
3. Das, R.; Pluskota, E.; Plow, E.F. Plasminogen and Its Receptors as Regulators of Cardiovascular Inflammatory Responses. *Trends Cardiovasc. Med.* **2010**, *20*, 120–124. [CrossRef] [PubMed]
4. Gong, Y.; Kim, S.-O.; Felez, J.; Grella, D.K.; Castellino, F.J.; Miles, L.A. Conversion of Glu-Plasminogen to Lys-Plasminogen Is Necessary for Optimal Stimulation of Plasminogen Activation on the Endothelial Cell Surface. *J. Biol. Chem.* **2001**, *276*, 19078–19083. [CrossRef]
5. Lawrence, D.; Strandberg, L.; Grundström, T.; Ny, T. Purification of Active Human Plasminogen Activator Inhibitor 1 from Escherichia Coli. Comparison with Natural and Recombinant Forms Purified from Eucaryotic Cells. *Eur. J. Biochem.* **1989**, *186*, 523–533. [CrossRef]
6. Draxler, D.F.; Sashindranath, M.; Medcalf, R.L. Plasmin: A Modulator of Immune Function. *Semin. Thromb. Hemost.* **2016**, *43*, 143–153. [CrossRef]
7. Muoio, V.; Persson, P.B.; Sendeski, M.M. The Neurovascular Unit—Concept Review. *Acta Physiol.* **2014**, *210*, 790–798. [CrossRef]
8. Pennica, D.; Holmes, W.E.; . Kohr, W.J.; Harkins, R.N.; Vehar, G.A.; Ward, C.A.; Bennett, W.F.; Yelverton, E.; Seeburg, P.H.; Heyneker, H.L.; et al. Cloning and Expression of Human Tissue-Type Plasminogen-Activator Cdna in Escherichia-Coli. *Nature* **1983**, *301*, 214–221. [CrossRef]
9. Yepes, M.; Roussel, B.D.; Ali, C.; Vivien, D. Tissue-Type Plasminogen Activator in the Ischemic Brain: More than a Thrombolytic. *Trends Neurosci.* **2009**, *32*, 48–55. [CrossRef]
10. Kim, J.A.; Tran, N.D.; Wang, S.-J.; Fisher, M.J. Astrocyte Regulation of Human Brain Capillary Endothelial Fibrinolysis. *Thromb. Res.* **2003**, *112*, 159–165. [CrossRef]
11. Huber, D.; Cramer, E.M.; Kaufmann, J.E.; Meda, P.; Massé, J.-M.; Kruithof, E.K.O.; Vischer, U.M. Tissue-Type Plasminogen Activator (t-PA) is Stored in Weibel-Palade Bodies in Human Endothelial Cells Both in Vitro and in Vivo. *Blood* **2002**, *99*, 3637–3645. [CrossRef]
12. Hanss, M.; Collen, D. Secretion of Tissue-Type Plasminogen Activator and Plasminogen Activator Inhibitor by Cultured Human Endothelial Cells: Modulation by Thrombin, Endotoxin, and Histamine. *J. Lab. Clin. Med.* **1987**, *109*, 97–104.
13. Pepper, M.S.; Rosnoblet, C.; Di Sanza, C.; Kruithof, E.K. Synergistic Induction of t-PA by VASCULAR endothelial Growth Factor and Basic Fibroblast Growth Factor and Localization of t-PA to Weibel-Palade Bodies in Bovine Microvascular Endothelial Cells. *Thromb. Haemost.* **2001**, *86*, 702–709.
14. Kawai, Y.; Matsumoto, Y.; Watanabe, K.; Yamamoto, H.; Satoh, K.; Murata, M.; Handa, M.; Ikeda, Y. Hemodynamic Forces Modulate the Effects of Cytokines on Fibrinolytic Activity of Endothelial Cells. *Blood* **1996**, *87*, 2314–2321. [CrossRef] [PubMed]
15. Chandler, W.L.; Levy, W.C.; Stratton, J.R. The Circulatory Regulation of TPA and UPA Secretion, Clearance, and Inhibition During Exercise and During the Infusion of Isoproterenol and Phenylephrine. *Circulation* **1995**, *92*, 2984–2994. [CrossRef]
16. Stein, C.; Brown, N.; E Vaughan, D.; Lang, C.C.; Wood, A.J. Regulation of Local Tissue-Type Plasminogen Activator Release by Endothelium-Dependent and Endothelium-Independent Agonists in Human Vasculature. *J. Am. Coll. Cardiol.* **1998**, *32*, 117–122. [CrossRef]
17. Levin, E.G.; Del Zoppo, G.J. Localization of Tissue Plasminogen Activator in the Endothelium of a Limited Number of Vessels. *Am. J. Pathol.* **1994**, *144*, 855–861.
18. Yang, F.; Liu, S.; Wang, S.-J.; Yu, C.; Paganini-Hill, A.; Fisher, M.J. Tissue Plasminogen Activator Expression and Barrier Properties of Human Brain Microvascular Endothelial Cells. *Cell Physiol. Biochem.* **2011**, *28*, 631–638. [CrossRef] [PubMed]
19. Cho, M.-K.; Sun, E.-S.; Kim, Y.-H. Zinc-Triggered Induction of Tissue Plasminogen Activator and Plasminogen in Endothelial Cells and Pericytes. *Exp. Neurobiol.* **2013**, *22*, 315–321. [CrossRef] [PubMed]
20. Kim, J.A.; Tran, N.D.; Li, Z.; Yang, F.; Zhou, W.; Fisher, M.J. Brain Endothelial Hemostasis Regulation by Pericytes. *J. Cereb. Blood Flow Metab.* **2006**, *26*, 209–217. [CrossRef]

21. Polavarapu, R.; Gongora, M.C.; Yi, H.; Ranganthan, S.; Lawrence, D.A.; Strickland, D.; Yepes, M. Tissue-Type Plasminogen Activator–Mediated Shedding of Astrocytic Low-Density Lipoprotein Receptor–Related Protein Increases the Permeability of the Neurovascular Unit. *Blood* **2006**, *109*, 3270–3278. [CrossRef]

22. Tang, H.; Fu, W.Y.; Ip, N.Y. Altered Expression of Tissue-Type Plasminogen Activator and Type 1 Inhibitor in Astrocytes of Mouse Cortex Following Scratch Injury in Culture. *Neurosci. Lett.* **2000**, *285*, 143–146. [CrossRef]

23. Zhang, C.; An, J.; Haile, W.B.; Echeverry, R.; Strickland, D.K.; Yepes, M. Microglial Low-Density Lipoprotein Receptor-Related Protein 1 Mediates the Effect of Tissue-Type Plasminogen Activator on Matrix Metalloproteinase-9 Activity in the Ischemic Brain. *J. Cereb. Blood Flow Metab.* **2009**, *29*, 1946–1954. [CrossRef]

24. Hu, K.; Yang, J.; Tanaka, S.; Gonias, S.L.; Mars, W.M.; Liu, Y. Tissue-Type Plasminogen Activator Acts as a Cytokine That Triggers Intracellular Signal Transduction and Induces Matrix Metalloproteinase-9 Gene Expression. *J. Biol. Chem.* **2006**, *281*, 2120–2127. [CrossRef]

25. Polavarapu, R.; Gongora, M.C.; Winkles, J.A.; Yepes, M. Tumor Necrosis Factor-Like Weak Inducer of Apoptosis Increases the Permeability of the Neurovascular Unit through Nuclear Factor-Kappa B Pathway Activation. *J. Neurosci.* **2005**, *25*, 10094–10100. [CrossRef]

26. Lin, L.; Wu, C.; Hu, K. Tissue Plasminogen Activator Activates NF-kappaB through a Pathway Involving Annexin A2/CD11b and Integrin-Linked Kinase. *J. Am. Soc. Nephrol.* **2012**, *23*, 1329–1338. [CrossRef]

27. Xin, H.; Li, Y.; Shen, L.H.; Liu, X.; Hozeska-Solgot, A.; Zhang, R.L.; Zhang, Z.G.; Chopp, M. Multipotent Mesenchymal Stromal Cells Increase tPA Expression and Concomitantly Decrease PAI-1 Expression in Astrocytes through the Sonic Hedgehog Signaling Pathway after Stroke (In Vitro Study). *J. Cereb. Blood Flow Metab.* **2011**, *31*, 2181–2188. [CrossRef]

28. Cassé, F.; Bardou, I.; Danglot, L.; Briens, A.; Montagne, A.; Parcq, J.; Alahari, A.; Galli, T.; Vivien, D.; Docagne, F. Glutamate Controls tPA Recycling by Astrocytes, Which in Turn Influences Glutamatergic Signals. *J. Neurosci.* **2012**, *32*, 5186–5199. [CrossRef]

29. Vincent, V.A.; Löwik, C.W.; Verheijen, J.H.; De Bart, A.C.; Tilders, F.J.; Van Dam, A. Role of Astrocyte-Derived Tissue-Type Plasminogen Activator in the Regulation of Endotoxin-Stimulated Nitric Oxide Production by Microglial Cells. *Glia* **1998**, *22*, 130–137. [CrossRef]

30. Siao, C.H.-J.; Susana, R.F.; Stella, E.T. Cell Type-Specific Roles for Tissue Plasminogen Activator Released by Neurons or Microglia after Excitotoxic Injury. *J. Neurosci.* **2003**, *23*, 3224–3242. [CrossRef]

31. Siao, C.J.; Tsirka, S.E. Tissue Plasminogen Activator Mediates Microglial Activation via its Finger Domain through Annexin II. *J. Neurosci.* **2002**, *22*, 3352–3358. [CrossRef]

32. Rogove, A.D.; Siao, C.; Keyt, B.; Strickland, S.; E Tsirka, S. Activation of Microglia Reveals a Non-Proteolytic Cytokine Function for Tissue Plasminogen Activator in the Central Nervous System. *J. Cell Sci.* **1999**, *112*, 4007–4016.

33. Flavin, M.P.; Zhao, G. Tissue Plasminogen Activator Protects Hippocampal Neurons from Oxygen-Glucose Deprivation Injury. *J. Neurosci. Res.* **2001**, *63*, 388–394. [CrossRef]

34. Krystosek, A.; Seeds, N.W. Plasminogen Activator Release at the Neuronal Growth Cone. *Science* **1981**, *213*, 1532–1534. [CrossRef]

35. Seeds, N.W.; Basham, M.E.; Haffke, S.P. Neuronal Migration is Retarded in Mice Lacking the Tissue Plasminogen Activator Gene. *Proc. Natl. Acad. Sci. USA* **1999**, *96*, 14118–14123. [CrossRef]

36. Lee, S.H.; Ko, H.M.; Kwon, K.J.; Lee, J.; Han, S.H.; Han, D.W.; Cheong, J.H.; Ryu, J.H.; Shin, C.Y. tPA Regulates Neurite Outgrowth by Phosphorylation of LRP5/6 in Neural Progenitor Cells. *Mol. Neurobiol.* **2014**, *49*, 199–215. [CrossRef]

37. Muller, C.M.; Griesinger, C.B. Tissue Plasminogen Activator Mediates Reverse Occlusion Plasticity in Visual Cortex. *Nat. Neurosci.* **1998**, *1*, 47–53. [CrossRef] [PubMed]

38. Shen, L.H.; Xin, H.; Li, Y.; Zhang, R.L.; Cui, Y.; Zhang, L.; Lü, M.; Zhang, Z.G.; Chopp, M. Endogenous Tissue Plasminogen Activator Mediates Bone Marrow Stromal Cell-Induced Neurite Remodeling after Stroke in Mice. *Stroke* **2011**, *42*, 459–464. [CrossRef] [PubMed]

39. Sappino, A.P.; Madani, R.; Huarte, J.; Belin, D.; Kiss, J.Z.; Wohlwend, A.; Vassalli, J.D. Extracellular Proteolysis in the Adult Murine Brain. *J. Clin. Investig.* **1993**, *92*, 679–685. [CrossRef]

40. Nicole, O.; Docagne, F.; Ali, C.; Margaill, I.; Carmeliet, P.; MacKenzie, E.T.; Vivien, D.; Buisson, A. The Proteolytic Activity of Tissue-Plasminogen Activator Enhances NMDA Receptor-Mediated Signaling. *Nat. Med.* **2001**, *7*, 59–64. [CrossRef] [PubMed]

41. Samson, A.L.; Medcalf, R.L. Tissue-Type Plasminogen Activator: A Multifaceted Modulator of Neurotransmission and Synaptic Plasticity. *Neuron* **2006**, *50*, 673–678. [CrossRef] [PubMed]

42. Qian, Z.; Gilbert, M.E.; Colicos, M.A.; Kandel, E.R.; Kuhl, D. Tissue-Plasminogen Activator is Induced as an Immediate-Early Gene during Seizure, Kindling and Long-Term Potentiation. *Nature* **1993**, *361*, 453–457. [CrossRef] [PubMed]

43. Seeds, N.W.; Basham, M.E.; Ferguson, J.E. Absence of Tissue Plasminogen Activator Gene or Activity Impairs Mouse Cerebellar Motor Learning. *J. Neurosci.* **2003**, *23*, 7368–7375. [CrossRef] [PubMed]

44. Seeds, N.W.; Williams, B.L.; Bickford, P.C. Tissue Plasminogen Activator Induction in Purkinje Neurons after Cerebellar Motor Learning. *Science* **1995**, *270*, 1992–1994. [CrossRef]

45. Pawlak, R.; Magarinos, A.M.; Melchor, J.; McEwen, B.; Strickland, S. Tissue Plasminogen Activator in the Amygdala is Critical for Stress-Induced Anxiety-Like Behavior. *Nat. Neurosci.* **2003**, *6*, 168–174. [CrossRef]

46. Stepanova, V.V.; Tkachuk, V.A. Urokinase as a Multidomain Protein and Polyfunctional Cell Regulator. *Biochemistry* **2002**, *67*, 109–118.

47. Danø, K.; Andreasen, P.; Grøndahl-Hansen, J.; Kristensen, P.; Nielsen, L.; Skriver, L. Plasminogen Activators, Tissue Degradation, and Cancer. *Adv. Cancer Res.* **1985**, *44*, 139–266.

48. Stoppelli, M.; Tacchetti, C.; Cubellis, M.; Corti, A.; Hearings, V.J.; Cassani, G.; Appella, E.; Blasi, F. Autocrine Saturation of Pro-urokinase Receptors on Human A431 Cells. *Cell* **1986**, *45*, 675–684. [CrossRef]

49. Cubellis, M.V.; Wun, T.C.; Blasi, F. Receptor-Mediated Internalization and Degradation of Urokinase is Caused by Its Specific Inhibitor PAI-1. *EMBO J.* **1990**, *9*, 1079–1085. [CrossRef]

50. Cao, D.J.; Guo, Y.L.; Colman, R.W. Urokinase-Type Plasminogen Activator Receptor is Involved in Mediating the Apoptotic Effect of Cleaved High Molecular Weight Kininogen in Human Endothelial Cells. *Circ. Res.* **2004**, *94*, 1227–1234. [CrossRef]

51. Breuss, J.M.; Uhrin, P. VEGF-Initiated Angiogenesis and the uPA/uPAR System. *Cell Adh. Migr.* **2012**, *6*, 535–615. [CrossRef]

52. Prager, G.W.; Breuss, J.M.; Steurer, S.; Mihaly, J.; Binder, B.R. Vascular Endothelial Growth Factor (VEGF) Induces Rapid Prourokinase (pro-uPA) Activation on the Surface of Endothelial Cells. *Blood* **2004**, *103*, 955–962. [CrossRef]

53. Pepper, M.S.; Vassalli, J.D.; Montesano, R.; Orci, L. Urokinase-Type Plasminogen Activator is Induced in Migrating Capillary Endothelial Cells. *J. Cell Biol.* **1987**, *105*, 2535–2541. [CrossRef]

54. Stie, J.; Fox, D. Induction of Brain Microvascular Endothelial Cell Urokinase Expression by Cryptococcus Neoformans Facilitates Blood-Brain Barrier Invasion. *PLoS ONE* **2012**, *7*, e49402. [CrossRef]

55. Benton, R.L.; Maddie, M.A.; Dincman, T.A.; Hagg, T.; Whittemore, S.R. Transcriptional Activation of Endothelial Cells by TGFbeta Coincides with Acute Microvascular Plasticity Following Focal Spinal Cord Ischaemia/Reperfusion Injury. *ASN Neuro.* **2009**, *1*, AN20090008. [CrossRef]

56. Yamamoto, M.; Sawaya, R.; Mohanam, S.; Bindal, A.K.; Bruner, J.M.; Oka, K.; Rao, V.H.; Tomonaga, M.; Nicolson, G.L.; Rao, J.S. Expression and Localization of Urokinase-Type Plasminogen Activator in Human Astrocytomas in Vivo. *Cancer Res.* **1994**, *54*, 3656–3661.

57. Diaz, A.; Merino, P.; Manrique, L.G.; Ospina, J.P.; Cheng, L.; Wu, F.; Jeanneret, V.; Yepes, M. A Cross-talk Between Neuronal Urokinase-type Plasminogen Activator (uPA) and Astrocytic uPA Receptor (uPAR) Promotes Astrocytic Activation and Synaptic Recovery in the Ischemic Brain. *J. Neurosci.* **2017**, *37*, 10310–10322. [CrossRef]

58. Diaz, A.; Merino, P.; Manrique, L.G.; Cheng, L.; Yepes, M. Urokinase-Type Plasminogen Activator (uPA) Protects the Tripartite Synapse in the Ischemic Brain via Ezrin-Mediated Formation of Peripheral Astrocytic Processes. *J. Cereb. Blood Flow Metab.* **2018**, *39*, 2157–2171. [CrossRef]

59. Rudlin, C.R. A Three-Dimensional Representation of Linear Growth and Skeletal Maturation. *Acta. Paediatr. Scand. Suppl.* **1989**, *356*, 46–50. [CrossRef]

60. Merino, P.; Diaz, A.; Jeanneret, V.; Wu, F.; Torre, E.; Cheng, L.; Yepes, M. Urokinase-Type Plasminogen Activator (uPA) Binding to the uPA Receptor (uPAR) Promotes Axonal Regeneration in the Central Nervous System. *J. Biol. Chem.* **2017**, *292*, 2741–2753. [CrossRef]

61. Wu, F.; Catano, M.; Echeverry, R.; Torre, E.; Haile, W.B.; An, J.; Chen, C.; Cheng, L.; Nicholson, A.; Tong, F.C.; et al. Urokinase-Type Plasminogen Activator Promotes Dendritic Spine Recovery and Improves Neurological Outcome Following Ischemic Stroke. *J. Neurosci.* **2014**, *34*, 14219–14232. [CrossRef] [PubMed]

62. Cho, E.; Lee, K.J.; Seo, J.-W.; Byun, C.J.; Chung, S.-J.; Suh, D.C.; Carmeliet, P.; Koh, J.-Y.; Kim, J.S.; Lee, J.-Y. Neuroprotection by Urokinase Plasminogen Activator in the Hippocampus. *Neurobiol. Dis.* **2012**, *46*, 215–224. [CrossRef] [PubMed]

63. Lino, N.; Fiore, L.; Rapacioli, M.; Teruel, L.; Flores, V.; Scicolone, G.; Sánchez, V. uPA-uPAR Molecular Complex is Involved in Cell Signaling during Neuronal Migration and Neuritogenesis. *Dev. Dyn.* **2014**, *243*, 676–689. [CrossRef] [PubMed]

64. Farias-Eisner, R.; Vician, L.; Silver, A.; Reddy, S.; Rabbani, S.A.; Herschman, H.R. The Urokinase Plasminogen Activator Receptor (UPAR) is Preferentially Induced by Nerve Growth Factor in PC12 Pheochromocytoma Cells and is Required for NGF-Driven Differentiation. *J. Neurosci.* **2000**, *20*, 230–239. [CrossRef]

65. Gilder, A.S.; Jones, K.A.; Hu, J.; Wang, L.; Chen, C.C.; Carter, B.S.; Gonias, S.L. Soluble Urokinase Receptor Is Released Selectively by Glioblastoma Cells That Express Epidermal Growth Factor Receptor Variant III and Promotes Tumor Cell Migration and Invasion. *J. Biol. Chem.* **2015**, *2902*, 14798–14809. [CrossRef]

66. Semina, E.; Rubina, K.; Sysoeva, V.; Rysenkova, K.; Klimovich, P.; Plekhanova, O.; Tkachuk, V. Urokinase and Urokinase Receptor Participate in Regulation of Neuronal Migration, Axon Growth and Branching. *Eur. J. Cell Biol.* **2016**, *95*, 295–310. [CrossRef]

67. Rivellini, C.; Dina, G.; Porrello, E.; Cerri, F.; Scarlato, M.; Domi, T.; Ungaro, D.; Del Carro, U.; Bolino, A.; Quattrini, A.; et al. Urokinase Plasminogen Receptor and the Fibrinolytic Complex Play a Role in Nerve Repair after Nerve Crush in Mice, and in Human Neuropathies. *PLoS ONE* **2012**, *7*, e32059. [CrossRef]

68. Diaz, A.; Merino, P.; Guo, J.D.; Yepes, M.A.; McCann, P.; Katta, T.; Tong, E.M.; Torre, E.; Rangaraju, S.; Yepes, M. Urokinase-Type Plasminogen Activator Protects Cerebral Cortical Neurons from Soluble Abeta-Induced Synaptic Damage. *J. Neurosci.* **2020**, *40*, 4251–4263. [CrossRef]

69. Feigin, V.L.; Norrving, B.; Mensah, G.A. Global Burden of Stroke. *Circ. Res.* **2017**, *120*, 439–448. [CrossRef]

70. Yepes, M.; Sandkvist, M.; Wong, M.K.; Coleman, T.A.; Smith, E.; Cohan, S.L.; Lawrence, D.A. Neuroserpin Reduces Cerebral Infarct Volume and Protects Neurons from Ischemia-Induced Apoptosis. *Blood* **2000**, *96*, 569–576. [CrossRef]

71. Rossetti, L.; Hu, M. Skeletal Muscle Glycogenolysis is More Sensitive to Insulin than is Glucose Transport/Phosphorylation. Relation to the Insulin-Mediated Inhibition of Hepatic Glucose Production. *J. Clin. Investig.* **1993**, *92*, 2963–2974. [CrossRef]

72. Gris, J.; Schved, J.; Brun, S.; Brunschwig, C.; Petris, I.; Lassonnery, M.; Martinez, P.; Sarlat, C. Venous Occlusion and Chronic Cigarette Smoking: Dose-Dependent Decrease in the Measurable Release of Tissue-Type Plasminogen Activator and Von Willebrand Factor. *Atherosclerosis* **1991**, *91*, 247–255. [CrossRef]

73. Gong, P.; Li, M.; Zou, C.; Tian, Q.; Xu, Z. Tissue Plasminogen Activator Causes Brain Microvascular Endothelial Cell Injury After Oxygen Glucose Deprivation by Inhibiting Sonic Hedgehog Signaling. *Neurochem. Res.* **2019**, *44*, 441–449. [CrossRef]

74. Lindgren, A.; Lindoff, C.; Norrving, B.; Åstedt, B.; Johansson, B.B. Tissue Plasminogen Activator and Plasminogen Activator Inhibitor-1 in Stroke Patients. *Stroke* **1996**, *27*, 1066–1071. [CrossRef]

75. The National Institute of Neurological Disorders and Stroke rt-PA Stroke Study Group. Stroke rt. Tissue Plasminogen Activator for Acute Ischemic Stroke. *N. Engl. J. Med.* **1995**, *333*, 1581–1587. [CrossRef]

76. Hacke, W.; Kaste, M.; Bluhmki, E.; Brozman, M.; Dávalos, A.; Guidetti, D.; Larrue, V.; Lees, K.R.; Medeghri, Z.; Machnig, T.; et al. Thrombolysis with Alteplase 3 to 4.5 Hours after Acute Ischemic Stroke. *N. Engl. J. Med.* **2008**, *359*, 1317–1329. [CrossRef]

77. Van Geffen, M.; Cugno, M.; Lap, P.; Loof, A.; Cicardi, M.; Van Heerde, W. Alterations of Coagulation and Fibrinolysis in Patients with Angioedema Due to C1-Inhibitor Deficiency. *Clin. Exp. Immunol.* **2012**, *167*, 472–478. [CrossRef]

78. Baker, S.K.; Chen, Z.-L.; Norris, E.H.; Revenko, A.S.; MacLeod, A.R.; Strickland, S. Blood-Derived Plasminogen Drives Brain Inflammation and Plaque Deposition in a Mouse Model of Alzheimer's Disease. *Proc. Natl. Acad. Sci. USA* **2018**, *115*, 9687–9696. [CrossRef]

79. Chen, Z.-L.; Revenko, A.S.; Singh, P.; MacLeod, A.R.; Norris, E.H.; Strickland, S. Depletion of Coagulation Factor XII Ameliorates Brain Pathology and Cognitive Impairment in Alzheimer Disease Mice. *Blood* **2017**, *129*, 2547–2556. [CrossRef]

80. Draxler, D.F.; Lee, F.; Ho, H.; Keragala, C.B.; Medcalf, R.L.; Niego, B. t-PA Suppresses the Immune Response and Aggravates Neurological Deficit in a Murine Model of Ischemic Stroke. *Front Immunol.* **2019**, *10*, 591. [CrossRef]

81. Graham, C.H.; Fitzpatrick, T.E.; McCrae, K.R. Hypoxia Stimulates Urokinase Receptor Expression through a Heme Protein-Dependent Pathway. *Blood* **1998**, *91*, 3300–3307. [CrossRef] [PubMed]

82. E Kroon, M.; Koolwijk, P.; Van Der Vecht, B.; Van Hinsbergh, V.W. Urokinase Receptor Expression on Human Microvascular Endothelial Cells is Increased by Hypoxia: Implications for Capillary-Like Tube Formation in a Fibrin Matrix. *Blood* **2000**, *96*, 2775–2783. [CrossRef]

83. Krock, B.L.; Skuli, N.; Simon, M.C. Hypoxia-Induced Angiogenesis: Good and Evil. *Genes Cancer* **2011**, *2*, 1117–1133. [CrossRef] [PubMed]

84. del Zoppo, G.J.; Higashida, R.T.; Furlan, A.J.; Pessin, M.S.; Rowley, H.A.; Gent, M. PROACT: A Phase II Randomized Trial of Recombinant Pro-Urokinase by Direct Arterial Delivery in Acute Middle Cerebral Artery Stroke. PROACT Investigators. Prolyse in Acute Cerebral Thromboembolism. *Stroke* **1998**, *29*, 4–11. [CrossRef]

85. Wardlaw, J.M. Overview of Cochrane Thrombolysis Meta-Analysis. *Neurology* **2001**, *57*, 69–76. [CrossRef]

86. Zhang, X.; Polavarapu, R.; She, H.; Mao, Z.; Yepes, M. Tissue-Type Plasminogen Activator and the Low-Density Lipoprotein Receptor-Related Protein Mediate Cerebral Ischemia-Induced Nuclear Factor-kappaB Pathway Activation. *Am. J. Pathol.* **2007**, *171*, 1281–1290. [CrossRef]

87. Yepes, M.; Sandkvist, M.; Moore, E.G.; Bugge, T.H.; Strickland, D.K.; Lawrence, D.A. Tissue-Type Plasminogen Activator Induces Opening of the Blood-Brain Barrier via the LDL Receptor-Related Protein. *J. Clin. Investig.* **2003**, *112*, 1533–1540. [CrossRef]

88. Kidwell, C.S.; Latour, L.; Saver, J.L.; Alger, J.R.; Starkman, S.; Duckwiler, G.; Jahan, R.; Viñuela, F.; Kang, D.-W.; Warach, S. Thrombolytic Toxicity: Blood Brain Barrier Disruption in Human Ischemic Stroke. *Cerebrovasc. Dis.* **2008**, *25*, 338–343. [CrossRef]

89. Zhang, L.; Li, Y.; Zhang, C.; Chopp, M.; Gosiewska, A.; Hong, K. Delayed Administration of Human Umbilical Tissue-Derived Cells Improved Neurological Functional Recovery in a Rodent Model of Focal Ischemia. *Stroke* **2011**, *42*, 1437–1444. [CrossRef]

90. Bretscher, A.; A Edwards, K.; Fehon, R.G. ERM Proteins and Merlin: Integrators at the Cell Cortex. *Nat. Rev. Mol. Cell Biol.* **2002**, *3*, 586–599. [CrossRef]

91. Tsukita, S.; Yonemura, S. Cortical Actin Organization: Lessons from ERM (Ezrin/Radixin/Moesin) Proteins. *J. Biol. Chem.* **1999**, *274*, 34507–34510. [CrossRef]

92. Fehon, R.G.; McClatchey, A.I.; Bretscher, A. Organizing the Cell Cortex: The Role of ERM Proteins. *Nat. Rev. Mol. Cell Biol.* **2010**, *11*, 276–287. [CrossRef]

93. Reichenbach, A.; Derouiche, A.; Kirchhoff, F. Morphology and Dynamics of Perisynaptic Glia. *Brain Res. Rev.* **2010**, *63*, 11–25. [CrossRef]

94. Lavialle, M.; Aumann, G.; Anlauf, E.; Pröls, F.; Arpin, M.; Derouiche, A. Structural Plasticity of Perisynaptic Astrocyte Processes Involves Ezrin and Metabotropic Glutamate Receptors. *Proc. Natl. Acad. Sci. USA* **2011**, *108*, 12915–12919. [CrossRef]

95. Taylor, R.A.; Sansing, L.H. Microglial Responses after Ischemic Stroke and Intracerebral Hemorrhage. *Clin. Dev. Immunol.* **2013**, *2013*, 1–10. [CrossRef]

96. Gravanis, I.; Tsirka, S.E. Tissue Plasminogen Activator and Glial Function. *Glia* **2004**, *49*, 177–183. [CrossRef]

97. Zhang, C.; An, J.; Strickland, D.K.; Yepes, M. The Low-Density Lipoprotein Receptor-Related Protein 1 Mediates Tissue-Type Plasminogen Activator-Induced Microglial Activation in the Ischemic Brain. *Am. J. Pathol.* **2009**, *174*, 586–594. [CrossRef]

98. Su, E.J.; Fredriksson, L.; Geyer, M.; Folestad, E.; Cale, J.; Andrae, J.; Gao, Y.; Pietras, K.; Mann, K.; Yepes, M.; et al. Activation of PDGF-CC by Tissue Plasminogen Activator Impairs Blood-Brain Barrier Integrity during Ischemic Stroke. *Nat. Med.* **2008**, *14*, 731–737. [CrossRef]

99. Kuo, P.C.; Weng, W.T.; Scofield, B.A.; Furnas, D.; Paraiso, H.C.; Intriago, A.J.; Bosi, K.D.; Yu, I.C.; Yen, J.H. Interferon-Beta Alleviates Delayed tPA-Induced Adverse Effects via Modulation of MMP3/9 Production in Ischemic Stroke. *Blood Adv.* **2020**, *4*, 4366–4381. [CrossRef]

100. Washington, R.A.; Becher, B.; Balabanov, R.; Antel, J.; Dore-Duffy, P. Expression of the Activation Marker Urokinase Plasminogen-Activator Receptor in Cultured Human Central Nervous System Microglia. *J. Neurosci. Res.* **1996**, *45*, 392–399. [CrossRef]

101. Cunningham, O.; Campion, S.; Perry, V.H.; Murray, C.; Sidenius, N.; Docagne, F.; Cunningham, C. Microglia and the urokinase plasminogen activator receptor/uPA system in innate brain inflammation. *Glia* **2009**, *57*, 1802–1814. [CrossRef] [PubMed]

102. Shin, S.M.; Cho, K.S.; Choi, M.S.; Lee, S.H.; Han, S.-H.; Kang, Y.-S.; Kim, H.J.; Cheong, J.H.; Shin, C.Y.; Ko, K.H. Urokinase-Type Plasminogen Activator Induces BV-2 Microglial Cell Migration Through Activation of Matrix Metalloproteinase-9. *Neurochem. Res.* **2010**, *35*, 976–985. [CrossRef] [PubMed]

103. Wang, Y.F.; Tsirka, S.E.; Strickland, S.; Stieg, P.E.; Soriano, S.G.; Lipton, S.A. Tissue plasminogen activator (tPA) increase neuronal damage after focal cerebral ischemia in wild-type and tPA-deficient mice. *Nat. Med.* **1998**, *4*, 228–231. [CrossRef] [PubMed]

104. Echeverry, R.; Wu, J.; Haile, W.B.; Guzman, J.; Yepes, M. Tissue-type plasminogen activator is a neuroprotectant in the mouse hippocampus. *J. Clin. Investig.* **2010**, *120*, 2194–2205. [CrossRef] [PubMed]

105. Nagai, N.; De Mol, M.; Lijnen, H.R.; Carmeliet, P.; Collen, D. Role of Plasminogen System Components in Focal Cerebral Ischemic Infarction: A Gene Targeting and Gene Transfer Study in Mice. *Circulation* **1999**, *99*, 2440–2444. [CrossRef] [PubMed]

106. Tabrizi, P.; Wang, L.; Seeds, N.; McComb, J.G.; Yamada, S.; Griffin, J.H.; Carmeliet, P.; Weiss, M.H.; Zlokovic, B.V. Tissue Plasminogen Activator (tPA) Deficiency Exacerbates Cerebrovascular Fibrin Deposition and Brain Injury in a Murine Stroke Model: Studies in tPA-Deficient Mice and Wild-Type Mice on a Matched Genetic Background. *Arterioscler. Thromb. Vasc. Biol.* **1999**, *19*, 2801–2806. [CrossRef]

107. Wu, F.; Wu, J.; Nicholson, A.D.; Echeverry, R.; Haile, W.B.; Catano, M.; An, J.; Lee, A.K.; Duong, D.; Dammer, E.B.; et al. Tissue-Type Plasminogen Activator Regulates the Neuronal Uptake of Glucose in the Ischemic Brain. *J. Neurosci.* **2012**, *32*, 9848–9858. [CrossRef]

108. Klein, G.M.; Li, H.; Sun, P.; Buchan, A.M. Tissue Plasminogen Activator does not Increase Neuronal Damage in Rat Models of Global and Focal Ischemia. *Neurology* **1999**, *52*, 1381–1384. [CrossRef]

109. Hacke, W.; ATLANTIS Trials Investigators; ECASS Trials Investigators; NINDS rt-PA Study Group Investigators. Association of Outcome with Early Stroke Treatment: Pooled Analysis of ATLANTIS, ECASS, and NINDS rt-PA Stroke Trials. *Lancet* **2004**, *363*, 768–774.

110. Adams, H.P., Jr.; Del Zoppo, G.; Alberts, M.J.; Bhatt, D.L.; Brass, L.; Furlan, A.; Grubb, R.L.; Higashida, R.T.; Jauch, E.C.; Kidwell, C.; et al. Guidelines for the Early Management of Adults with Ischemic Stroke: A Guideline from the American Heart Association/American Stroke Association Stroke Council, Clinical Cardiology Council, Cardiovascular Radiology and Intervention Council, and the Atherosclerotic Peripheral Vascular Disease and Quality of Care Outcomes in Research Interdisciplinary Working Groups: The American Academy of Neurology affirms the value of this guideline as an educational tool for neurologists. *Circulation* **2007**, *115*, 478–534.

111. Haile, W.B.; Wu, J.; Echeverry, R.; Wu, F.; An, J.; Yepes, M. Tissue-Type Plasminogen Activator has a Neuroprotective Effect in the Ischemic Brain Mediated by Neuronal TNF-α. *J. Cereb. Blood Flow Metab.* **2012**, *32*, 57–69. [CrossRef]

112. Tsirka, S.E.; Gualandris, A.; Amaral, D.G.; Strickland, S. Excitotoxin-induced neuronal degeneration and seizure are mediated by tissue plasminogen activator. *Nat. Cell Biol.* **1995**, *377*, 340–344. [CrossRef]

113. Chen, Z.-L.; Strickland, S. Neuronal Death in the Hippocampus Is Promoted by Plasmin-Catalyzed Degradation of Laminin. *Cell* **1997**, *91*, 917–925. [CrossRef]

114. Reddrop, C.; Moldrich, R.X.; Beart, P.M.; Farso, M.; Liberatore, G.T.; Howells, D.W.; Petersen, K.-U.; Schleuning, W.-D.; Medcalf, R.L. Vampire Bat Salivary Plasminogen Activator (Desmoteplase) Inhibits Tissue-Type Plasminogen Activator-Induced Potentiation of Excitotoxic Injury. *Stroke* **2005**, *36*, 1241–1246. [CrossRef]

115. Wu, F.; Echeverry, R.; Wu, J.; An, J.; Haile, W.B.; Cooper, D.S.; Catano, M.; Yepes, M. Tissue-Type Plasminogen Activator Protects Neurons from Excitotoxin-Induced Cell Death via Activation of the ERK1/2-CREB-ATF3 Signaling Pathway. *Mol. Cell Neurosci.* **2013**, *52*, 9–19. [CrossRef]

116. Salles, F.J.; Strickland, S. Localization and Regulation of the Tissue Plasminogen Activator-Plasmin System in the Hippocampus. *J. Neurosci.* **2002**, *22*, 2125–2134. [CrossRef]

117. Kim, Y.H. Nonproteolytic Neuroprotection by Human Recombinant Tissue Plasminogen Activator. *Science* **1999**, *284*, 647–650. [CrossRef]

118. Parcq, J.; Bertrand, T.; Montagne, A.; Baron, A.F.; Macrez, R.; Billard, J.M.; Briens, A.; Hommet, Y.; Wu, J.; Yepes, M.; et al. Unveiling an Exceptional Zymogen: The Single-Chain Form of tPA is a Selective Activator of NMDA Receptor-Dependent Signaling and Neurotoxicity. *Cell Death Differ.* **2012**, *19*, 1983–1991. [CrossRef]

119. Anfray, A.; Brodin, C.; Drieu, A.; Potzeha, F.; Dalarun, B.; Agin, V.; Vivien, D.; Orset, C. Single- and Two- Chain Tissue Type Plasminogen Activator Treatments Differentially Influence Cerebral Recovery after Stroke. *Exp. Neurol.* **2021**, *338*, 113606. [CrossRef]

120. Nagai, N.; Okada, K.; Kawao, N.; Ishida, C.; Ueshima, S.; Collen, D.; Matsuo, O. Urokinase-Type Plasminogen Activator receptor (uPAR) Augments Brain Damage in a Murine Model of Ischemic Stroke. *Neurosci. Lett.* **2008**, *432*, 46–49. [CrossRef]

121. Morales, D.; McIntosh, T.; Conte, V.; Fujimoto, S.; Graham, D.; Grady, M.S.; Stein, S.C. Impaired Fibrinolysis and Traumatic Brain Injury in Mice. *J. Neurotrauma* **2006**, *23*, 976–984. [CrossRef] [PubMed]

122. Yu, X.; Ji, C.; Shao, A. Neurovascular Unit Dysfunction and Neurodegenerative Disorders. *Front. Neurosci.* **2020**, *14*, 334. [CrossRef] [PubMed]
123. Zlokovic, B.V. The Blood-Brain Barrier in Health and Chronic Neurodegenerative Disorders. *Neuron* **2008**, *57*, 178–201. [CrossRef] [PubMed]
124. Sloane, P.D.; Zimmerman, S.; Suchindran, C.; Reed, P.; Wang, L.; Boustani, M.; Sudha, S. The Public Health Impact of Alzheimer's Disease, 2000–2050: Potential Implication of Treatment Advances. *Annu. Rev. Public Health* **2002**, *23*, 213–231. [CrossRef] [PubMed]
125. Kelleher, R.J.; Soiza, R.L. Evidence of Endothelial Dysfunction in the Development of Alzheimer's Disease: Is Alzheimer's a Vascular Disorder? *Am. J. Cardiovasc. Dis.* **2013**, *3*, 197–226. [PubMed]
126. Kisler, K.; Nelson, A.R.; Montagne, A.; Zlokovic, B.V. Cerebral Blood Flow Regulation and Neurovascular Dysfunction in Alzheimer Disease. *Nat. Rev. Neurosci.* **2017**, *18*, 419–434. [CrossRef] [PubMed]
127. Sweeney, M.D.; Kisler, K.; Montagne, A.; Toga, A.W.; Zlokovic, B.V. The role of brain vasculature in neurodegenerative disorders. *Nat. Neurosci.* **2018**, *21*, 1318–1331. [CrossRef]
128. Park, L.; Zhou, J.; Koizumi, K.; Wang, G.; Anfray, A.; Ahn, S.J.; Seo, J.; Zhou, P.; Zhao, L.; Paul, S.; et al. tPA Deficiency Underlies Neurovascular Coupling Dysfunction by Amyloid-beta. *J. Neurosci.* **2020**, *40*, 8160–8173. [CrossRef]
129. Davis, J.; Wagner, M.R.; Zhang, W.; Xu, F.; Van Nostrand, W.E. Amyloid Beta-Protein Stimulates the Expression of Urokinase-Type Plasminogen Activator (uPA) and Its Receptor (uPAR) in Human Cerebrovascular Smooth Muscle Cells. *J. Biol. Chem.* **2003**, *278*, 19054–19061. [CrossRef]
130. Storck, S.E.; Meister, S.; Nahrath, J.; Meißner, J.N.; Schubert, N.; Di Spiezio, A.; Baches, S.; Vandenbroucke, R.E.; Bouter, Y.; Prikulis, I.; et al. Endothelial LRP1 Transports Amyloid-Beta(1-42) Across the Blood-Brain Barrier. *J. Clin. Investig.* **2016**, *126*, 123–136. [CrossRef]
131. Lillis, A.P.; Mikhailenko, I.; Strickland, D.K. Beyond Endocytosis: LRP Function in Cell Migration, Proliferation and Vascular Permeability. *J. Thromb. Haemost.* **2005**, *3*, 1884–1893. [CrossRef]
132. Rodríguez-Arellano, J.; Parpura, V.; Zorec, R.; Verkhratsky, A. Astrocytes in Physiological Aging and Alzheimer's Disease. *Neurosci.* **2016**, *323*, 170–182. [CrossRef]
133. A Bates, K.; Fonte, J.; A Robertson, T.; Martins, R.N.; Harvey, A.R. Chronic Gliosis Triggers Alzheimer's Disease-Like Processing of Amyloid Precursor Protein. *Neurosci.* **2002**, *113*, 785–796. [CrossRef]
134. Habib, N.; McCabe, C.; Medina, S.; Varshavsky, M.; Kitsberg, D.; Dvir-Szternfeld, R.; Green, G.; Dionne, D.; Nguyen, L.; Marshall, J.L.; et al. Disease-Associated Astrocytes in Alzheimer's Disease and Aging. *Nat. Neurosci.* **2020**, *23*, 701–706. [CrossRef]
135. Ballabh, P.; Braun, A.; Nedergaard, M. The Blood-Brain Barrier: An Overview: Structure, Regulation, and Clinical Implications. *Neurobiol. Dis.* **2004**, *16*, 1–13. [CrossRef]
136. Montagne, A.; Barnes, S.R.; Sweeney, M.D.; Halliday, M.R.; Sagare, A.P.; Zhao, Z.; Toga, A.W.; Jacobs, R.E.; Liu, C.Y.; Amezcua, L.; et al. Blood-Brain Barrier Breakdown in the Aging Human Hippocampus. *Neuron* **2015**, *85*, 296–302. [CrossRef]
137. Kuchibhotla, K.V.; Lattarulo, C.R.; Hyman, B.T.; Bacskai, B.J. Synchronous Hyperactivity and Intercellular Calcium Waves in Astrocytes in Alzheimer Mice. *Science* **2009**, *323*, 1211–1215. [CrossRef]
138. Hemonnot, A.-L.; Hua, J.; Ulmann, L.; Hirbec, H. Microglia in Alzheimer Disease: Well-Known Targets and New Opportunities. *Front. Aging Neurosci.* **2019**, *11*, 233. [CrossRef] [PubMed]
139. Griffin, W.S.; Stanley, L.C.; Ling, C.; White, L.; MacLeod, V.; Perrot, L.J.; White, C.L.; Araoz, C. Brain interleukin 1 and S-100 immunoreactivity are elevated in Down syndrome and Alzheimer disease. *Proc. Natl. Acad. Sci. USA* **1989**, *86*, 7611–7615. [CrossRef] [PubMed]
140. Hansen, D.V.; Hanson, J.E.; Sheng, M. Microglia in Alzheimer's Disease. *J. Cell Biol.* **2018**, *217*, 459–472. [CrossRef]
141. Mehra, A.; Ali, C.; Parcq, J.; Vivien, D.; Docagne, F. The Plasminogen Activation System in Neuroinflammation. *Biochim. et Biophys. Acta (BBA) Mol. Basis Dis.* **2016**, *1862*, 395–402. [CrossRef]
142. ElAli, A.; Bordeleau, M.; Thériault, P.; Filali, M.; Lampron, A.; Rivest, S. Tissue-Plasminogen Activator Attenuates Alzheimer's Disease-Related Pathology Development in APPswe/PS1 Mice. *Neuropsychopharmacology* **2016**, *41*, 1297–1307. [CrossRef] [PubMed]
143. Walker, D.G.; Lue, L.F.; Beach, T.G. Increased Expression of the Urokinase Plasminogen-Activator Receptor in Amyloid Beta Peptide-Treated Human Brain Microglia and in AD Brains. *Brain Res.* **2002**, *926*, 69–79. [CrossRef]
144. Hopperton, K.E.; Mohammad, D.; O Trépanier, M.; Giuliano, V.; Bazinet, R.P. Markers of Microglia in Post-Mortem Brain Samples from Patients with Alzheimer's Disease: A Systematic Review. *Mol. Psychiatry* **2018**, *23*, 177–198. [CrossRef]
145. Hawkins, B.T.; Davis, T.P. The Blood-Brain Barrier/Neurovascular Unit in Health and Disease. *Pharmacol. Rev.* **2005**, *57*, 173–185. [CrossRef]
146. Muller, U.C.; Deller, T.; Korte, M. Not Just Amyloid: Physiological Functions of the Amyloid Precursor Protein Family. *Nat. Rev. Neurosci.* **2017**, *18*, 281–298. [CrossRef]
147. Selkoe, D.J. Toward a Comprehensive Theory for Alzheimer's Disease. Hypothesis: Alzheimer's Disease is Caused by the Cerebral Accumulation and Cytotoxicity of Amyloid Beta-Protein. *Ann. NY Acad. Sci.* **2000**, *924*, 17–25. [CrossRef]
148. Selkoe, D.J.; Yamazaki, T.; Citron, M.; Podlisny, M.B.; Koo, E.H.; Teplow, D.B.; Haass, C. The Role of APP Processing and Trafficking Pathways in the Formation of Amyloid Beta-Protein. *Ann. NY Acad. Sci.* **1996**, *777*, 57–64. [CrossRef]

149. Kamenetz, F.; Tomita, T.; Hsieh, H.; Seabrook, G.; Borchelt, D.; Iwatsubo, T.; Sisodia, S.; Malinow, R. APP Processing and Synaptic Function. *Neuron* **2003**, *37*, 925–937. [CrossRef]

150. Blennow, K.; de Leon, M.J.; Zetterberg, H. Alzheimer's Disease. *Lancet* **2006**, *368*, 387–403. [CrossRef]

151. Collaborators, G.B.D.D. Global, Regional, and National Burden of Alzheimer's Disease and Other Dementias, 1990–2016: A Systematic Analysis for the Global Burden of Disease Study 2016. *Lancet Neurol.* **2019**, *18*, 88–106.

152. Almeida, C.G.; Tampellini, D.; Takahashi, R.H.; Greengard, P.; Lin, M.T.; Snyder, E.M.; Gouras, G.K. Beta-Amyloid Accumulation in APP Mutant Neurons Reduces PSD-95 and GluR1 in Synapses. *Neurobiol. Dis.* **2005**, *20*, 187–198. [CrossRef] [PubMed]

153. Aizenstein, H.J.; Nebes, R.D.; Saxton, J.A.; Price, J.C.; Mathis, C.A.; Tsopelas, N.D.; Ziolko, S.K.; James, J.A.; Snitz, B.E.; Houck, P.R.; et al. Frequent Amyloid Deposition Without Significant Cognitive Impairment Among the Elderly. *Arch. Neurol.* **2008**, *65*, 1509–1517. [CrossRef] [PubMed]

154. Lowe, S.L.; Willis, B.A.; Hawdon, A.; Natanegara, F.; Chua, L.; Foster, J.; Shcherbinin, S.; Ardayfio, P.; Sims, J.R. Donanemab (LY3002813) Dose-Escalation Study in Alzheimer's Disease. *Alzheimers Dement (NY)* **2021**, *7*, e12112.

155. Ledesma, M.D.; Da Silva, J.S.; Crassaerts, K.; Delacourte, A.; De Strooper, B.; Dotti, C.G. Brain Plasmin Enhances APP Alpha-Cleavage and Abeta Degradation and is Reduced in Alzheimer's Disease Brains. *EMBO Rep.* **2000**, *1*, 530–535. [CrossRef]

156. Tucker, H.M.; Kihiko, M.; Caldwell, J.N.; Wright, S.; Kawarabayashi, T.; Price, D.; Walker, D.; Scheff, S.; McGillis, J.P.; Rydel, R.E.; et al. The Plasmin System is Induced by and Degrades Amyloid-Beta Aggregates. *J. Neurosci.* **2000**, *20*, 3937–3946. [CrossRef]

157. Tucker, H.M.; Kihiko-Ehmann, M.; Wright, S.; Rydel, R.E.; Estus, S. Tissue Plasminogen Activator Requires Plasminogen to Modulate Amyloid-Beta Neurotoxicity and Deposition. *J. Neurochem.* **2000**, *75*, 2172–2177. [CrossRef]

158. Fabbro, S.; Seeds, N.W. Plasminogen activator activity is inhibited while neuroserpin is up-regulated in the Alzheimer disease brain. *J. Neurochem.* **2009**, *109*, 303–315. [CrossRef]

159. Barker, R.; Love, S.; Kehoe, P.G. Plasminogen and Plasmin in Alzheimer's Disease. *Brain Res.* **2010**, *1355*, 7–15. [CrossRef]

160. Dotti, C.G.; Galvan, C.; Ledesma, M.D. Plasmin Deficiency in Alzheimer's Disease Brains: Causal or Casual? *Neurodegener. Dis.* **2004**, *1*, 205–212. [CrossRef]

161. Kingston, I.B.; Castro, M.J.; Anderson, S. In Vitro Stimulation of Tissue-Type Plasminogen Activator by Alzheimer Amyloid Beta-Peptide Analogues. *Nat. Med.* **1995**, *1*, 138–142. [CrossRef] [PubMed]

162. Melchor, J.P.; Pawlak, R.; Strickland, S. The Tissue Plasminogen Activator-Plasminogen Proteolytic Cascade Accelerates Amyloid-Beta (Abeta) Degradation and Inhibits Abeta-Induced Neurodegeneration. *J. Neurosci.* **2003**, *23*, 8867–8871. [CrossRef] [PubMed]

163. Sutton, R.; Keohane, M.E.; Vandenberg, S.R.; Gonias, S.L. Plasminogen activator inhibitor-1 in the cerebrospinal fluid as an index of neurological disease. *Blood Coagul. Fibrinolysis* **1994**, *5*, 167–172. [CrossRef] [PubMed]

164. Liu, R.M.; Van Groen, T.; Katre, A.; Cao, D.; Kadisha, I.; Ballinger, C.; Wang, L.; Carroll, S.L.; Li, L. Knockout of Plasminogen Activator Inhibitor 1 Gene Reduces Amyloid Beta Peptide Burden in a Mouse Model of Alzheimer's Disease. *Neurobiol. Aging* **2011**, *32*, 1079–1089. [CrossRef]

165. Medina, M.G.; Ledesma, M.D.; Domínguez, J.E.; Medina, M.; Zafra, D.; Alameda, F.; Dotti, C.G.; Navarro, P. Tissue Plasminogen Activator Mediates Amyloid-Induced Neurotoxicity via Erk1/2 Activation. *EMBO J.* **2005**, *24*, 1706–1716. [CrossRef]

166. Tucker, H.M.; Kihiko-Ehmann, M.; Estus, S. Urokinase-Type Plasminogen Activator Inhibits Amyloid-Beta Neurotoxicity and Fibrillogenesis via Plasminogen. *J. Neurosci. Res.* **2002**, *70*, 249–255. [CrossRef]

167. Price, J.L.; Morris, J.C. Tangles and Plaques in Nondemented Aging and "Preclinical" Alzheimer's Disease. *Ann. Neurol.* **1999**, *45*, 358–368. [CrossRef]

168. Terry, R.D.; Masliah, E.; Salmon, D.P.; Butters, N.; DeTeresa, R.; Hill, R.; Hansen, L.A.; Katzman, R. Physical Basis of Cognitive Alterations in Alzheimer's Disease: Synapse Loss is the Major Correlate of Cognitive Impairment. *Ann. Neurol.* **1991**, *30*, 572–580. [CrossRef]

169. Selkoe, D.J. Alzheimer's Disease is A Synaptic Failure. *Science* **2002**, *298*, 789–791. [CrossRef]

170. Palop, J.J.; Mucke, L. Amyloid-Beta-Induced Neuronal Dysfunction in Alzheimer's Disease: From Synapses toward Neural Networks. *Nat Neurosci.* **2010**, *13*, 812–818. [CrossRef]

171. Hsieh, H.; Boehm, J.; Sato, C.; Iwatsubo, T.; Tomita, T.; Sisodia, S.; Malinow, R. AMPAR Removal Underlies Abeta-Induced Synaptic Depression and Dendritic Spine Loss. *Neuron* **2006**, *52*, 831–843. [CrossRef]

172. Li, S.; Hong, S.; Shepardson, N.E.; Walsh, D.M.; Shankar, G.M.; Selkoe, D. Soluble Oligomers of Amyloid Beta Protein Facilitate Hippocampal Long-Term Depression by Disrupting Neuronal Glutamate Uptake. *Neuron* **2009**, *62*, 788–801. [CrossRef]

173. Cleary, J.P.; Walsh, D.M.; Hofmeister, J.J.; Shankar, G.M.; Kuskowski, M.A.; Selkoe, D.J.; Ashe, K.H. Natural Oligomers of the Amyloid-Beta Protein Specifically Disrupt Cognitive Function. *Nat. Neurosci.* **2005**, *8*, 79–84. [CrossRef]

174. Stratton, J.R.; Chandler, W.L.; Schwartz, R.S.; Cerqueira, M.D.; Levy, W.C.; E Kahn, S.; Larson, V.G.; Cain, K.C.; Beard, J.C.; Abrass, I.B. Effects of physical conditioning on fibrinolytic variables and fibrinogen in young and old healthy adults. *Circulation* **1991**, *83*, 1692–1697. [CrossRef]

175. Yaffe, K.; Barnes, D.; Nevitt, M.; Lui, L.Y.; Covinsky, K. A Prospective Study of Physical Activity and Cognitive Decline in Elderly Women: Women Who Walk. *Arch. Intern. Med.* **2001**, *161*, 1703–1708. [CrossRef]

176. Smith, P.J.; Blumenthal, J.A.; Hoffman, B.M.; Cooper, H.; Strauman, T.A.; Welsh-Bohmer, K.; Browndyke, J.N.; Sherwood, A. Aerobic Exercise and Neurocognitive Performance: A Meta-Analytic Review of Randomized Controlled Trials. *Psychosom. Med.* **2010**, *72*, 239–252. [CrossRef]

177. Poewe, W.; Seppi, K.; Tanner, C.M.; Halliday, G.M.; Brundin, P.; Volkmann, J.; Schrag, A.E.; Lang, A.E. Parkinson Disease. *Nat. Rev. Dis. Primers.* **2017**, *3*, 17013. [CrossRef]
178. Kim, K.S.; Choi, Y.R.; Park, J.Y.; Lee, J.H.; Kim, D.K.; Lee, S.J.; Paik, S.R.; Jou, I.; Park, S.M. Proteolytic Cleavage of Extracellular Alpha-Synuclein by Plasmin: Implications for Parkinson Disease. *J. Biol. Chem.* **2012**, *287*, 24862–24872. [CrossRef]
179. Reuland, C.J.; Church, F.C. Synergy between plasminogen activator inhibitor-1, α-synuclein, and neuroinflammation in Parkinson's disease. *Med. Hypotheses* **2020**, *138*, 109602. [CrossRef]
180. Pan, H.; Zhao, Y.; Zhai, Z.; Zheng, J.; Zhou, Y.; Zhai, Q.; Cao, X.; Tian, J.; Zhao, L. Role of Plasminogen Activator Inhibitor-1 in the Diagnosis and Prognosis of Patients with Parkinson's Disease. *Exp. Ther. Med.* **2018**, *15*, 5517–5522. [CrossRef]
181. Demestre, M.; Howard, R.S.; Orrell, R.W.; Pullen, A.H. Serine proteases purified from sera of patients with amyotrophic lateral sclerosis (ALS) induce contrasting cytopathology in murine motoneurones to IgG. *Neuropathol. Appl. Neurobiol.* **2006**, *32*, 141–156. [CrossRef] [PubMed]
182. Glas, M.; Popp, B.; Angele, B.; Koedel, U.; Chahli, C.; Schmalix, W.; Anneser, J.; Pfister, H.; Lorenzl, S. A role for the urokinase-type plasminogen activator system in amyotrophic lateral sclerosis. *Exp. Neurol.* **2007**, *207*, 350–356. [CrossRef] [PubMed]

International Journal of
Molecular Sciences

Review

Fibrinolytic System and Cancer: Diagnostic and Therapeutic Applications

Niaz Mahmood [1,2] and Shafaat A. Rabbani [1,2,*]

[1] Department of Medicine, McGill University, Montréal, QC H4A3J1, Canada; niaz.mahmood@mail.mcgill.ca
[2] Department of Medicine, McGill University Health Centre, Montréal, QC H4A3J1, Canada
* Correspondence: shafaat.rabbani@mcgill.ca

Abstract: Fibrinolysis is a crucial physiological process that helps to maintain a hemostatic balance by counteracting excessive thrombosis. The components of the fibrinolytic system are well established and are associated with a wide array of physiological and pathophysiological processes. The aberrant expression of several components, especially urokinase-type plasminogen activator (uPA), its cognate receptor uPAR, and plasminogen activator inhibitor-1 (PAI-1), has shown a direct correlation with increased tumor growth, invasiveness, and metastasis. As a result, targeting the fibrinolytic system has been of great interest in the field of cancer biology. Even though there is a plethora of encouraging preclinical evidence on the potential therapeutic benefits of targeting the key oncogenic components of the fibrinolytic system, none of them made it from "bench to bedside" due to a limited number of clinical trials on them. This review summarizes our existing understanding of the various diagnostic and therapeutic strategies targeting the fibrinolytic system during cancer.

Keywords: uPA; uPAR; PAI-1; PA system; cancer

Citation: Mahmood, N.; Rabbani, S.A. Fibrinolytic System and Cancer: Diagnostic and Therapeutic Applications. *Int. J. Mol. Sci.* **2021**, *22*, 4358. https://doi.org/10.3390/ijms22094358

Academic Editor: Hau C. Kwaan

Received: 28 February 2021
Accepted: 19 April 2021
Published: 22 April 2021

Publisher's Note: MDPI stays neutral with regard to jurisdictional claims in published maps and institutional affiliations.

1. Fibrinolysis and Fibrinolytic System: An Introduction

The association between fibrinolysis and cancer has been known for more than a hundred years. In the early part of the 20th century, it was reported that tumor tissues possess fibrinolytic properties [1,2]. Fischer observed that explants of tumors obtained from viral sarcomas in chickens could cause the breakdown of blood clots, whereas the explants obtained from normal connective tissue lacked such activity [3]. However, the importance of fibrinolysis during cancer progression remained underappreciated until the early 1970s, when the involvement of a proteolytic enzyme that increases fibrinolysis during oncogenic transformation was described [4,5]. Since then, understanding the biology of fibrinolysis during cancer progression as well as its therapeutic targeting has gained much attention. The increase in fibrinolytic activity in the tumor is primarily attributed to the plasminogen activators (PA) secreted by the tumors [6], and therefore the fibrinolytic system is interchangeably designated as the PA system in cancer. It is now well established that several components belonging to the fibrinolytic system are deregulated in cancer [7].

In normal physiological conditions, fibrinolysis refers to the enzymatic degradation process of the fibrin mesh of blood clots and is facilitated by the components of the PA system including the key protease plasmin, its precursor inactive plasminogen, and activators tissue- and urokinase-type plasminogen activators (tPA and uPA) [7]. Plasmin is primarily derived from inactive plasminogen by the action of either tPA or uPA. Once activated, plasmin degrades the accumulated fibrin into soluble fibrin degradation products (FDP). Plasmin activities are balanced by plasmin inhibitors α2-antiplasmin (α2-AP) and α2-macroglobulin to counteract the free plasmin concentration [8,9]. On the other hand, plasminogen activator inhibitor-1 (PAI-1) and -2 (PAI-2) and activated protein C inhibitor (PAI-3) modulate the function of plasminogen activators (tPA or uPA) [7,8,10,11]. The components of the fibrinolytic system play crucial roles in various other biological processes that include cell migration, tissue remodeling, modulation of various growth factors and cytokines, regulation of immune response, and chemotaxis [12–16].

Plasmin can directly deregulate the fibrinolytic system in various pathological conditions including cancer; however, therapeutic targeting of plasmin is not always a feasible option in clinical settings. This is due to the role of plasmin in crucial physiological processes such as tissue remodeling and thrombolysis [10], and blocking plasmin activity may lead to systemic disorders. Ploplis et al. demonstrated that animals with homozygous deletion of the plasminogen ($Plg^{-/-}$) gene showed growth retardation and reduced fertility and survival compared to the wildtype ($Plg^{+/+}$) group [17]. However, targeting the molecular factors (tPA, uPA, uPAR, and PAI-1) that are upstream of plasminogen did not impair normal growth characteristics of the animals. While there is strong evidence suggesting the involvement of uPA, uPA receptor (uPAR), and PAI-1 in various stages of cancer growth and progression, tPA is less commonly associated with cancer [7]. For example, Shapiro et al. showed that depletion of uPA inhibits tumorigenesis in a rodent model of melanoma by limiting the availability of critical growth-promoting factors such as the basic fibroblast growth factor (bFGF) in the tumor microenvironment [18]. Loss-of-function assays against uPA and uPAR genes increased apoptosis of cancer cells [19,20]. Furthermore, the uPA/uPAR/PAI-1 axis has been shown to play a critical role in mediating angiogenesis and metastasis in different types of cancer via the modulation of several well-known cancer-related signaling pathways such as the Ras/Raf/MEK/ERK and PI3K/AKT pathways [20–23]. Therefore, attempts to pharmacologically target uPA, uPAR, and PAI-1 have been made in the case of almost all types of cancer. For the rest of the review, we will therefore focus on these three components (uPA, uPAR, and PAI-1) of the fibrinolytic system and describe the advances made to target them in cancer therapeutics and diagnostics.

2. Components of the Fibrinolytic System as Diagnostic Biomarkers

Due to the extensively described multifaceted role of the components of the fibrinolytic system during different stages of cancer progression, various efforts have been conducted for their use in diagnostic approaches [9]. Some key examples of the potential use of the fibrinolytic components in cancer diagnosis are described below.

2.1. uPA and PAI-1 as Cancer Biomarkers

Around three decades ago, Duffy et al. first described that breast cancer patients showing a higher activity of uPA had a significantly shorter disease-free interval compared to those with lower uPA activity [24]. Later on, Jänicke et al., in two separate studies, reported on the elevated levels of uPA and PAI-1 proteins in primary breast tumors, which correlated with the poor prognosis of the patients [25,26]. Using data from 8377 breast cancer patients, it was further confirmed that the higher levels of uPA and/or PAI-1 in breast tumors correlate with the aggressiveness of cancer and poor relapse-free and overall survival of the cancer patients [27]. The clinical utility of uPA and PAI-1 as biomarkers has been demonstrated by two separate level-of-evidence-1 studies [28,29]. Several commercially available enzyme-linked immunosorbent assay (ELISA) kits have been developed for detecting the levels of uPA and PAI-1 proteins [30]. The clinical utility of one of the commercially available ELISA kits, FEMTELLE®, was further validated by a multicenter external quality assessment (EQA) program [31]. The results from six independent laboratories showed that the coefficient of variation (CV) to detect uPA and PAI-1 by using the FEMTELLE® kit ranged between 6.2–8.2% and 13.2–16.6%, respectively [31,32]. The American Society of Clinical Oncology (ASCO) recommended using an ELISA test to measure uPA and PAI-1 levels as prognostic markers for assessing the risk of breast cancer and a predictive marker to determine the suitable adjuvant therapies for the patients [33]. One of the major caveats of ELISA-based assays is the requirement of either fresh or fresh-frozen tissue samples, which is logistically challenging [34]. Therefore, the use of alternative methods without the need for freshly processed samples to determine the uPA and PAI-1 levels has been explored. The utilization of formalin-fixed paraffin-embedded tissues to assess uPA and

PAI-1 seems to be the most straightforward solution; however, the presence of uPA and PAI-1 antigens in both the tumor and stroma cells makes consistent immunohistochemical scoring very challenging [34]. With the emerging use of machine learning algorithms in various aspects of biology, it would be interesting if artificial intelligence (AI) technology can be used in this regard to automatically distinguish the tumor and surrounding stromal tissue and thereby solve a long-standing biological problem.

Several groups have attempted to measure uPA (also known as the *PLAU* gene) and PAI-1 (also known as *SERPINE1*) mRNAs in cancer through quantitative reverse transcription-polymerase chain reaction (qRT-PCR) [35,36] and nucleic acid sequence-based amplification (NASBA) assays [37]. Even though the use of the mRNA-based approach does not require fresh/fresh-frozen tissues, the incongruities between the studies to quantify and use the uPA and PAI-1 mRNAs as biomarkers for cancer patients prevented their use in clinical settings [34]. The recent advancements in sequencing technologies may overcome the cross-laboratory discrepancies in mRNA measurement, and targeted sequencing of uPA and PAI-1 mRNA may provide concrete evidence as to whether they can be used as biomarkers for cancer patients.

Epigenetic modification through DNA methylation provides an additional layer of transcriptional regulation of gene expression [38]. Anomalous DNA methylation is a hallmark of cancer [38], and DNA methylation-based biomarkers are emerging as promising frontiers for cancer diagnosis [39]. The higher stability of DNA, as well as its methylation marks and the fact that it can be efficiently isolated from formalin-fixed paraffin-embedded tissue samples, makes it well suited for use as diagnostic biomarkers [34]. Our group was the first to test and report the alteration in the status of uPA promoter methylation in cancer cells [40]. We further demonstrated an inverse association between uPA promoter DNA methylation and its mRNA expression as the tumor progresses to a more clinically aggressive grade [41]. uPA promoter methylation is decreased as the cancer becomes more aggressive, which indicates that the assessment of uPA promoter methylation may serve as an early diagnostic marker. A similar inverse relationship between mRNA expression and promoter DNA methylation has been demonstrated in the case of the PAI-1 gene [42]. DNA methylation assays are relatively simple to develop [39], and the advent of targeted sequencing technologies made it easier to assess the methylation sites on a specific location of the genome. Further studies on uPA and PAI-1 promoter methylation using larger cohorts of cancer patients belonging to different demographics are warranted to confirm their prognostic significance in cancer diagnosis.

2.2. uPAR as a Diagnostic Biomarker

Among the various members of the fibrinolytic system, uPAR holds a dominant position in terms of its applicability in cancer diagnosis and therapeutics. This is because of the fact that uPAR is scarcely present in healthy tissues, but its levels are elevated in malignancies where it is often associated with the aggressiveness of the cancer [9,43]. These characteristics of uPAR make it an ideal candidate for non-invasive imaging for cancer diagnosis and response to therapy. Almost two decades ago, we had shown that when an anti-rat uPAR antibody radiolabeled with ^{125}I was inoculated into animals with pre-established prostate and mammary tumors, an increase in radioactivity was determined in the primary tumors as well as various metastatic sites [44]. However, the rats receiving control IgG antibody radiolabeled with ^{125}I showed minimum radioactivity. This study suggested that uPAR imaging can be used for cancer diagnosis.

Various attempts have been made by different groups to utilize uPAR for cancer diagnosis. A small molecule uPAR binding peptide called AE105 in conjugation with ^{64}Cu-labeled DOTA was evaluated for positron-emission tomography (PET)-based molecular imaging [45,46]. In 2015, Persson et al. reported on the utilization of AE105 for the first ever uPAR PET scanning in humans [47]. Several other groups are using a uPAR-based imaging strategy to determine the aggressiveness of cancer in humans, and there are more than ten clinical trials that are either ongoing or have recently been completed. Some of the ongoing clinical trials include NCT02965001 (for head and neck cancer), NCT03307460 (for prostate cancer), NCT03278275 (neuroendocrine tumors), and NCT02755675 (malignant pleural mesothelioma, non-small cell lung cancer, and large cell neuroendocrine carcinoma of the lung).

Antibodies targeting uPAR (such as ATN-658, 2G10) and uPA (ATN-291) were also utilized for cancer imaging [48–50]. One of the major advantages of using antibodies for oncological imaging is that they possess a relatively longer half-life in the serum compared to peptide-based agents, thereby prolonging the timeframes for cancer imaging up to days [47,51]. On the other hand, the half-life of peptide-based agents such as AE105 is shorter, and the imaging timeframes may last several hours only [47]. Yang et al. targeted uPAR imaging through conjugation of the amino-terminal fragment (ATF) of uPA with magnetic iron oxide nanoparticles (ATF-IO) for imaging mammary tumors in vivo [52]. In summary, regardless of the strategies used, uPAR-based oncological imaging holds great promise for cancer diagnosis.

2.3. suPAR as a Cancer Biomarker

Another important avenue that holds great promise but needs more exploration is using soluble urokinase plasminogen activator receptor (suPAR) from body fluids as a biomarker of cancer. The plasma levels of suPAR are elevated in different types of pathological conditions such as inflammation, autoimmune diseases, virus infection, and chronic kidney diseases, where they serve as plasma biomarkers [53–56]. Rovina et al. demonstrated that suPAR can be potentially used as an early marker of respiratory failure in patients suffering from COVID-19-related pneumonia [57]. Elevated levels of suPAR were found in different types of cancer including colon, lung, prostate, breast, and ovarian cancers [53,58–61].

There two main forms of suPAR in the circulation: (i) the full-length suPAR, and (ii) the cleaved soluble uPAR (containing the D2 and D3 domains of uPAR) [62]. Due to the lack of the N-terminal domain, the cleaved form of suPAR cannot bind to most of the uPAR ligands, with the exception of formyl peptide receptor (FPR)-like 1 and 2 that do not require the presence of the D1 domain [62,63]. Therefore, the full-length suPAR is also known as the active form of suPAR and is more suitable as a biomarker than the cleaved form [10]. Different types of ELISA-based commercial kits are available, but there are variabilities between the kits. More recently, Winnicki et al. compared the performance of two well-known ELISA kits (Quantikine Human uPAR ELISA and suPARnostic™ assay) for kidney disease [64]. They took samples from patients suffering from kidney disease and compared them to healthy controls and found that the cut-off values vary between the two ELISA kits. Similar studies are also warranted in the case of cancer in order to exploit the true diagnostic potential of suPAR.

3. Therapeutic Targeting of the Fibrinolytic System

Among the components of the fibrinolytic system, uPA was the first to be targeted for the treatment of cancer. Over the years, different classes of agents were used to target uPA. Some of the most important ones are summarized below and depicted in Figure 1.

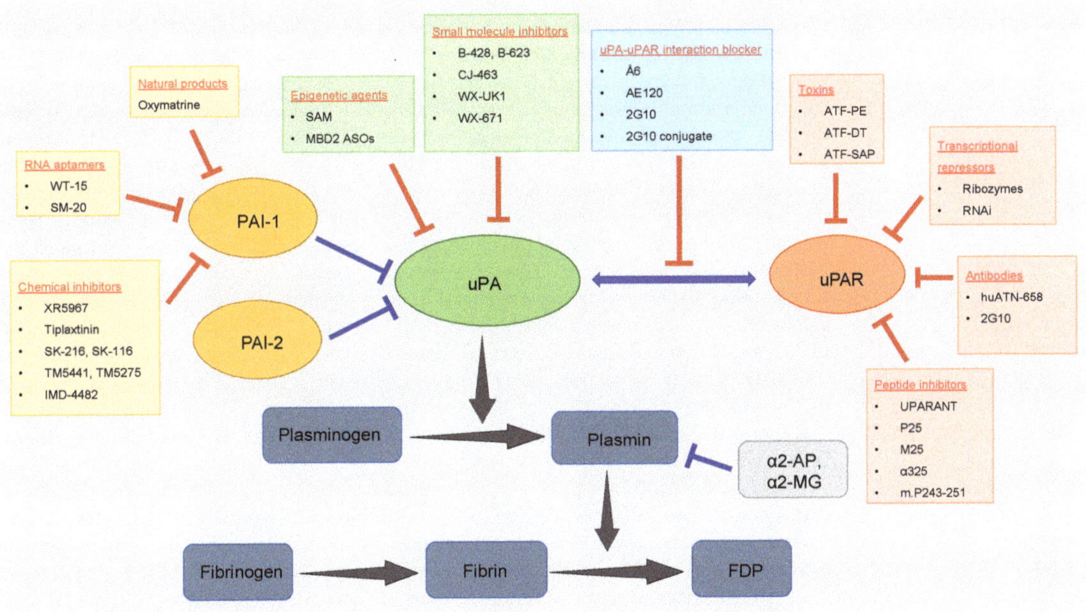

Figure 1. Components of the fibrinolytic system. The drug/inhibitors targeting the key members of the fibrinolytic system (uPA, uPAR, and PAI-1) that are deregulated in cancer are listed inside the colored boxes.

3.1. Small Molecule Inhibitors of uPA

In 1987, Vassalli et al. reported that amiloride could selectively block the catalytic activity of uPA [65]. Later on, by some modification in the amiloride structure, Towle et al. developed a novel class of small molecule inhibitors that interfered with the catalytic activity of uPA [66]. Further investigations of two compounds (B-428 and B-623) belonging to this class revealed that they possess higher inhibitory effects than amiloride to repress the uPA catalytic activity. Subsequently, work from our group demonstrated that B-428 administration significantly reduced prostate cancer growth and metastasis in vivo without causing any detrimental side effects [67]. We then checked whether single-agent treatment with B-428 could show similar anti-cancer effects in the case of other types of cancer and found that the inhibitor caused a statistically significant reduction in breast tumor volume and metastatic spread into the distant organ in preclinical settings [68]. Furthermore, B-428 in combination with tamoxifen (an approved drug for the treatment of hormone receptor-positive breast cancer) additively reduced mammary tumor volume and distant metastasis in vivo. CJ-463 is another small molecule inhibitor of uPA with an inhibitory constant (K_i) value of 20 nM [69] and significantly reduced tumor volume and metastasis in a murine model of lung cancer [70]. Wilex AG, a biopharmaceutical company, developed several potent uPA inhibitors that went to human clinical trials, where they showed some promising results. WX-UK1, a small molecule uPA inhibitor from Wilex AG, significantly reduced breast cancer growth and metastasis in rodent models [71]. Upamostat (also known as MESUPRON® or WX-671) is a second-generation serine protease inhibitor that targets uPA. Upamostat is a prodrug of WX-UK1 and has shown encouraging results in human clinical trials [72–74].

3.2. Repression of uPA Gene Expression by Epigenetic Agents

Epigenetic alterations are commonly seen during different malignancies [38], and targeting such abnormalities using epigenetic agents has become an area of immense interest over the last two decades, especially after the approval of the first epigenetic drug Vidaza against hematological cancers [75]. We have previously shown that *uPA* gene expression is increased in cancer because of DNA hypomethylation of its promoter region [41]. DNA methylation is a chemically reversible process [76], and the treatment of cancer cells with the global methyl donor S-adenosylmethionine (SAM) significantly reduces the expression of uPA through the reversal of the hypomethylated state in vitro [77–79]. More recently, we showed that oral administration of SAM reduces the tumor volume and metastasis of highly invasive triple-negative MDA-MB-231 breast cancer xenografts implanted into the fat pad of immunocompromised mice [80]. Further analysis of the primary tumors revealed that SAM treatment significantly reduces *uPA* gene expression compared to the vehicle-treated control arm. SAM is a naturally occurring agent with little to no documented toxicity, and such treatment holds promise for future combination studies with different potent anti-cancer agents. Moreover, microarray-based gene expression analyses of MDA-MB-231 breast cancer cells revealed a downregulation of genes involved in the uPA/uPAR pathway upon SAM treatment, suggesting the possible epigenetic modulation of the axis in cancer [80]. SAM is a pleiotropic molecule and utilized as a cofactor by many enzymes, some of which are involved in tumorigenesis. For example, nicotinamide N-methyltransferase (NNMT) uses endogenous SAM to mediate its enzymatic activity [81]. Overexpression of NNMT has been seen in many cancers [82–84]. However, there is no direct evidence showing increased NNMT activity upon exogenous SAM treatment to treat cancer. Regardless, more detailed studies are warranted to determine whether exogenous administration of SAM provides survival advantages to the cancer cells.

We have also shown that targeting methyl-CpG-binding domain protein 2 (MBD2), a protein that can read the DNA methylation marks and is frequently upregulated in many cancers [85], using a 20-mer antisense oligonucleotide (ASO) reduced uPA gene expression in breast and prostate cancer cells [77,78].

3.3. Transcriptional Repression of uPA and uPAR

Gene therapy-based strategies have been attempted to target the transcription of the uPA or uPAR gene. Karikó et al. synthesized a 37-mer hammerhead ribozyme to target the uPAR mRNA and delivered it into osteosarcoma cells using lipofectaime [86]. They found that the artificially synthesized ribozymes entered into the cytoplasm of cancer cells, cleaved the 1.4 kilobase uPAR mRNA, and thereby caused a dose-dependent decrease in its translation into a protein.

RNA interference (RNAi) against the uPA/uPAR genes has also been tested. Mohan et al. demonstrated that adenovirus-mediated delivery of an antisense construct targeting the elevated uPAR expression in high-grade glioma markedly reduced tumor growth in vivo [87]. Margheri et al. used an 18-mer ASO against uPAR known as "uPAR aODNs" that significantly reduced uPAR levels and subsequently decreased prostate cancer bone metastases when PC3 cells were injected into immunocompromised animals via intracardiac injections [88]. In SNB19 glioma cells, RNAi-mediated repression of uPA and uPAR gene expression retarded the oncogenic PI3K/AKT/mTOR axis and increased Fas ligand-mediated apoptosis [89]. Even though various attempts have been made to repress uPA and uPAR transcriptionally, none of them made it to clinical trials. One possible reason is that there has been a general reluctance to use "nucleic acid-based therapies" over the years. However, with the global use of mRNA-based vaccines against SARS-CoV-2, "nucleic acid-based therapies" have finally become more acceptable. Therefore, attempts to transcriptionally repress uPA, uPAR, and many other known oncogenes may see a boost in clinical settings in the near future.

3.4. Blocking the uPA–uPAR Interaction

Several approaches to block the interaction between uPA and its receptor uPAR have been tested over the years. In 1993, Crowley et al. used an enzymatically inactive uPA analog by mutating the 356th residue of uPA from serine to alanine and found a significant decrease in prostate cancer metastasis [90]. We have previously shown that Å6, a non-competitive inhibitor of the uPA–uPAR interaction, can cause a significant reduction in breast tumor growth and distant metastasis in vivo [91]. Other groups have reported similar anti-cancer therapeutic benefits of Å6 in other types of cancer as a single-agent monotherapy [92,93]. Moreover, when used in combination with other drugs, Å6 could significantly enhance the anti-cancer effects of tamoxifen [94] and cisplatin [95].

Using a phage display library, Duriseti et al. identified the 2G10 antibody that stably binds to the uPAR protein, blocks its interaction with uPA, and significantly suppresses the invasiveness of cancer cells in vitro [96]. The 2G10 antibody has shown promising therapeutic and diagnostic benefits in animal models of breast cancer [50,97]. More recently, Harel et al. conjugated the 2G10 antibody to an antimitotic cytotoxic payload called monomethylauristatin E (MMAE) via a cathepsin B-cleavable linker, RED-244 [98]. The 2G10 conjugate (2G10-RED-244-MMAE) showed an enhanced anti-cancer therapeutic potential to reduce breast tumors compared to the vehicle-treated control and 2G10 monotherapy-treated groups in vivo. Further studies using the 2G10 antibody alone or the conjugate in different types of cancer are warranted. AE120 is a peptide-based inhibitor of uPAR that has the ability to block uPA binding to uPAR and reduce the invasiveness of HEp-3 human epidermoid carcinoma cells [99].

3.5. Peptide Inhibitors against uPAR

uPAR is susceptible to cleavage by proteases such as plasmin, uPA, and metalloproteases. The most susceptible cleavage site of uPAR is located in the linker regions between the D1 and D2 domains. The uPAR protein lacking the D1 domain cannot bind to its most canonical binding partners uPA and vitronectin; however, it can bind to the FPRs and functions in cell migration [100]. Several small molecule peptide inhibitors have been developed that block the interaction between uPAR and FPRs to inhibit their internalization. UPARANT (also known as cenupatide) is the most well-known peptide inhibitor of uPAR that competes with N-formyl-Met-Leu-Phe (fMLF) for binding to the FPRs and thereby blocks the VEGF-directed angiogenesis [100,101]. In addition, several other peptide inhibitors against uPAR have been reported which include P25 [102], M25 [103], α325 [104], and m.P243-251 [105].

3.6. Antibodies against uPAR

Antibody-based targeted therapies have shown great promise over the last two decades. We have previously shown that a polyclonal rat anti-uPAR antibody causes a significant reduction in primary breast tumor growth and metastasis in preclinical settings [44]. Thereafter, a monoclonal antibody called ATN-658 was developed to target the human uPAR protein, and administration of the ATN-658 antibody significantly reduced the growth, invasiveness, and metastatic ability of prostate cancer cells both in vitro and in vivo [106]. The ATN-658 antibody is now fully humanized, and our recent studies demonstrated that treatment with the humanized ATN-658 (huATN-658) caused a significant reduction in primary breast tumor growth in vivo [107]. Furthermore, when human MDA-MB-231 and bone metastatic variant MDA-BoM-1833 breast cancer cells were implanted into the tibia of immunocompromised animals, the huATN-658 antibody significantly decreased breast tumor-induced skeletal lesions as well as the growth of the tumor cells within the bone microenvironment. Importantly, the anti-cancer effects showed a further enhancement in the group of animals treated with huATN-658 in combination with the Food and Drug Administration (FDA)-approved bisphosphonate zoledronic acid, suggesting the clinical utility of the antibody for human breast cancer patients. The anti-cancer therapeutic potential of ATN-658 has been evaluated in different types of cancer, where the antibody showed equally promising results [108,109]. As mentioned before, 2G10 is another uPAR antibody that has shown anti-cancer therapeutic potential in vivo [97].

3.7. Toxin Conjugation to Deliver the Drugs Targeting uPA/uPAR System

The use of toxins to treat cancer dates back to the early nineteenth century when Vautier described the spontaneous regression of tumors in patients with concurrent *Clostridium* infection [110]. Later on, William B. Coley, a physician based in New York, noticed a curative effect of erysipelas (a bacterial infection of the skin) on patients with sarcoma [111]. He later developed a vaccine from the cocktail of two killed bacteria (*Serratia marcescens* and *Streptococcus pyogenes*) known as "Coley's toxins" for the treatment of various types of cancer [112–114]. The earlier success of Dr. Coley's treatment strategy led to the development of recombinant toxins for use in cancer treatment [115]. The major advantage of toxin-based cancer therapeutics is that the bacteria can be manipulated for a more selective delivery system [110]. When used in combination with conventional standard-of-care cancer therapies, the bacterial toxins may enhance therapeutic response [115,116]. Several toxins targeting the uPA/uPAR system have been assessed for the treatment of different malignancies in the past two decades. The most prominent strategy in this regard has been the conjugation of ATF with a suitable toxin for targeting uPAR-expressing cells. For example, ATF conjugated with a truncated *Pseudomonas* exotoxin (PE) showed significant cytotoxic effects in a panel of well-established cancer cell lines belonging to various malignancies [117]. Vallera et al. demonstrated that conjugation of the catalytic portion of diphtheria toxin (DT) with ATF caused a significant reduction in the proliferation of glioblastoma cells in vitro and reduced tumor volumes in vivo [118]. In another study, a bispecific immunotoxin, DTATEGF, targeting the EGF/EGFR and uPA/uPAR axes showed a potent cytotoxic effect in human metastatic non-small cell lung cancer (NSCLC) brain tumor xenografts [119]. More recently, Zuppone et al. showed that conjugation of a ribosome-inactivating protein called saporin (SAP) with ATF significantly reduced the viability of breast and bladder cancer cell lines [120]. Furthermore, the anti-cancer effects of the ATF–SAP conjugate were selective towards cancer cells with no discernable effect on the growth of non-tumorigenic fibroblast cells expressing high levels of uPAR. Even though the bacterial toxins showed great promise for targeting the uPA/uPAR system, more research is needed before their successful translation to human clinical trials. Bacterial toxins may elicit unwanted immunogenicity and septic shock to the host, which is a major concern regarding their use in humans [115].

3.8. Chemical Inhibitors of PAI-1

Even though PAI-1 levels are markedly elevated in many types of cancer, therapeutic agents targeting PAI-1 have not been developed to the same extent as uPA/uPAR. The first class of PAI-1 inhibitors was reported in the 1990s [121]; however, they were mostly used for clot lysis rather than cancer therapeutics. XR5967, a diketopiperazine that can block the activity of human and murine PAI-1, significantly reduced cancer cell invasion, migration, and angiogenesis in vitro [122]. Tiplaxtinin (also known as PAI-039) inhibits cell proliferation, colony formation, and angiogenesis and increases apoptosis by blocking PAI-1 expression in several types of cancer [123–125]. Oral administration of a specific PAI-1 inhibitor, SK-216, inhibited tumor progression in a murine model of Lewis lung carcinoma [126]. However, in the same study, SK-216 administration did not show any significant effect in reducing B16 melanoma tumor volume in vivo, suggesting a possible cancer-type specificity of PAI-1 inhibition. Mutoh et al. showed that two PAI-1 inhibitors, SK-216 and SK-116, could individually reduce the number of intestinal polyps and thereby function as chemopreventive agents for colorectal cancer [127]. Two orally available small molecule anti-PAI-1 agents, TM5441 and TM5275, inhibited the proliferation of different types of cancer cell lines (MDA-MB-231 (breast cancer), HCT116 (colorectal cancer), HT1080 (fibrosarcoma), Jurkat (acute T cell leukemia), Daoy (medulloblastoma)) with an IC50 range between 9.7 and 60.3 μM [128]. However, the anti-cancer effects were not common for all the cancer types as some cell lines did not respond to TM5441 and TM5275 treatments in vitro [128]. In another study, TM5275 inhibited ovarian cancer cell proliferation in vitro [129]. Another PAI-1 inhibitor, IMD-4482, caused decreased ovarian cancer cell proliferation and invasion and induced apoptosis in vitro [130]. Moreover, IMD-4482 administration could reduce tumor volume and distant metastasis by inhibiting the phosphorylation of focal adhesion kinase (FAK) in vivo.

3.9. RNA Aptamers as PAI-1 Inhibitors

RNA aptamers are single-stranded nucleic acids that can tightly bind to specific targets and are used for various diagnostic and therapeutic applications [131]. By using combinatorial chemistry techniques, Blake et al. identified the RNA aptamers SM-20 and WT-15 that bind to PAI-1 with high affinity and specificity and thereby disrupt the interaction of PAI-1 with vitronectin and heparin [132]. The disruption in the PAI-1–vitronectin interaction shows anti-metastatic potential. WT-15 and SM-20 could also reduce the invasiveness and migratory properties of the highly invasive MDA-MB-231 breast cancer cell line in vitro [133]. Further studies are warranted to know whether similar anti-metastatic properties of the RNA aptamers can be replicated in the case of other types of cancer. In addition, preclinical studies using RNA aptamers are also needed to understand whether they show similar anti-cancer effects in vivo.

3.10. Natural Products as PAI-1 Inhibitors

Wang et al. showed that treatment of colorectal cancer cells with oxymatrine, a quinolizidine alkaloid derived from the Chinese herb *Sophora flavescens*, caused a significant reduction in cell proliferation and migration [134]. At the molecular level, oxymatrine treatment reduced the expression of PAI-1 and components of the TGF-β signaling. Moreover, oxymatrine induced the expression of the epithelial cell marker (E-cadherin) while decreasing the expression of the mesenchymal marker (α-Smooth muscle actin), which reversed the epithelial-to-mesenchymal (EMT) state and thereby reduced migration. However, the exact mechanism by which oxymatrine downregulates PAI-1 expression is not known and warrants further exploration.

4. Conclusions and Future Perspectives

Even though several key components (uPA, uPAR, and PAI-1) of the fibrinolytic system are clearly deregulated in almost all cancers and are potential diagnostic and therapeutic targets, their clinical translation into human cancer patients has been relatively less explored. Some of the earlier attempts to inhibit components of the fibrinolytic system showed a modest response in preclinical settings as the drugs/inhibitors used were not entirely specific against the component. For example, amiloride can target several factors other than uPA, notably, the epithelial sodium channel (ENaC), acid-sensitive ionic channel (ASIC), and ornithine decarboxylase [135,136]. It is possible that the amiloride-mediated anti-cancer effects were seen because of the combined inhibition of several targets. Moreover, it is possible that the tumor cells activate alternative pathways to interfere with the efficacy of the drugs targeting the fibrinolytic system. Such activation of compensatory pathways has been observed for several approved anti-cancer drugs [137,138]. Therefore, detailed studies on the interplay between the various inhibitors of the fibrinolytic system during tumor progression are warranted in the future. With the advent of high-throughput screening technologies, it is now straightforward to assess the genome-wide effects of a particular drug. It will be interesting to utilize such screening methods to evaluate the collateral effects of targeting the various components of the fibrinolytic system. This will also help to design better combination strategies for treating different types of cancer.

One of the encouraging frontiers that has recently drawn much attention is the utility of uPAR-based oncological imaging. More than ten phase 2 trials assessing the effectiveness of uPAR-based imaging are either ongoing or have been completed so far. suPAR is emerging as a new candidate biomarker for several pathological conditions; however, the studies related to the use of suPAR as a potential biomarker are still at an infancy. More research is warranted in this regard. It remains to be seen whether the strong preclinical evidence of some of the antibodies and small molecule inhibitors against the components of the fibrinolytic system could still be translated in human patients through appropriate double-blinded randomized clinical trials.

Author Contributions: Conceptualization, S.A.R. and N.M.; writing—original draft preparation, N.M.; writing—review and editing, S.A.R. and N.M.; visualization, N.M.; supervision, S.A.R.; project administration and funding acquisition, S.A.R. All authors have read and agreed to the published version of the manuscript.

Funding: This work was supported by a grant from the Canadian Institutes for Health Research, PJT-156225, to S.A.R., N.M. is the recipient of the Fonds de la recherche en santé du Québec (FRQS) Doctoral fellowship from the Government of Quebec, Canada.

Institutional Review Board Statement: Not applicable.

Informed Consent Statement: Not applicable.

Data Availability Statement: Not applicable.

Conflicts of Interest: The authors declare no conflict of interest.

References

1. Carrel, A.; Burrows, M.T. Cultivation in vitro of malignant tumors. *J. Exp. Med.* **1911**, *13*, 571. [CrossRef]
2. Fischer, A. The cultivation of malignant tumor cells indefinitely outside the body. *J. Cancer Res.* **1925**, *9*, 62–70. [CrossRef]
3. Fischer, A. Beitrag zur Biologie der Gewebezellen. Eine vergleichend biologische Studie 92 PROTEINASES SECRETED BY BREAST TUMC) RS der normalen und malignen Gewebezellen in vitro. *Arch. Mikrosk. Anat. Entwickl.* **1925**, *104*, 210–261.
4. Unkeless, J.; Tobia, A.; Ossowski, L.; Quigley, J.; Rifkin, D.; Reich, E. An enzymatic function associated with transformation of fibroblasts by oncogenic viruses: I. Chick embryo fibroblast cultures transformed by avian RNA tumor viruses. *J. Exp. Med.* **1973**, *137*, 85–111. [CrossRef] [PubMed]
5. Ossowski, L.; Unkeless, J.; Tobia, A.; Quigley, N.P.; Rifkin, D.; Reich, E. An enzymatic function associated with transformation of fibroblasts by oncogenic viruses: II. Mammalian fibroblast cultures transformed by DNA and RNA tumor viruses. *J. Exp. Med.* **1973**, *137*, 112–126. [CrossRef] [PubMed]
6. Südhoff, T.; Schneider, W. Fibrinolytic mechanisms in tumor growth and spreading. *Clin. Investig.* **1992**, *70*, 631–636. [CrossRef] [PubMed]
7. Kwaan, H.C.; Lindholm, P.F. Seminars in thrombosis and hemostasis. In *Fibrin and Fibrinolysis in Cancer*; Thieme Medical Publishers: New York, NY, USA, 2019; pp. 413–422.
8. Bannish, B.E.; Chernysh, I.N.; Keener, J.P.; Fogelson, A.L.; Weisel, J.W. Molecular and physical mechanisms of fibrinolysis and thrombolysis from mathematical modeling and experiments. *Sci. Rep.* **2017**, *7*, 1–11. [CrossRef]
9. Mahmood, N.; Mihalcioiu, C.; Rabbani, S.A. Multifaceted role of the urokinase-type plasminogen activator (uPA) and its receptor (uPAR): Diagnostic, prognostic, and therapeutic applications. *Front. Oncol.* **2018**, *8*, 24. [CrossRef]
10. Lin, H.; Xu, L.; Yu, S.; Hong, W.; Huang, M.; Xu, P. Therapeutics targeting the fibrinolytic system. *Exp. Mol. Med.* **2020**, *52*, 367–379. [CrossRef] [PubMed]
11. Kwaan, H.C.; Mazar, A.P.; McMahon, B.J. Seminars in thrombosis and hemostasis. In *The Apparent uPA/PAI-1 Paradox in Cancer: More Than Meets the Eye*; Thieme Medical Publishers: New York, NY, USA, 2013; pp. 382–391.
12. Naldini, L.; Tamagnone, L.; Vigna, E.; Sachs, M.; Hartmann, G.; Birchmeier, W.; Daikuhara, Y.; Tsubouchi, H.; Blasi, F.; Comoglio, P.M. Extracellular proteolytic cleavage by urokinase is required for activation of hepatocyte growth factor/scatter factor. *Embo J.* **1992**, *11*, 4825–4833. [CrossRef] [PubMed]
13. Degryse, B.; Resnati, M.; Rabbani, S.A.; Villa, A.; Fazioli, F.; Blasi, F. Src-dependence and pertussis-toxin sensitivity of urokinase receptor-dependent chemotaxis and cytoskeleton reorganization in rat smooth muscle cells. *Blood J. Am. Soc. Hematol.* **1999**, *94*, 649–662.
14. Blasi, F. Urokinase and urokinase receptor: A paracrine/autocrine system regulating cell migration and invasiveness. *Bioessays* **1993**, *15*, 105–111. [CrossRef] [PubMed]
15. Mondino, A.; Blasi, F. uPA and uPAR in fibrinolysis, immunity and pathology. *Trends Immunol.* **2004**, *25*, 450–455. [CrossRef]
16. Sappino, A.P.; Huarte, J.; Belin, D.; Vassalli, J.D. Plasminogen activators in tissue remodeling and invasion: mRNA localization in mouse ova-ries and implanting embryos. *J Cell Biol.* **1989**, *109*, 2471–2479. [CrossRef] [PubMed]
17. Ploplis, V.A.; Carmeliet, P.; Vazirzadeh, S.; Van Vlaenderen, I.; Moons, L.; Plow, E.F.; Collen, D.S. Effects of disruption of the plasminogen gene on thrombosis, growth, and health in mice. *Circulation* **1995**, *92*, 2585–2593. [CrossRef] [PubMed]
18. Shapiro, R.L.; Duquette, J.G.; Roses, D.F.; Nunes, I.; Harris, M.N.; Kamino, H.; Wilson, E.L.; Rifkin, D.B. Induction of primary cutaneous melanocytic neoplasms in urokinase-type plasminogen activator (uPA)-deficient and wild-type mice: Cellular blue nevi invade but do not progress to malignant melanoma in uPA-deficient animals. *Cancer Res.* **1996**, *56*, 3597–3604.

19. Subramanian, R.; Gondi, C.S.; Lakka, S.S.; Jutla, A.; Rao, J.S. siRNA-mediated simultaneous downregulation of uPA and its receptor inhibits angiogenesis and invasiveness triggering apoptosis in breast cancer cells. *Int. J. Oncol.* **2006**, *28*, 831–839. [CrossRef]

20. Ma, Z.; Webb, D.J.; Jo, M.; Gonias, S.L. Endogenously produced urokinase-type plasminogen activator is a major determinant of the basal level of activated ERK/MAP kinase and prevents apoptosis in MDA-MB-231 breast cancer cells. *J. Cell Sci.* **2001**, *114*, 3387–3396.

21. Aguirre-Ghiso, J.A.; Estrada, Y.; Liu, D.; Ossowski, L. ERKMAPK activity as a determinant of tumor growth and dormancy; regulation by p38SAPK. *Cancer Res.* **2003**, *63*, 1684–1695.

22. LaRusch, G.A.; Mahdi, F.; Shariat-Madar, Z.; Adams, G.; Sitrin, R.G.; Zhang, W.M.; McCrae, K.R.; Schmaier, A.H. Factor XII stimulates ERK1/2 and Akt through uPAR, integrins, and the EGFR to initiate angiogenesis. *Blood* **2010**, *115*, 5111–5120. [CrossRef]

23. Zhang, W.; Xu, J.; Fang, H.; Tang, L.; Chen, W.; Sun, Q.; Zhang, Q.; Yang, F.; Sun, Z.; Cao, L. Endothelial cells promote triple-negative breast cancer cell metastasis via PAI-1 and CCL5 signaling. *FASEB J.* **2018**, *32*, 276–288. [CrossRef] [PubMed]

24. Duffy, M.J.; O'siorain, L.; O'grady, P.; Devaney, D.; Fennelly, J.J.; Lijnen, H.J. Urokinase-plasminogen activator, a marker for aggressive breast carcinomas. Preliminary report. *Cancer* **1988**, *62*, 531–533. [CrossRef]

25. Janicke, F.; Schmitt, M.; Ulm, K.; Gössner, W.; Graeff, H. Urokinase-type plasminogen activator antigen and early relapse in breast cancer. *Lancet* **1989**, *334*, 1049. [CrossRef]

26. Jänicke, F.; Schmitt, M.; Graeff, H. Seminars in Thrombosis and Hemostasis. In *Clinical Relevance of the Urokinase-Type and Tissue-Type Plasminogen Activators and of Their Type 1 Inhibitor in Breast Cancer*; Copyright© 2021 by Thieme Medical Publishers, Inc.: New York, NY, USA, 1991; pp. 303–312.

27. Look, M.P.; van Putten, W.L.; Duffy, M.J.; Harbeck, N.; Christensen, I.J.; Thomssen, C.; Kates, R.; Spyratos, F.; Fernö, M.R.; Eppenberger-Castori, S. Pooled analysis of prognostic impact of urokinase-type plasminogen activator and its inhibitor PAI-1 in 8377 breast cancer patients. *J. Natl. Cancer Inst.* **2002**, *94*, 116–128. [CrossRef] [PubMed]

28. Hayes, D.F.; Bast, R.C.; Desch, C.E.; Fritsche, H., Jr.; Kemeny, N.E.; Jessup, J.M.; Locker, G.Y.; Macdonald, J.S.; Mennel, R.G.; Norton, L.; et al. Tumor marker utility grading system: A framework to evaluate clinical utility of tumor markers. *J. Natl. Cancer Inst.* **1996**, *88*, 1456–1466. [CrossRef] [PubMed]

29. Simon, R.M.; Paik, S.; Hayes, D.F. Use of archived specimens in evaluation of prognostic and predictive biomarkers. *J. Natl. Cancer Inst.* **2009**, *101*, 1446–1452. [CrossRef]

30. Benraad, T.J.; Geurts-Moespot, J.; Grøndahl-Hansen, J.; Schmitt, M.; Heuvel, J.; De Witte, J.; Foekens, J.; Leake, R.; Brünner, N.; Sweep, C. Immunoassays (ELISA) of urokinase-type plasminogen activator (uPA): Report of an EORTC/BIOMED-1 workshop. *Eur. J. Cancer* **1996**, *32*, 1371–1381. [CrossRef]

31. Sweep, C.G.; Geurts-Moespot, J.; Grebenschikov, N.; de Witte, J.H.; Heuvel, J.J.; Schmitt, M.; Duffy, M.J.; Jänicke, F.; Kramer, M.D.; Foekens, J.A.; et al. External quality assessment of trans-European multicentre antigen determinations (enzyme-linked immunosorbent assay) of urokinase-type plasminogen activator (uPA) and its type 1 inhibitor (PAI-1) in human breast cancer tissue extracts. *Br. J. Cancer* **1998**, *78*, 1434–1441. [CrossRef]

32. Duffy, M.J.; McGowan, P.M.; Harbeck, N.; Thomssen, C.; Schmitt, M. uPA and PAI-1 as biomarkers in breast cancer: Validated for clinical use in level-of-evidence-1 studies. *Breast Cancer Res.* **2014**, *16*, 1–10. [CrossRef]

33. Harris, L.; Fritsche, H.; Mennel, R.; Norton, L.; Ravdin, P.; Taube, S.; Somerfield, M.R.; Hayes, D.F.; Bast, R.C., Jr. American Society of Clinical Oncology 2007 update of recommendations for the use of tumor markers in breast cancer. *J. Clin. Oncol.* **2007**, *25*, 5287–5312. [CrossRef]

34. Schmitt, M.; Mengele, K.; Gkazepis, A.; Napieralski, R.; Magdolen, V.; Reuning, U.; Harbeck, N. Assessment of urokinase-type plasminogen activator and its inhibitor PAI-1 in breast cancer tissue: Historical aspects and future prospects. *Breast Care* **2008**, *3*, 3. [CrossRef]

35. Spyratos, F.; Bouchet, C.; Tozlu, S.; Labroquere, M.; Vignaud, S.; Becette, V.; Lidereau, R.; Bieche, I. Prognostic value of uPA, PAI-1 and PAI-2 mRNA expression in primary breast cancer. *Anticancer Res.* **2002**, *22*, 2997–3003. [PubMed]

36. Holzscheiter, L.; Kotzsch, M.; Luther, T.; Kiechle-Bahat, M.; Sweep, F.C.; Span, P.N.; Schmitt, M.; Magdolen, V. Quantitative RT-PCR assays for the determination of urokinase-type plasminogen activator and plasminogen activator inhibitor type 1 mRNA in primary tumor tissue of breast cancer patients: Comparison to antigen quantification by ELISA. *Int. J. Mol. Med.* **2008**, *21*, 251–259.

37. Lamy, P.-J.; Verjat, T.; Servanton, A.-C.; Paye, M.; Leissner, P.; Mougin, B. Urokinase-type plasminogen activator and plasminogen activator inhibitor type-1 mRNA assessment in breast cancer by means of NASBA: Correlation with protein expression. *Am. J. Clin. Pathol.* **2007**, *128*, 404–413. [CrossRef] [PubMed]

38. Mahmood, N.; Rabbani, S.A. DNA methylation and breast cancer: Mechanistic and therapeutic applications. *Trends Cancer Res.* **2017**, *12*, 1–18.

39. Locke, W.J.; Guanzon, D.; Ma, C.; Liew, Y.J.; Duesing, K.R.; Fung, K.Y.C.; Ross, J.P. DNA Methylation Cancer Biomarkers: Translation to the Clinic. *Front. Genet.* **2019**, *10*, 1150. [CrossRef]

40. Xing, R.H.; Rabbani, S.A. Transcriptional regulation of urokinase (uPA) gene expression in breast cancer cells: Role of DNA methylation. *Int. J. Cancer* **1999**, *81*, 443–450. [CrossRef]

41. Pakneshan, P.; Têtu, B.; Rabbani, S.A. Demethylation of urokinase promoter as a prognostic marker in patients with breast carcinoma. *Clin. Cancer Res.* **2004**, *10*, 3035–3041. [CrossRef] [PubMed]
42. Gao, S.; Skeldal, S.; Krogdahl, A.; Sørensen, J.A.; Andreasen, P.A. CpG methylation of the PAI-1 gene 5′-flanking region is inversely correlated with PAI-1 mRNA levels in human cell lines. *Thromb. Haemost.* **2005**, *94*, 651–660.
43. Boonstra, C.M.; Verspaget, W.H.; Ganesh, S.; Kubben, J.G.M.F.L.; Vahrmeijer, A.; van de Velde, J.H.C.; Kuppen, J.K.P.; Quax, H.A.P.; Sier, F.M.C. Clinical applications of the urokinase receptor (uPAR) for cancer patients. *Curr. Pharm. Des.* **2011**, *17*, 1890–1910. [CrossRef]
44. Rabbani, S.A.; Gladu, J. Urokinase receptor antibody can reduce tumor volume and detect the presence of occult tumor metastases in vivo. *Cancer Res.* **2002**, *62*, 2390–2397.
45. Persson, M.; Madsen, J.; Østergaard, S.; Jensen, M.M.; Jørgensen, J.T.; Juhl, K.; Lehmann, C.; Ploug, M.; Kjaer, A. Quantitative PET of human urokinase-type plasminogen activator receptor with 64Cu-DOTA-AE105: Implications for visualizing cancer invasion. *J. Nucl. Med.* **2012**, *53*, 138–145. [CrossRef]
46. Persson, M.; Hosseini, M.; Madsen, J.; Jørgensen, T.J.; Jensen, K.J.; Kjaer, A.; Ploug, M. Improved PET imaging of uPAR expression using new 64Cu-labeled cross-bridged peptide ligands: Comparative in vitro and in vivo studies. *Theranostics* **2013**, *3*, 618. [CrossRef] [PubMed]
47. Persson, M.; Skovgaard, D.; Brandt-Larsen, M.; Christensen, C.; Madsen, J.; Nielsen, C.H.; Thurison, T.; Klausen, T.L.; Holm, S.; Loft, A. First-in-human uPAR PET: Imaging of cancer aggressiveness. *Theranostics* **2015**, *5*, 1303. [CrossRef]
48. Yang, D.; Severin, G.W.; Dougherty, C.A.; Lombardi, R.; Chen, D.; Van Dort, M.E.; Barnhart, T.E.; Ross, B.D.; Mazar, A.P.; Hong, H. Antibody-based PET of uPA/uPAR signaling with broad applicability for cancer imaging. *Oncotarget* **2016**, *7*, 73912–73924. [CrossRef] [PubMed]
49. Boonstra, M.C.; van Driel, P.B.; van Willigen, D.M.; Stammes, M.A.; Prevoo, H.A.; Tummers, Q.R.; Mazar, A.P.; Beekman, F.J.; Kuppen, P.J.; van de Velde, C.J. uPAR-targeted multimodal tracer for pre-and intraoperative imaging in cancer surgery. *Oncotarget* **2015**, *6*, 14260. [CrossRef] [PubMed]
50. LeBeau, A.M.; Sevillano, N.; King, M.L.; Duriseti, S.; Murphy, S.T.; Craik, C.S.; Murphy, L.L.; VanBrocklin, H.F. Imaging the urokinase plasminongen activator receptor in preclinical breast cancer models of acquired drug resistance. *Theranostics* **2014**, *4*, 267. [CrossRef]
51. Baart, V.M.; Boonstra, M.C.; Sier, C.F. uPAR directed-imaging of head-and-neck cancer. *Oncotarget* **2017**, *8*, 20519. [CrossRef]
52. Yang, L.; Peng, X.-H.; Wang, Y.A.; Wang, X.; Cao, Z.; Ni, C.; Karna, P.; Zhang, X.; Wood, W.C.; Gao, X. Receptor-targeted nanoparticles for in vivo imaging of breast cancer. *Clin. Cancer Res.* **2009**, *15*, 4722–4732. [CrossRef]
53. Langkilde, A.; Hansen, T.W.; Ladelund, S.; Linneberg, A.; Andersen, O.; Haugaard, S.B.; Jeppesen, J.; Eugen-Olsen, J. Increased plasma soluble uPAR level is a risk marker of respiratory cancer in initially cancer-free individuals. *Cancer Epidemiol. Prev. Biomark.* **2011**, *20*, 609–618. [CrossRef] [PubMed]
54. Backes, Y.; van der Sluijs, K.F.; Mackie, D.P.; Tacke, F.; Koch, A.; Tenhunen, J.J.; Schultz, M.J. Usefulness of suPAR as a biological marker in patients with systemic inflammation or infection: A systematic review. *Intensive Care Med.* **2012**, *38*, 1418–1428. [PubMed]
55. Andersen, O.; Eugen-Olsen, J.; Kofoed, K.; Iversen, J.; Haugaard, S.B. Soluble urokinase plasminogen activator receptor is a marker of dysmetabolism in HIV-infected patients receiving highly active antiretroviral therapy. *J. Med Virol.* **2008**, *80*, 209–216. [CrossRef] [PubMed]
56. Hayek, S.S.; Sever, S.; Ko, Y.-A.; Trachtman, H.; Awad, M.; Wadhwani, S.; Altintas, M.M.; Wei, C.; Hotton, A.L.; French, A.L. Soluble urokinase receptor and chronic kidney disease. *N. Engl. J. Med.* **2015**, *373*, 1916–1925. [CrossRef] [PubMed]
57. Rovina, N.; Akinosoglou, K.; Eugen-Olsen, J.; Hayek, S.; Reiser, J.; Giamarellos-Bourboulis, E.J. Soluble urokinase plasminogen activator receptor (suPAR) as an early predictor of severe respiratory failure in patients with COVID-19 pneumonia. *Crit. Care* **2020**, *24*, 187. [CrossRef]
58. Brünner, N.; Nielsen, H.J.; Hamers, M.; Christensen, I.J.; Thorlacius-Ussing, O.; Stephens, R.W. The urokinase plasminogen activator receptor in blood from healthy individuals and patients with cancer. *Apmis* **1999**, *107*, 160–167. [CrossRef]
59. Riisbro, R.; Christensen, I.J.; Piironen, T.; Greenall, M.; Larsen, B.; Stephens, R.W.; Han, C.; Høyer-Hansen, G.; Smith, K.; Brünner, N. Prognostic significance of soluble urokinase plasminogen activator receptor in serum and cytosol of tumor tissue from patients with primary breast cancer. *Clin. Cancer Res.* **2002**, *8*, 1132–1141.
60. Almasi, C.E.; Brasso, K.; Iversen, P.; Pappot, H.; Høyer-Hansen, G.; Danø, K.; Christensen, I.J. Prognostic and predictive value of intact and cleaved forms of the urokinase plasminogen activator receptor in metastatic prostate cancer. *Prostate* **2011**, *71*, 899–907. [CrossRef]
61. Almasi, C.E.; Drivsholm, L.; Pappot, H.; Høyer-Hansen, G.; Christensen, I.J. The liberated domain I of urokinase plasminogen activator receptor–a new tumour marker in small cell lung cancer. *Apmis* **2013**, *121*, 189–196. [CrossRef]
62. Montuori, N.; Ragno, P. Multiple activities of a multifaceted receptor: Roles of cleaved and soluble uPAR. *Front. Biosci.* **2009**, *14*, 2494–2503. [CrossRef]
63. Iribarren, P.; Zhou, Y.; Hu, J.; Le, Y.; Wang, J.M. Role of formyl peptide receptor-like 1 (FPRL1/FPR2) in mononuclear phagocyte responses in Alzheimer disease. *Immunol. Res.* **2005**, *31*, 165–176. [CrossRef]

64. Winnicki, W.; Sunder-Plassmann, G.; Sengölge, G.; Handisurya, A.; Herkner, H.; Kornauth, C.; Bielesz, B.; Wagner, L.; Kikić, Ž.; Pajenda, S.; et al. Diagnostic and Prognostic Value of Soluble Urokinase-type Plasminogen Activator Receptor (suPAR) in Focal Segmental Glomerulosclerosis and Impact of Detection Method. *Sci. Rep.* **2019**, *9*, 13783. [CrossRef]

65. Vassalli, J.-D.; Belin, D. Amiloride selectively inhibits the urokinase-type plasminogen activator. *Febs Lett.* **1987**, *214*, 187–191. [CrossRef]

66. Towle, M.J.; Lee, A.; Maduakor, E.C.; Schwartz, C.E.; Bridges, A.J.; Littlefield, B.A. Inhibition of urokinase by 4-substituted benzo [b] thiophene-2-carboxamidines: An important new class of selective synthetic urokinase inhibitor. *Cancer Res.* **1993**, *53*, 2553–2559.

67. Rabbani, S.; Harakidas, P.; Davidson, D.J.; Henkin, J.; Mazar, A.P. Prevention of prostate-cancer metastasis in vivo by a novel synthetic inhibitor of urokinase-type plasminogen activator (uPA). *Int. J. Cancer* **1995**, *63*, 840–845. [CrossRef] [PubMed]

68. Xing, R.H.; Mazar, A.; Henkin, J.; Rabbani, S.A. Prevention of breast cancer growth, invasion, and metastasis by antiestrogen tamoxifen alone or in combination with urokinase inhibitor B-428. *Cancer Res.* **1997**, *57*, 3585–3593.

69. Schweinitz, A.; Steinmetzer, T.; Banke, I.J.; Arlt, M.J.; Stürzebecher, A.; Schuster, O.; Geissler, A.; Giersiefen, H.; Zeslawska, E.; Jacob, U. Design of novel and selective inhibitors of urokinase-type plasminogen activator with improved pharmacokinetic properties for use as antimetastatic agents. *J. Biol. Chem.* **2004**, *279*, 33613–33622. [CrossRef] [PubMed]

70. Henneke, I.; Greschus, S.; Savai, R.; Korfei, M.; Markart, P.; Mahavadi, P.; Schermuly, R.T.; Wygrecka, M.; Stürzebecher, J.R.; Seeger, W. Inhibition of urokinase activity reduces primary tumor growth and metastasis formation in a murine lung carcinoma model. *Am. J. Respir. Crit. Care Med.* **2010**, *181*, 611–619. [CrossRef] [PubMed]

71. Setyono-Han, B.; Stürzebecher, J.; Schmalix, W.A.; Muehlenweg, B.; Sieuwerts, A.M.; Timmermans, M.; Magdolen, V.; Schmitt, M.; Klijn, J.G.; Foekens, J.A. Suppression of rat breast cancer metastasis and reduction of primary tumour growth by the small synthetic urokinase inhibitor WX-UK1. *Thromb. Haemost.* **2005**, *93*, 779–786. [CrossRef]

72. Heinemann, V.; Ebert, M.P.; Pinter, T.; Bevan, P.; Neville, N.G.; Mala, C. Randomized phase II trial with an uPA inhibitor (WX-671) in patients with locally advanced nonmetastatic pancreatic cancer. *J. Clin. Oncol.* **2010**, *28* (Suppl. 15), 4060. [CrossRef]

73. Heinemann, V.; Ebert, M.P.; Laubender, R.P.; Bevan, P.; Mala, C.; Boeck, S. Phase II randomised proof-of-concept study of the urokinase inhibitor upamostat (WX-671) in combination with gemcitabine compared with gemcitabine alone in patients with non-resectable, locally advanced pancreatic cancer. *Br. J. Cancer* **2013**, *108*, 766–770. [CrossRef]

74. Meyer, J.E.; Brocks, C.; Graefe, H.; Mala, C.; Thäns, N.; Bürgle, M.; Rempel, A.; Rotter, N.; Wollenberg, B.; Lang, S. The oral serine protease inhibitor WX-671–first experience in patients with advanced head and neck carcinoma. *Breast Care* **2008**, *3* (Suppl. 2), 20. [CrossRef]

75. Lübbert, M.; Suciu, S.; Baila, L.; Rüter, B.H.; Platzbecker, U.; Giagounidis, A.; Selleslag, D.; Labar, B.; Germing, U.; Salih, H.R. Low-dose decitabine versus best supportive care in elderly patients with intermediate-or high-risk myelodysplastic syndrome (MDS) ineligible for intensive chemotherapy: Final results of the randomized phase III study of the European Organisation for Research and Treatment of Cancer Leukemia Group and the German MDS Study Group. *J. Clin. Oncol.* **2011**, *29*, 1987–1996.

76. Mahmood, N.; Rabbani, S.A. Targeting DNA Hypomethylation in Malignancy by Epigenetic Therapies. In *Human Cell Transformation*; Springer: Berlin/Heidelberg, Germany, 2019; pp. 179–196.

77. Pakneshan, P.; Szyf, M.; Farias-Eisner, R.; Rabbani, S.A. Reversal of the hypomethylation status of urokinase (uPA) promoter blocks breast cancer growth and metastasis. *J. Biol. Chem.* **2004**, *279*, 31735–31744. [CrossRef]

78. Shukeir, N.; Pakneshan, P.; Chen, G.; Szyf, M.; Rabbani, S.A. Alteration of the methylation status of tumor-promoting genes decreases prostate cancer cell invasiveness and tumorigenesis in vitro and in vivo. *Cancer Res.* **2006**, *66*, 9202–9210. [CrossRef]

79. Parashar, S.; Cheishvili, D.; Arakelian, A.; Hussain, Z.; Tanvir, I.; Khan, H.A.; Szyf, M.; Rabbani, S.A. S-adenosylmethionine blocks osteosarcoma cells proliferation and invasion in vitro and tumor metastasis in vivo: Therapeutic and diagnostic clinical applications. *Cancer Med.* **2015**, *4*, 732–744. [CrossRef] [PubMed]

80. Mahmood, N.; Cheishvili, D.; Arakelian, A.; Tanvir, I.; Khan, H.A.; Pépin, A.-S.; Szyf, M.; Rabbani, S.A. Methyl donor S-adenosylmethionine (SAM) supplementation attenuates breast cancer growth, invasion, and metastasis in vivo; therapeutic and chemopreventive applications. *Oncotarget* **2018**, *9*, 5169. [CrossRef]

81. Roberti, A.; Fernández, A.F.; Fraga, M.F. Nicotinamide N-methyltransferase: At the crossroads between cellular metabolism and epigenetic regulation. *Mol. Metab.* **2021**, *45*, 101165. [CrossRef] [PubMed]

82. Wang, Y.; Zeng, J.; Wu, W.; Xie, S.; Yu, H.; Li, G.; Zhu, T.; Li, F.; Lu, J.; Wang, G.Y. Nicotinamide N-methyltransferase enhances chemoresistance in breast cancer through SIRT1 protein stabilization. *Breast Cancer Res.* **2019**, *21*, 1–17. [CrossRef] [PubMed]

83. Pozzi, V.; Sartini, D.; Morganti, S.; Giuliante, R.; Di Ruscio, G.; Santarelli, A.; Rocchetti, R.; Rubini, C.; Tomasetti, M.; Giannatempo, G. RNA-mediated gene silencing of nicotinamide N-methyltransferase is associated with decreased tumorigenicity in human oral carcinoma cells. *PLoS ONE* **2013**, *8*, e71272. [CrossRef]

84. Palanichamy, K.; Kanji, S.; Gordon, N.; Thirumoorthy, K.; Jacob, J.R.; Litzenberg, K.T.; Patel, D.; Chakravarti, A. NNMT silencing activates tumor suppressor PP2A, inactivates oncogenic STKs, and inhibits tumor forming ability. *Clin. Cancer Res.* **2017**, *23*, 2325–2334. [CrossRef]

85. Mahmood, N.; Rabbani, S.A. DNA methylation readers and cancer: Mechanistic and therapeutic applications. *Front. Oncol.* **2019**, *9*, 489. [CrossRef]

86. Karikó, K.; Megyeri, K.; Xiao, Q.; Barnathan, E.S. Lipofectin-aided cell delivery of ribozyme targeted to human urokinase receptor mRNA. *Febs Lett.* **1994**, *352*, 41–44. [CrossRef]

136. Koo, J.; Parekh, D.; Townsend, C., Jr.; Saydjari, R.; Evers, B.; Farre, A.; Ishizuka, J.; Thompson, J. Amiloride inhibits the growth of human colon cancer cells in vitro. *Surg. Oncol.* **1992**, *1*, 385–389. [CrossRef]
137. Fiszman, G.L.; Jasnis, M.A. Molecular Mechanisms of Trastuzumab Resistance in HER2 Overexpressing Breast Cancer. *Int. J. Breast Cancer* **2011**, *2011*, 352182. [CrossRef]
138. Chandarlapaty, S. Negative feedback and adaptive resistance to the targeted therapy of cancer. *Cancer Discov.* **2012**, *2*, 311–319. [CrossRef] [PubMed]

International Journal of
Molecular Sciences

MDPI

Review

Thrombin Activatable Fibrinolysis Inhibitor (TAFI): An Updated Narrative Review

Machteld Sillen and Paul J. Declerck *

Laboratory for Therapeutic and Diagnostic Antibodies, Department of Pharmaceutical and Pharmacological Sciences, KU Leuven, B-3000 Leuven, Belgium; machteld.sillen@kuleuven.be
* Correspondence: paul.declerck@kuleuven.be

Abstract: Thrombin activatable fibrinolysis inhibitor (TAFI), a proenzyme, is converted to a potent attenuator of the fibrinolytic system upon activation by thrombin, plasmin, or the thrombin/thrombomodulin complex. Since TAFI forms a molecular link between coagulation and fibrinolysis and plays a potential role in venous and arterial thrombotic diseases, much interest has been tied to the development of molecules that antagonize its function. This review aims at providing a general overview on the biochemical properties of TAFI, its (patho)physiologic function, and various strategies to stimulate the fibrinolytic system by interfering with (activated) TAFI functionality.

Keywords: thrombin activatable fibrinolysis inhibitor; TAFI; coagulation; fibrinolysis; proCPU; proCPB; proCPR; carboxypeptidase

Citation: Sillen, M.; Declerck, P.J. Thrombin Activatable Fibrinolysis Inhibitor (TAFI): An Updated Narrative Review. *Int. J. Mol. Sci.* **2021**, *22*, 3670. https://doi.org/10.3390/ijms22073670

Academic Editor: Hau C. Kwaan

Received: 3 March 2021
Accepted: 29 March 2021
Published: 1 April 2021

Publisher's Note: MDPI stays neutral with regard to jurisdictional claims in published maps and institutional affiliations.

1. Introduction

Hemostasis is an essential process to safeguard the patency of the vascular system and the surrounding tissues and requires a delicate balance between the formation (coagulation) and the dissolution (fibrinolysis) of blood clots. Upon vascular injury, the coagulatory response is initiated, ultimately resulting in a thrombin burst, which plays a key role in the formation and stabilization of the fibrin clot as well as in the protection of this clot from degradation through the activation of thrombin activatable fibrinolysis inhibitor (TAFI) [1,2]. Normally, the coagulatory response is balanced by the action of the plasminogen activator/plasmin system [3]. The key fibrinolytic enzyme, plasmin, dissolves the blood clot by degrading the fibrin meshwork into soluble fibrin degradation products and exposing new carboxy-terminal (C-terminal) lysines at the fibrin surface, which serve to mediate a positive feedback mechanism in the fibrinolytic process by (I) promoting the binding of plasminogen and therefore also its activation to plasmin by tissue-type plasminogen activator (tPA) [4] and (II) by binding plasmin and thus protecting it against its major plasma inhibitor α_2-antiplasmin [5]. To prevent hyperfibrinolysis, the action of plasmin is negatively modulated at different levels. Firstly, at the level of plasminogen activation by plasminogen activator inhibitor-1 (PAI-1), which is the main physiological inhibitor of tPA and urokinase-type plasminogen activator (uPA) (reviewed in [6,7]). Secondly, at the level of plasmin by α_2-antiplasmin (reviewed in [8]). Thirdly, at the level of the blood clot by activated TAFI (TAFIa), a zinc-dependent metallocarboxypeptidase that removes the C-terminal lysines from the partially degraded fibrin clot and thereby abrogates the fibrin cofactor function in plasminogen activation (reviewed in [9,10]). As TAFI is being activated by thrombin, the key component of the coagulatory system, and attenuates the fibrinolytic response, TAFI forms an important molecular link between coagulation and fibrinolysis. Since a variety of studies have demonstrated a role for TAFI in thrombotic disorders (reviewed in [11,12]), several small molecule-, peptide-, and antibody-based inhibitors have been designed in order to explore the potential benefit of pharmacological inhibition of TAFI. This narrative review aims at providing a general overview on the biochemical properties of TAFI/TAFIa, the (patho)physiologic role of TAFIa, and various strategies to stimulate the fibrinolytic system by interfering with TAFI functionality.

2. Discovery and Nomenclature

Thrombin activatable fibrinolysis inhibitor (TAFI) was first discovered more than three decades ago in 1989 as a novel unstable molecular form of arginine carboxypeptidase activity in fresh serum prepared from human blood. Because of its instability, it was first named unstable carboxypeptidase (CPU) [13]. Shortly after, another independent study reported the identification of an arginine-specific carboxypeptidase (CPR) generated in blood during coagulation or inflammation [14]. In 1991, a third study revealed a plasminogen-binding protein being present in plasma with a similar amino acid sequence to pancreatic carboxypeptidase B and was therefore named plasma procarboxypeptidase B (plasma proCPB) [15]. In 1995, the discovery of a 60-kDa carboxypeptidase zymogen was reported, that upon activation by thrombin exerted antifibrinolytic effects [16]. This protein was accordingly named thrombin activatable fibrinolysis inhibitor (TAFI). Subsequent amino-terminal sequencing revealed that CPU, CPR, plasma proCPB, and TAFI were identical [17]. To emphasize its main physiological function during fibrinolysis (antifibrinolytic) and its connection to the coagulation system (activatable by thrombin), the term thrombin activatable fibrinolysis inhibitor (TAFI) is widely accepted and used to refer to the zymogen.

3. TAFI Synthesis and Distribution

The human TAFI encoding gene, *CPB2*, was mapped to chromosome 13 (13q14.11) and contains 11 exons [17,18]. Two out of 19 identified single-nucleotide polymorphisms (SNPs), +505 G/A and +1040 C/T located in the coding region, result in amino acid substitutions 147 Ala/Thr and 325 Thr/Ile, respectively [19]. As a consequence, TAFI exists as four isoforms, i.e., TAFI-A147-T325, TAFI-A147-I325, TAFI-T147-T325, and TAFI-T147-I325, of which the 325 Thr/Ile polymorphism has an impact on TAFIa stability [20].

TAFI is predominantly synthesized in the liver as a preproenzyme containing 423 amino acids and, after removal of the 22-residue long signal peptide, is secreted into the blood circulation as a 56-kDa proenzyme [15]. TAFI circulates in plasma at concentrations ranging from 73 to 275 nM (corresponding to 4–15 µg/mL) [21,22], of which the apparently large interindividual variation can mainly be attributed to the differential reactivity of the 325 Thr/Ile isoforms of TAFI in some commercially available enzyme-linked immunosorbent assays (ELISAs) [23]. Using isoform-independent ELISAs, it was observed that less than 25% of the variation in plasma levels is due to TAFI gene polymorphisms (outside the encoding region) that may modulate gene expression or affect mRNA stability [19,24,25].

Another pool of TAFI is synthesized in the precursors of blood platelets, the megakaryocytes, and is released upon activation of platelets by thrombin, adenosine diphosphate, and collagen [26]. Despite the minute amounts of TAFI stored in platelets, representing 0.1% of total blood TAFI, it was suggested that platelet-derived TAFI may play an important antifibrinolytic role through a local boost of TAFIa activity owing to the high concentration of platelets within the blood clot [26]. Indeed, TAFI secreted from platelets may affect the resistance to fibrinolysis that is conferred upon platelet-rich clots; however, it was recently shown that activation of plasma-derived TAFI, but not platelet-derived TAFI, is essential for the attenuation of fibrinolysis [27]. Activated TAFI (TAFIa) exerts its antifibrinolytic properties through a threshold-dependent mechanism, with the threshold-value being proportional to the plasmin concentration in plasma, which in turn depends on the concentrations of tPA and α_2-antiplasmin [28,29]. Importantly, only small amounts of TAFIa, i.e., 1% of total TAFI protein, are required to attenuate fibrinolysis [28].

4. TAFI Activation and Instability

4.1. TAFI Is a Metallocarboxypeptidase

Activated TAFI (TAFIa), a member of the metallocarboxypeptidase family, is a zinc-dependent exopeptidase that cleaves carboxy-terminal peptide bonds. The metallocarboxypeptidases are divided into two subfamilies, A and B. TAFI belongs to subfamily A that is characterized by a high structural similarity, i.e., a globular proenzyme consisting of two separate moieties: The activation peptide and the catalytic domain. In this respect, TAFI is a proenzyme that contains a 92-residue long amino-terminal (N-terminal) activation peptide (Phe1-Arg92, 20 kDa, heavily glycosilated) and a catalytic domain of 309 residues (Ala93-Val401, 36 kDa) (Figure 1A).

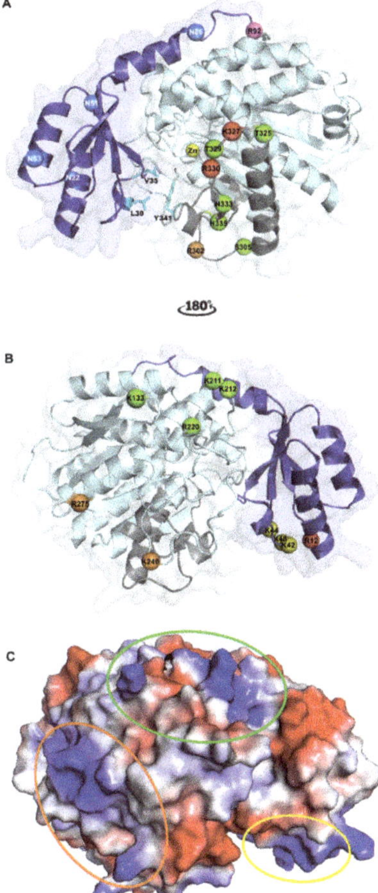

Figure 1. Crystallographic structure of human thrombin activatable fibrinolysis inhibitor (TAFI). (**A**) Cartoon representation of TAFI. The activation peptide (AP) and the catalytic moiety are colored in dark and light blue, respectively. The catalytic zinc-ion in the active center is shown as a yellow sphere. The four glycosylation sites in the AP (Asn22, Asn51, Asn63, and Asn86) are represented by blue spheres. TAFI can be activated through cleavage at Arg92 (shown as a magenta sphere) by thrombin, plasmin, or the thrombin/thrombomodulin complex. Upon the subsequent conformational change to inactivated TAFIa, (TAFIai) a cryptic cleavage site at Arg302 (shown as an orange sphere)

becomes exposed and can be cleaved by plasmin or thrombin. Two additional plasmin cleavage sites, Lys327 and Arg330, are indicated by red spheres. Five residues that have been mutated within the dynamic flap region (colored in grey) and result in the most stable TAFIa mutant are indicated by green spheres (Ser305Cys, Thr325Ile, Thr329Ile, His333Tyr, and His335Gln). The dynamic flap, of which the mobility leads to conformational changes that disrupt the catalytic site to form TAFIai, is stabilized by hydrophobic interactions between Val35 and Leu39 of the AP and Tyr341 in the dynamic flap (shown as cyan sticks). (**B**) Binding sites on TAFI for TAFI-activators after rotating panel A by 180° along the y-axis. The three putative thrombomodulin (TM) binding sites, Lys42/Lys43/Lys44, Lys133/Lys211/Lys212/Arg220, and Lys240/Arg275, are indicated by yellow, green, and orange spheres, respectively. Arginine at position 12, which plays an important role in TM-stimulated TAFI activation by thrombin, is indicated by a red sphere and may either constitute a potential cleavage site for thrombin or an exosite for TM. Furthermore, Lys133 may also be a part of the plasmin binding site on TAFI. (**C**) Charged surface representation of TAFI. The three putative TM binding sites are indicated by ovals in the same color as the spheres representing the binding sites in panel B. This figure was generated using the Protein Data Bank structure with PDB ID 3D66 [30]).

4.2. Three-Dimensional Strucures of TAFI and TAFIa

To date, 17 structures containing human, bovine, or porcine TAFI or TAFIa have been published online in the Protein Data Bank (PDB) (Table 1). The first crystal structure of TAFI was solved in 2008 using TAFI that was recombinantly expressed in a HEK293ES cell line and thus contained a homogenous N-linked glycan profile of the $(Man)_5(GlcNAc)_2$ type [30]. Five putative N-linked glycosylation sites were identified, of which four sites reside within the activation peptide (Asn22, Asn51, Asn63, and Asn86, Figure 1) and one within the TAFIa moiety (Asn219) [31]. However, only the activation peptide was shown to be heavily glycosylated, whereas glycosylation at the Asn219 site seems irreconcilable with the TAFI crystal structure as this residue is completely buried within the catalytic domain [30]. The first 76 residues of the activation peptide (Phe1-Val76) fold into four β-strands and two α-helices that form an open sandwich antiparallel α/β-fold, which is connected by a partially α-helical linker region (Glu77-Arg92) to the catalytic moiety. Both the structures of intact TAFI (PDB ID 3D66 [30]) and TAFIa (PDB ID 3LMS [32]) reveal that the catalytic moiety has a globular shape characterized by a typical α/β-hydrolase fold, comprising an eight-stranded mixed β-sheet flanked by nine α-helices.

Table 1. List of X-ray crystallographic structures containing TAFI or activated TAFI (TAFIa) in the Protein Data Bank (PDB).

Form	PDB ID	TAFI Variant [1]	Ligand [2]	Resolution (Å)	Ref.
TAFI	3D66	Human TAFI	-	3.1	[30]
	3D67	Human TAFI	GEMSA	3.4	[30]
	3D68	Human TAFI-IIYQ	-	2.8	[30]
	4P10	Human TAFI	Compound **5**	2.0	[33]
	3DGV	Bovine TAFI	-	2.5	-
	3OSL	Bovine TAFI	TCI	6.0	[34]
	5HVF	Human TAFI-CIIYQ	Nanobody VHH-i83	2.8	[35]
	5HVG	Human TAFI-CIIYQ	Nanobody VHH-a204	3.0	[35]
	5HVH	Human TAFI-CIIYQ	VHH-i83 + VHH-a204	3.0	[35]
TAFIa	3D4U	Bovine TAFIa	TCI	2.5	[36]
	3LMS	Human TAFIa	TCI	1.7	[32]
	4UIA	Porcine TAFIa	Compound **3a**	2.2	[37]
	4UIB	Porcine TAFIa	Compound **3p**	1.9	[37]
	5LYF	Porcine TAFIa	Urea lead of compounds **1** and **6–7a**	2.0	[38]
	5LYD	Porcine TAFIa	Compound **1**	2.0	[38]
	5LYI	Porcine TAFIa	Compound **7a**	1.6	[38]
	5LYL	Porcine TAFIa	Compound **6a**	1.8	[38]

[1] TAFI-IIYQ: TAFI-T325I-T329I-H333Y-H335Q; TAFI-CIIYQ: TAFI-S305C-T325I-T329I-H333Y-H335Q. [2] GEMSA: (2-guanidinoethylmercapto)-succinic acid; TCI: tick carboxypeptidase inhibitor.

4.3. TAFI Activation

Even though TAFI exerts low intrinsic carboxypeptidase activity, also referred to as zymogen activity [39], the antifibrinolytic property of TAFI relies on the carboxypeptidase activity of TAFIa [40]. TAFIa is generated upon proteolytic cleavage of the Arg92-Ala93 bond by trypsin-like proteases, such as thrombin or plasmin, resulting in the release of the activation peptide from the catalytic TAFIa moiety [15].

Thrombin is a weak activator of TAFI; however, by forming a complex with either soluble or membrane-bound thrombomodulin (TM), the catalytic efficiency of thrombin-mediated TAFI activation is increased 1250-fold [41,42]. Furthermore, the thrombin/thrombomodulin (T/TM) complex also efficiently generates activated protein C. Activated protein C is both a direct anticoagulant, i.e., by inactivating activated clotting factors V and VIII, as well as indirectly profibrinolytic, i.e., by attenuating prothrombin activation and thus the subsequent TAFI activation. Thus, whereas massive coagulation is prevented through the activation of protein C by T/TM, generation of TAFIa by T/TM results in a protection of the formed clot [43].

Alternatively, TAFI can be activated by plasmin, which is a stronger activator than thrombin [44]. Even though the efficiency of plasmin-mediated TAFI activation is enhanced by glycosaminoglycans such as heparin, the catalytic efficiency of plasmin/heparin remains 10-fold lower than that of the T/TM complex [44]. Therefore, the T/TM complex was postulated to be the main physiological activator of TAFI, as was also suggested by an in vivo study using a monoclonal antibody (mAb) that selectively inhibits T/TM-mediated TAFI activation [45]. In in vitro settings, a biphasic pattern in time associated with thrombin-induced (during thrombin formation) versus plasmin-induced TAFI activation (during the fibrinolytic phase) has been demonstrated [46]. In a consecutive study, the second TAFIa activity peak generated through plasmin-mediated activation could not be translated directly to the fibrinolytic rate [47]. However, other studies using mAbs that mainly impair plasmin-mediated TAFI activation revealed that plasmin contributes to TAFI activation both during clot formation and lysis in vitro [48], and showed to be a relevant physiological activator of TAFI in vivo as well [49]. This therefore indicates that plasmin-mediated TAFI activation may be of direct importance in vivo, where a more dynamic interplay exists between coagulation and fibrinolysis.

Using the crystal structure of thrombin in a complex with thrombomodulin fragments (epidermal growth factor (EGF)-like domains 4, 5, and 6 designated as TM-EGF456) (PDB ID 1DX5 [50]) and a homology model of TAFI that was built based on a crystal structure of human procarboxypeptidase B (PDB ID 1KWM [51]), a structural model of the ternary TAFI/thrombin/TM-EGF456 complex was built [52]. Based on this model and mutagenesis studies, three positively charged surface patches on TAFI, comprising residues Lys42/Lys43/Lys44, Lys133/Lys211/Lys212/Arg220, and Lys240/Arg275, have been suggested as binding sites for the C-loop of the TM-EGF-like domain 3 [52,53]. Furthermore, Arg12, which is located in close proximity to Lys42/Lys43/Lys44, is a potential thrombin cleavage site and plays an important role in TM-stimulated TAFI activation by thrombin [54]. However, at that point, it remained unclear whether cleavage at Arg12 accelerates TM-mediated TAFI activation or whether Arg12 belongs to an exosite for TM. A mutagenesis study later confirmed that both Arg12 and the triple lysine cluster (Lys42/Lys43/Lys44) were critical for the interaction of TAFI with TM as well as for the antifibrinolytic potential of TAFI [55]. Another study employing a deletion mutant of TAFI lacking the first 73 residues of the activation peptide (TAFI-S305C-T325I-T329I-H333Y-H335Q-Δ1–73 or TAFI-CIIYQ-Δ1–73), and thus also lacking Arg12 and the triple lysine cluster, further demonstrated that indeed this N-terminal part of the activation peptide is essential for the co-factor function of TM in accelerating TAFI-activation by thrombin [56]. In contrast, another TAFI mutant TAFI-K133A could be activated by thrombin and the T/TM complex but not by plasmin, indicating that Lys133 may be a part of the plasmin binding site on TAFI [49]. Together, this demonstrates that even though activation of TAFI by thrombin, the T/TM complex, or plasmin involves the same cleavage site at Arg92, these activators bind different residues or regions in the TAFI molecule.

4.4. TAFI Instability

Activated TAFI is thermally unstable, and spontaneously converts to an inactive conformation (designated as TAFIai) with a half-life of 8 min (Thr325 polymorphism) or 15 min (Ile325 polymorphism) at 37 °C [57,58]. As no physiological inhibitors of TAFI have been described, this intrinsic instability of TAFIa is thought to play a role in autoregulation of its antifibrinolytic activity in vivo. The crystal structure of intact TAFI (PDB ID 3D66) revealed poor electron density levels and increased B-factors for a segment comprising residues Phe296-Trp350 in the catalytic domain, which is part of the catalytic cleft wall [30]. Because reduced electron density levels are mostly caused by a higher mobility, this region is therefore referred to as the dynamic flap region (Figure 1). In intact TAFI, the activation peptide shields the catalytic pocket, which contains the catalytic zinc ion coordinated by His159, Glu162, and His288, and stabilizes the dynamic flap region through hydrophobic interactions involving residues Val35 and Leu39 of the activation peptide and Tyr341 of the catalytic domain. In a later study, the TAFI deletion mutant TAFI-CIIYQ-Δ1–73 showed similar stability to that of intact TAFI-CIIYQ, whereas cleavage of the Arg92-Ala93 bond leads to the formation of a less stable activated TAFI-CIIYQ. This suggests that, apart from the interactions between the activation peptide and the dynamic flap, the segment Ala74-Arg92 may also contribute to the role of the activation peptide in stabilizing regions in the catalytic domain outside the dynamic flap region in intact TAFI [56]. Activation of TAFI, however, leads to the dissociation of the activation peptide and thus a disruption of these stabilizing interactions, resulting in an increased mobility in the dynamic flap, eventually leading to conformational changes that disrupt the catalytic site.

Interestingly, comparison of the crystal structure of intact TAFI (PDB ID 3D66 [30]) with the crystal structure of TAFI in complex with a carboxypeptidase inhibitor, (2-guanidinoethylmercapto)-succinic acid (GEMSA) (PDB ID 3D67 [30]) revealed that the dynamic flap is stabilized upon binding of GEMSA within the active site. Apart from being more stable at lower temperature [57], several TAFI mutants have been reported that result in a remarkable stabilization of TAFIa [59–61]. These mutants contain either four (TAFI-T325I-T329I-H333Y-H335Q or TAFI-IIYQ, PDB ID 3D68 [30]) or five (TAFI-S305C-T325I-T329I-H333Y-H335Q or TAFI-CIIYQ, [61]) stabilizing mutations in the dynamic flap region, stabilizing TAFIa through more extensive interactions between the dynamic flap and the stable core of the catalytic moiety, indicating the important role of this region in TAFIa instability. Importantly, the antifibrinolytic effects of these TAFIa mutants correlate with their increased stability, underscoring the importance of the intrinsic instability in limiting TAFIa activity. Furthermore, upon the conformational change of TAFIa to TAFIai, a cryptic cleavage site at Arg302 becomes exposed [30], resulting in a subsequent irreversible degradation of TAFIai through proteolytic cleavage by thrombin or plasmin [62]. In addition, cleavage of the proenzyme TAFI at Lys327 and Arg330 by plasmin results in an inactive 45 kDa fragment [63].

5. (Patho)Physiological Role of TAFI

Cardiovascular disease and thrombotic disorders are often caused by an increased coagulatory or an impaired fibrinolytic response. Due to the antifibrinolytic activity of TAFIa, elevated levels of TAFI/TAFIa are expected to generate a hypofibrinolytic state and constitute a potential risk factor for various thrombotic diseases. Furthermore, studies have shown that the SNPs in the TAFI gene contribute to plasma TAFI concentrations and may thus also contribute to a higher risk for these diseases [64–66]. Even though mice engineered to be completely TAFI deficient by gene targeting did not display any observable phenotype [67], another study demonstrated that a significant reduction in thrombus formation was observed in TAFI-deficient mice upon FeCl$_3$-induced vena cava thrombosis, indicating that TAFI may still play an important physiological role [68]. In this respect, several studies investigated the role of TAFI levels or the TAFI gene polymorphism as risk factors for the development of cardiovascular disease (extensively reviewed in [12]).

Even though several studies could provide a link between TAFI gene SNPs and cerebral venous thrombosis [69], venous thromboembolic disease [70,71], myocardial infarction [72,73], stroke [74], and coronary heart disease [75,76], a clear link remains controversial as it could not be established by many other studies [77–82]. Similarly, independent of the contribution from TAFI gene SNPs, ample evidence has been provided of a link between elevated TAFI levels and venous thromboembolic disease [83,84], deep vein thrombosis [85], stroke [82,86,87], and coronary heart disease [88]. On the other hand, whereas in France carriers of the 505A allele (i.e., the Thr147 isoform), which is associated with higher TAFI levels, showed an increased risk of coronary heart disease; this risk was decreased in carriers of the 505 A allele from Northern Ireland [89]. Similar controversial results were reported in studies in which carriers of SNPs resulting in lower TAFI levels showed an increased risk of deep vein thrombosis [78] and myocardial infarction [90], suggesting a complex relationship between TAFI and thrombotic disease.

Apart from having a profound role as an antifibrinolytic protein, an anti-inflammatory role has been described for TAFIa, as it is able to directly inactivate several inflammatory proteins, such as bradykinin, anaphylatoxins C3a and C5a, thrombin-cleaved osteopontin, and plasmin-cleaved chemerin (reviewed in [91]). Because bradykinin has vasodilating properties, TAFIa may also have a function in blood pressure regulation; however, the physiological relevance of this link is not completely understood as several studies reported conflicting data [92–94]. Moreover, TAFIa attenuates the formation of plasmin and it has also been reported that C-terminal lysines and arginines from cellular plasminogen receptors are also substrates of TAFIa [95], therefore suggesting a role for TAFIa in cellular processes involving wound healing, cell migration, and angiogenesis, which also contributes to its anti-inflammatory activity [96–98]. However, the exact role of TAFI in inflammation seems to be very complex, since TAFI deficiency in mice has shown to either worsen, have no effect on, or to improve the outcomes in diverse models of inflammatory disease [99]. Moreover, inflammation has been shown to affect hemostasis and may therefore contribute to the atherosclerotic and thrombotic components of cardiovascular disease [100]. Indeed, the high-grade systemic inflammation, which is observed in patients with inflammatory diseases, such as rheumatoid arthritis and inflammatory bowel disease, puts them at greater risk for developing cardiovascular disease [101,102]. Because of the potentially protective role of TAFI in inflammation-related disorders, as demonstrated in mice models of alveolitis, arthritis, and hepatic inflammation, caution must be taken when inhibiting TAFI [103–105].

6. Inhibition of TAFI Functionality

To date, no physiological TAFIa inhibitors have been found in plasma. However, several small molecules, peptides, and antibody-based inhibitors have been designed and characterized. Owing to its anti-fibrinolytic effect and association with thrombotic tendencies and risk for cardiovascular disease, TAFIa remains a putative drug target. Prevention of TAFI activation and direct inhibition of TAFIa are two potential pharmacological strategies in the development of profibrinolytic drugs.

6.1. Synthetic Peptides

Even though no physiological inhibitors of TAFIa have been identified, protein inhibitors that naturally occur in potatoes (potato tuber carboxypeptidase inhibitor, PTCI) [106], leeches (leech carboxypeptidase inhibitor, LCI) [107], and ticks (tick carboxypeptidase inhibitor, TCI) [108] have been described. They are competitive inhibitors of TAFIa, and the crystal structures of TAFIa in complex with TCI revealed their inhibitory mechanism, i.e., binding across the flexible surface segments that form the rim of the active site cleft and penetrating the active site, thereby blocking access of substrates to the active site [32]. Remarkably, a biphasic effect, i.e., prolonging clot lysis at low concentrations and enhancing clot lysis at high concentrations, has been observed for PTCI [106,109]. This phenomenon can be explained by the stabilizing effect of TAFIa inhibitors and the equilibrium between free and inhibitor-bound TAFIa. While free TAFIa is irreversibly inactivated by its thermal instability, inhibitor-bound TAFIa is stabilized by preventing conformational changes that cause inactivation of TAFIa. However, when the TAFIa-inhibitor complex slowly dissociates, TAFIa is released to replenish the free pool. As long as free TAFIa concentrations stay above the tPA-dependent threshold value, fibrinolysis will be attenuated and remains in its initial phase [106].

6.2. Small Molecule Inhibitors

Since TAFIa is a zinc-dependent metallocarboxypeptidase, its catalytic activity can be inhibited by chelating agents such as o-phenanthroline and ethylenediaminetetraacetic acid (EDTA) that chelate the essential zinc-ion in the active site [13,15,110]. On the other hand, reducing agents, such as 2-mercaptoethanol and dithiothreitol, can inhibit TAFIa by disrupting the disulfide bonds in the active site of TAFIa (Cys156-Cys169, Cys228-Cys252, and Cys243-Cys257) [13,15,110]. TAFIa is also sensitive to inhibition by small synthetic substrate analogs including organic arginine analogs such as 2-mercaptomethyl-3-guanidinoethylthiopropanoic acid (MERGETPA) and GEMSA or organic lysine analogs such as ε-aminocaproic acid (ε-ACA) [57,110,111]. Even though these inhibitors are most widely used both in in vitro and in vivo studies, the major drawback of these inhibitors is that they also show inhibitory capacity towards other plasma-circulating carboxypeptidases such as carboxypeptidase N (CPN). Apart from their lack of specificity, they are extremely polar, which may limit their oral availability when using them in an in vivo setting. As a consequence, from a drug discovery point of view, many efforts have been devoted to obtain more selective inhibitors of TAFIa with a favorable pharmacokinetic profile.

Several low molecular weight (LMW) inhibitors of TAFIa have been patented (extensively reviewed elsewhere [12]). Most of these inhibitors display a consensus structure consisting of three characteristic groups, i.e., (I) a basic group that mimics the lysine side chain to bind Asp256 at the bottom of the S1′ specificity pocket of TAFIa, (II) a carboxylic acid that corresponds to the C-terminal carboxylic acid of the lysine that it is replacing, and (III) a functional group to coordinate the catalytic zinc ion [112]. These synthetic inhibitors can best be categorized based on the zinc-coordinating functional group, which often contains an imidazole, thiol, phosphonic or phosphinic acid, sulfonamide, or selenium group. Even though several of these LMW inhibitors have entered phase I and phase II clinical studies and showed an excellent safety profile and selectivity towards TAFI, further development was often discontinued due to various reasons, e.g., unfavorable pharmacokinetic properties (low oral bioavailability, short elimination half-life), no superiority over the standard treatment, or for unknown reasons.

6.3. Antibodies and Antibody Fragments

Since small synthetic inhibitors often deal with specificity issues, antibodies have become a key tool in drug discovery owing to their specific binding characteristics and amenability to protein engineering. Several monoclonal antibodies (mAbs) have been raised against TAFI in mice and antibody fragments thereof, such as single-chain variable fragments (scFv), have been generated to circumvent immunogenicity problems that are frequently encountered with the murine parental mAb. However, these smaller derivatives often encounter other difficulties such as a reduced binding affinity or a decreased stability. In contrast, variable antigen-binding domains of camelid antibodies, called nanobodies, share a high degree of sequence identity with human variable domains, indicating lower immunogenicity in human, and show excellent binding affinities, a remarkable stability, and solubility in various conditions. The panels of mAbs and nanobodies that were generated to target TAFI can interfere with TAFI or TAFIa activity, TAFI activation, or use a combination of both inhibitory mechanisms (Table 2).

Table 2. Non-exhaustive list [1] of monoclonal antibodies (mAbs) and nanobodies (Nbs) that target TAFI or activated TAFI (TAFIa).

		Mechanism	Epitope Residues	Ref.
mAbs	MA-RT36A3F5	Destabilizes TAFIa	Arg188, His192	[113]
	MA-RT13B2	Destabilizes TAFIa; Directly inhibits TAFIa	Arg227, Ser251	[113]
	MA-RT30D8	Directly inhibits TAFIa	Arg227, Ser249, Ser251, Tyr260	[113]
	MA-RT82F12	Directly inhibits TAFIa	Arg227, Ser249, Ser251, Tyr260	[113]
	MA-TCK27A4	Inhibits P-, T-, and T/TM-mediated TAFI activation	Phe113	[48]
	MA-T12D11	Inhibits T/TM-mediated TAFI activation	Gly66	[114]
	MA-T94H3	Inhibits T/TM- and P-mediated TAFI activation	Val41	[114]
	MA-T1C10	Inhibits T/TM- and P-mediated TAFI activation	Gln45	[114]
	MA-TCK22G2	Inhibits P- and T-mediated TAFI activation	Thr147, Ala148	[48]
	MA-TCK11A9	Inhibits P-mediated TAFI activation; Directly inhibits TAFIa activity on fibrin	Lys268, Ser272, Arg276	[48,115]
	MA-TCK26D6	Inhibits P-mediated, and to a lesser extent, T-mediated TAFI activation; Directly inhibits TAFIa activity on fibrin	Asp87, Thr88	[49,115]
Nbs	VHH-mTAFI-i49	Stimulates TAFI activity Destabilizes TAFI	Arg227, Lys212	[116]
	VHH-a204	Inhibits P-, T-, and T/TM-mediated TAFI activation	Arg117, His118, His175, Gln178, Ile182, Gln184, Tyr186, Arg384	[35,117]
	VHH-i83	Inhibits P- and T/TM-mediated TAFI activation; Directly inhibits TAFIa	Arg12, Gln33, Gln45, Ser70, Val71, Gln292, Arg330, Thr367, Thr369	[117]

[1] This list only includes mAbs and Nbs for which the epitope has been mapped my mutagenesis or X-ray crystallographic studies.

Within the panel of monoclonal antibodies generated towards rat TAFIa, two mAbs, MA-RT36A3F5 and MA-RT13B2, were found to have a destabilizing effect on TAFIa, thereby shortening its functional half-life [113]. Mutagenesis studies suggested an important role for residues Arg188 and His192 on α-helix 6 in the epitope for MA-RT36A3F5 (Figure 2A). Since α-helix 6 is connected to the active site through α-helix 5, it was hypothesized that binding of MA-RT36A3F5 induces a conformational change, leading to a disruption of the zinc-binding motive in the active site, thereby destabilizing TAFIa. On the other hand, MA-RT13B2 binds on the opposite side of the TAFI molecule with respect to MA-RT36A3F5 and makes interactions with Arg227 and Ser251 located on the loops connecting α-helix 7 with α-helix 8 via the active site residue Arg217 and substrate binding sites Asn234 and Arg235. Apart from destabilizing TAFIa, MA-RT13B2 was also shown to directly interfere with TAFIa activity by reducing the hydrolysis rate of a chromogenic TAFIa substrate. Indeed, by binding to this region in TAFI, MA-RT13B2 may induce a conformational change in the aforementioned loop, which could impact both the stability of TAFIa as well as the accessibility of the active site. The latter is in agreement with the observation that the binding site of MA-RT13B2 was shown to partially overlap with those of two other mAbs within the panel, MA-RT30D8 and MA-RT82F12, which directly inhibit TAFIa activity [113]. Residues Arg227, Ser249, Ser251, and Tyr260 were involved in binding of both mAbs.

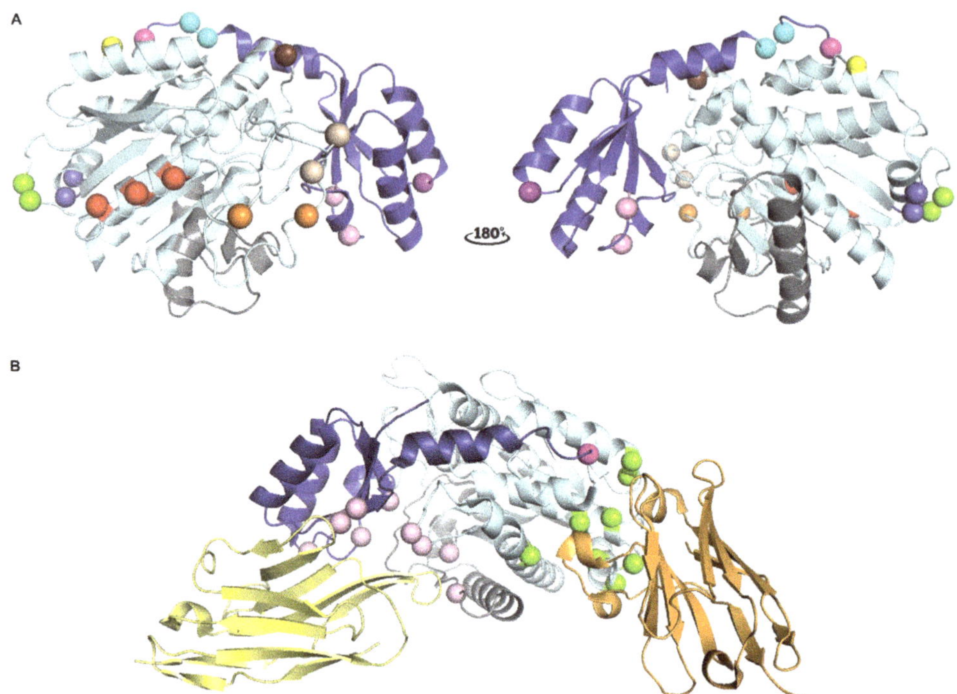

Figure 2. Localization of different epitopes in the structure of TAFI. (**A**) Localization of the epitopes of monoclonal antibodies (mAbs) that interfere with TAFI or TAFIa as determined by mutagenesis studies. The activation peptide (AP) and the catalytic moiety are colored in dark and light blue, respectively. The dynamic flap region is colored in grey. Major determinants of the epitopes are indicated as spheres. The cleavage site at Arg92 is indicated as a magenta sphere. Epitope residues for MA-RT36A3F5 and MA-RT13B2 are indicated by dark blue and light orange spheres, respectively. Epitope residues for MA-RT30D8 and MA-RT82F12 are indicated by the light and dark orange spheres. Epitope residues for nanobody VHH-mTAFI-i49, which destabilizes the TAFI proenzyme, are indicated by the brown and light orange sphere indicated by an asterisk. The epitope residue identified for MA-TCK27A4, which interferes with all modes of TAFI activation, is indicated by a yellow sphere. The epitope residue for MA-T12D11, which selectively inhibits T/TM-mediated TAFI activation, is indicated by a purple sphere. Epitope residues for MA-T94H3 and MA-T1C10, which interfere with both T/TM- and plasmin-mediated TAFI activation, are indicated by pink spheres. Epitope residues for MA-TCK22G2, which interferes with plasmin- and thrombin-mediated TAFI activation, are represented by green spheres. Epitope residues for MA-TCK11A9 and MA-TCK26D6, which mainly inhibit plasmin-mediated TAFI activation, are indicated by red and cyan spheres, respectively. Panel A was generated using the structure of intact human TAFI (PDB ID 3D66). (**B**) Cartoon representation of the crystal structure of TAFI in complex with nanobodies VHH-i83 (yellow) and VHH-a204 (orange) (PDB ID 5HVH). Residues that are engaged in polar interactions with VHH-i83 and VHH-a204 are indicated by pink and green spheres, respectively. The cleavage site at Arg12 is indicated as a magenta sphere.

Apart from modulating TAFIa activity and destabilizing TAFIa, another concept of TAFI inhibition was discovered with a nanobody, VHH-mTAFI-i49, which transiently stimulates the intrinsic activity of TAFI and simultaneously destabilizes the proenzyme, depleting the pool of activatable TAFI [116]. Epitope mapping revealed that Arg227 and Lys212 belong to the epitope and suggested that binding of the nanobody destabilizes TAFI by disrupting the stabilizing interactions between the activation peptide and the catalytic moiety of PAI-1. Importantly, this hypothesis is in line with the observation that the activation peptide stabilizes regions both within and outside of the dynamic flap of the catalytic moiety [56].

Antibodies that can inhibit T/TM-mediated TAFI activation, either exclusively or in combination with the inhibition of plasmin-mediated TAFI activation, have been shown to bind different regions that do not comprise the cleavage site (Figure 2A) [114]. Binding studies using a human/murine TAFI chimer revealed that the binding sites for these activation-inhibiting mAbs reside in the N-terminal region of the activation peptide of TAFI. Residue Gly66 was identified as a major determinant of the epitope of mAbs, such as MA-T12D11, that exclusively inhibit the T/TM-mediated TAFI activation [114]. Importantly, Gly66 is located on the surface of the protein in close proximity to Arg12 and the Lys42/Lys43/Lys44 region, which was proposed to be important for TM-stimulated TAFI activation by thrombin [52,53]. More recently, structures of TAFI in complex with a nanobody that specifically interferes with TM-dependent TAFI activation, VHH-i83, revealed that this nanobody directly interacts with Arg12, and thereby sterically blocks the binding of TM to the triple lysine cluster within the activation peptide (Figure 2B) [35]. Interestingly, this nanobody also has a direct inhibitory effect on TAFIa; however, only in the presence of the activation peptide [35]. Indeed, the structure of the TAFI-CIIYQ/VHH-i83 complex revealed a previously undescribed mechanism of TAFIa inhibition, i.e., tightly bridging the activation peptide with the catalytic moiety, forming a ternary complex that resembles the inactive proenzyme in which the active site is shielded. On the other hand, important binding residues for mAbs that can interfere with both T/TM- and plasmin-mediated TAFI activation, Val41 (MA-T94H3) and Gln45 (MA-T1C10), are located adjacent to the triple lysine cluster within the activation peptide but at a distance from Gly66 [114].

Antibodies that only inhibit plasmin and/or thrombin-mediated TAFI activation have been shown to bind regions outside the activation peptide (Figure 2A). Monoclonal Ab MA-TCK22G2 was shown to inhibit both plasmin- and thrombin-mediated TAFI activation and presumably binds to residues Thr147 and Ala148 in the loop connecting β-strand 2 and 3 within the catalytic moiety [48]. Monoclonal Ab MA-TCK11A9 was shown to selectively inhibit plasmin-mediated TAFI activation. The major determinants of the MA-TCK11A9 epitope were shown to reside in the α_4 helix (Lys268, Ser272, and Arg276) in the catalytic moiety of TAFI [48]. Since the epitope residues for MA-TCK22G2 and MA-TCK11A9 are located at a distance from Arg92, it was hypothesized that binding of these mAbs might induce a conformational change or allosteric modulation in the TAFI molecule, preventing plasmin and/or thrombin to activate TAFI. In the case of MA-TCK26D6, which mainly inhibits plasmin-mediated TAFI activation, residues Asp87 and Thr88 located within the activation peptide were shown to contribute to the epitope [49]. It should be noted that this binding site is located close to the Arg92 cleavage site, which is in line with the ability of MA-TCK26D6 to also inhibit thrombin-mediated TAFI activation; however, only to a lesser extent. Apart from the inhibition of plasmin-mediated TAFI activation, a supplemental inhibitory mechanism was revealed, as MA-TCK11A9 and MA-TCK26D6 also have a direct inhibitory effect against TAFIa activity on fibrin [115]. Most interestingly, in the presence of MA-TCK11A9 and MA-TCK26D6, TAFIa was still able to exert its anti-inflammatory role, through inactivation of pro-inflammatory mediators such as thrombin-cleaved osteopontin and C5a. This concept of TAFIa inhibition, leading to a profibrinolytic effect without compromising the strongly intertwined anti-inflammatory role, might therefore be of interest for the development of TAFI inhibitors to treat thrombotic diseases. Notably, due to the in vitro and in vivo potency and cross-reactivity toward rodent TAFI, the scFv fragment of MA-TCK26D6 (scFv-TCK26D6) was developed into a bispecific diabody format together with PAI-1-inhibiting scFv-33H1F7 [118]. Further in vivo evaluation and comparison with the standard thrombolytic therapy showed that the diabody, Db-TCK26D6x33H1F7, holds great promise in both the prevention and treatment of thrombotic disease [119,120]. Alternatively, taking into consideration the numerous advantages of nanobodies over conventional antibody formats, such as higher stability, lower immunogenicity, and better tissue and clot penetration, it might be of interest to pursue a similar dual-targeting strategy using bispecific nanobody constructs comprising one anti-TAFI and one

anti-PAI-1 nanobody.

Author Contributions: Conceptualization, M.S. and P.J.D.; writing—original draft preparation, M.S.; writing—review and editing, M.S. and P.J.D.; visualization, M.S. All authors have read and agreed to the published version of the manuscript.

Funding: This research received no external funding.

Institutional Review Board Statement: Not applicable.

Informed Consent Statement: Not applicable.

Data Availability Statement: Not applicable.

Conflicts of Interest: The authors declare no conflict of interest.

Abbreviations

C-terminal	Carboxy-terminal
CPN	Carboxypeptidase N
CPR	Arginine-specific carboxypeptidase
CPU	Unstable carboxypeptidase
EGF	Epidermal growth factor
ELISA	Enzyme-linked immunosorbent assay
GEMSA	(2-guanidinoethylmercapto)-succinic acid
LCI	Leech carboxypeptidase inhibitor
LMW	Low molecular weight
mAb	Monoclonal antibody
N-terminal	Amino-terminal
PAI-1	Plasminogen activator inhibitor-1
PDB	Protein Data Bank
proCPB	Procarboxypeptidase B
PTCI	Potato tuber carboxypeptidase inhibitor
scFv	Single-chain variable fragment
SNPs	Single-nucleotide polymorphisms
TAFI	Thrombin activatable fibrinolysis inhibitor
TAFIa	Activated thrombin activatable fibrinolysis inhibitor
TAFIai	Inactivated conformation of activated thrombin activatable fibrinolysis inhibitor
TAFI-CIIYQ	TAFI-S305C-T325I-T329I-H333Y-H335Q
TCI	Tick carboxypeptidase inhibitor
TM	thrombomodulin
tPA	Tissue-type plasminogen activator
T/TM	Thrombin/thrombomodulin
uPA	Urokinase-type plasminogen activator

References

1. Lord, S.T. Molecular mechanisms affecting fibrin structure and stability. *Arterioscler. Thromb Vasc. Biol.* **2011**, *31*, 494–499. [CrossRef]
2. Foley, J.H.; Kim, P.Y.; Mutch, N.J.; Gils, A. Insights into thrombin activatable fibrinolysis inhibitor function and regulation. *J. Thromb. Haemost.* **2013**, *11*, 306–315. [CrossRef]
3. Chapin, J.C.; Hajjar, K.A. Fibrinolysis and the control of blood coagulation. *Blood Rev.* **2015**, *29*, 17–24. [CrossRef]
4. Silva, M.M.; Thelwell, C.; Williams, S.C.; Longstaff, C. Regulation of fibrinolysis by C-terminal lysines operates through plasminogen and plasmin but not tissue-type plasminogen activator. *J. Thromb. Haemost.* **2012**, *10*, 2354–2360. [CrossRef] [PubMed]
5. Schneider, M.; Nesheim, M. A study of the protection of plasmin from antiplasmin inhibition within an intact fibrin clot during the course of clot lysis. *J. Biol. Chem.* **2004**, *279*, 13333–13339. [CrossRef]
6. Declerck, P.J.; Gils, A. Three decades of research on plasminogen activator inhibitor-1: A multifaceted serpin. *Semin. Thromb. Hemost.* **2013**, *39*, 356–364. [CrossRef] [PubMed]

7.   Sillen, M.; Declerck, P.J. Targeting PAI-1 in Cardiovascular Disease: Structural Insights Into PAI-1 Functionality and Inhibition. *Front. Cardiovasc. Med.* **2020**, *7*, 622473. [CrossRef]
8.   Singh, S.; Saleem, S.; Reed, G.L. Alpha2-Antiplasmin: The Devil You Don't Know in Cerebrovascular and Cardiovascular Disease. *Front. Cardiovasc. Med.* **2020**, *7*, 608899. [CrossRef] [PubMed]
9.   Declerck, P.J. Thrombin activatable fibrinolysis inhibitor. *Hamostaseologie* **2011**, *31*, 168–173. [CrossRef]
10.   Vercauteren, E.; Gils, A.; Declerck, P.J. Thrombin activatable fibrinolysis inhibitor: A putative target to enhance fibrinolysis. *Semin. Thromb. Hemost.* **2013**, *39*, 365–372. [CrossRef]
11.   Leurs, J.; Hendriks, D. Carboxypeptidase U (TAFIa): A metallocarboxypeptidase with a distinct role in haemostasis and a possible risk factor for thrombotic disease. *Thromb. Haemost.* **2005**, *94*, 471–487. [CrossRef] [PubMed]
12.   Claesen, K.; Mertens, J.C.; Leenaerts, D.; Hendriks, D. Carboxypeptidase U (CPU, TAFIa, CPB2) in Thromboembolic Disease: What Do We Know Three Decades after Its Discovery? *Int. J. Mol. Sci.* **2021**, *22*, 883. [CrossRef] [PubMed]
13.   Hendriks, D.; Scharpé, S.; van Sande, M.; Lommaert, M.P. Characterisation of a carboxypeptidase in human serum distinct from carboxypeptidase N. *J. Clin. Chem. Clin. Biochem.* **1989**, *27*, 277–285. [CrossRef] [PubMed]
14.   Campbell, W.; Okada, H. An arginine specific carboxypeptidase generated in blood during coagulation or inflammation which is unrelated to carboxypeptidase N or its subunits. *Biochem. Biophys. Res. Commun.* **1989**, *162*, 933–939. [CrossRef]
15.   Eaton, D.L.; Malloy, B.E.; Tsai, S.P.; Henzel, W.; Drayna, D. Isolation, molecular cloning, and partial characterization of a novel carboxypeptidase B from human plasma. *J. Biol. Chem.* **1991**, *266*, 21833–21838. [CrossRef]
16.   Côté, H.C.; Stevens, W.K.; Bajzar, L.; Banfield, D.K.; Nesheim, M.E.; MacGillivray, R.T. Characterization of a stable form of human meizothrombin derived from recombinant prothrombin (R155A, R271A, and R284A). *J. Biol. Chem.* **1994**, *269*, 11374–11380. [CrossRef]
17.   Vanhoof, G.; Wauters, J.; Schatteman, K.; Hendriks, D.; Goossens, F.; Bossuyt, P.; Scharpé, S. The gene for human carboxypeptidase U (CPU)—A proposed novel regulator of plasminogen activation–maps to 13q14.11. *Genomics* **1996**, *38*, 454–455. [CrossRef]
18.   Tsai, S.P.; Drayna, D. The gene encoding human plasma carboxypeptidase B (CPB2) resides on chromosome 13. *Genomics* **1992**, *14*, 549–550. [CrossRef]
19.   Boffa, M.B.; Maret, D.; Hamill, J.D.; Bastajian, N.; Crainich, P.; Jenny, N.S.; Tang, Z.; Macy, E.M.; Tracy, R.P.; Franco, R.F.; et al. Effect of single nucleotide polymorphisms on expression of the gene encoding thrombin-activatable fibrinolysis inhibitor: A functional analysis. *Blood* **2008**, *111*, 183–189. [CrossRef] [PubMed]
20.   Brouwers, G.J.; Vos, H.L.; Leebeek, F.W.; Bulk, S.; Schneider, M.; Boffa, M.; Koschinsky, M.; van Tilburg, N.H.; Nesheim, M.E.; Bertina, R.M.; et al. A novel, possibly functional, single nucleotide polymorphism in the coding region of the thrombin-activatable fibrinolysis inhibitor (TAFI) gene is also associated with TAFI levels. *Blood* **2001**, *98*, 1992–1993. [CrossRef] [PubMed]
21.   Bajzar, L.; Manuel, R.; Nesheim, M.E. Purification and characterization of TAFI, a thrombin-activable fibrinolysis inhibitor. *J. Biol. Chem.* **1995**, *270*, 14477–14484. [CrossRef] [PubMed]
22.   Mosnier, L.O.; von dem Borne, P.A.; Meijers, J.C.; Bouma, B.N. Plasma TAFI levels influence the clot lysis time in healthy individuals in the presence of an intact intrinsic pathway of coagulation. *Thromb. Haemost.* **1998**, *80*, 829–835. [PubMed]
23.   Gils, A.; Alessi, M.C.; Brouwers, E.; Peeters, M.; Marx, P.; Leurs, J.; Bouma, B.; Hendriks, D.; Juhan-Vague, I.; Declerck, P.J. Development of a genotype 325-specific proCPU/TAFI ELISA. *Arterioscler. Thromb. Vasc. Biol.* **2003**, *23*, 1122–1127. [CrossRef] [PubMed]
24.   Frère, C.; Morange, P.E.; Saut, N.; Tregouet, D.A.; Grosley, M.; Beltran, J.; Juhan-Vague, I.; Alessi, M.C. Quantification of thrombin activatable fibrinolysis inhibitor (TAFI) gene polymorphism effects on plasma levels of TAFI measured with assays insensitive to isoform-dependent artefact. *Thromb. Haemost.* **2005**, *94*, 373–379. [CrossRef]
25.   Frère, C.; Tregouet, D.A.; Morange, P.E.; Saut, N.; Kouassi, D.; Juhan-Vague, I.; Tiret, L.; Alessi, M.C. Fine mapping of quantitative trait nucleotides underlying thrombin-activatable fibrinolysis inhibitor antigen levels by a transethnic study. *Blood* **2006**, *108*, 1562–1568. [CrossRef]
26.   Mosnier, L.O.; Buijtenhuijs, P.; Marx, P.F.; Meijers, J.C.; Bouma, B.N. Identification of thrombin activatable fibrinolysis inhibitor (TAFI) in human platelets. *Blood* **2003**, *101*, 4844–4846. [CrossRef]
27.   Suzuki, Y.; Sano, H.; Mochizuki, L.; Honkura, N.; Urano, T. Activated platelet-based inhibition of fibrinolysis via thrombin-activatable fibrinolysis inhibitor activation system. *Blood Adv.* **2020**, *4*, 5501–5511. [CrossRef]
28.   Leurs, J.; Nerme, V.; Sim, Y.; Hendriks, D. Carboxypeptidase U (TAFIa) prevents lysis from proceeding into the propagation phase through a threshold-dependent mechanism. *J. Thromb. Haemost.* **2004**, *2*, 416–423. [CrossRef]
29.   Walker, J.B.; Bajzar, L. The intrinsic threshold of the fibrinolytic system is modulated by basic carboxypeptidases, but the magnitude of the antifibrinolytic effect of activated thrombin-activable fibrinolysis inhibitor is masked by its instability. *J. Biol. Chem.* **2004**, *279*, 27896–27904. [CrossRef]
30.   Marx, P.F.; Brondijk, T.H.; Plug, T.; Romijn, R.A.; Hemrika, W.; Meijers, J.C.; Huizinga, E.G. Crystal structures of TAFI elucidate the inactivation mechanism of activated TAFI: A novel mechanism for enzyme autoregulation. *Blood* **2008**, *112*, 2803–2809. [CrossRef]
31.   Valnickova, Z.; Christensen, T.; Skottrup, P.; Thøgersen, I.B.; Højrup, P.; Enghild, J.J. Post-translational modifications of human thrombin-activatable fibrinolysis inhibitor (TAFI): Evidence for a large shift in the isoelectric point and reduced solubility upon activation. *Biochemistry* **2006**, *45*, 1525–1535. [CrossRef]

32. Sanglas, L.; Arolas, J.L.; Valnickova, Z.; Aviles, F.X.; Enghild, J.J.; Gomis-Rüth, F.X. Insights into the molecular inactivation mechanism of human activated thrombin-activatable fibrinolysis inhibitor. *J. Thromb. Haemost.* **2010**, *8*, 1056–1065. [CrossRef] [PubMed]

33. Brink, M.; Dahlén, A.; Olsson, T.; Polla, M.; Svensson, T. Design and synthesis of conformationally restricted inhibitors of active thrombin activatable fibrinolysis inhibitor (TAFIa). *Bioorg. Med. Chem.* **2014**, *22*, 2261–2268. [CrossRef] [PubMed]

34. Valnickova, Z.; Sanglas, L.; Arolas, J.L.; Petersen, S.V.; Schar, C.; Otzen, D.; Aviles, F.X.; Gomis-Rüth, F.X.; Enghild, J.J. Flexibility of the thrombin-activatable fibrinolysis inhibitor pro-domain enables productive binding of protein substrates. *J. Biol. Chem.* **2010**, *285*, 38243–38250. [CrossRef]

35. Zhou, X.; Weeks, S.D.; Ameloot, P.; Callewaert, N.; Strelkov, S.V.; Declerck, P.J. Elucidation of the molecular mechanisms of two nanobodies that inhibit thrombin-activatable fibrinolysis inhibitor activation and activated thrombin-activatable fibrinolysis inhibitor activity. *J. Thromb. Haemost.* **2016**, *14*, 1629–1638. [CrossRef]

36. Sanglas, L.; Valnickova, Z.; Arolas, J.L.; Pallarés, I.; Guevara, T.; Solà, M.; Kristensen, T.; Enghild, J.J.; Aviles, F.X.; Gomis-Rüth, F.X. Structure of activated thrombin-activatable fibrinolysis inhibitor, a molecular link between coagulation and fibrinolysis. *Mol. Cell* **2008**, *31*, 598–606. [CrossRef]

37. Halland, N.; Brönstrup, M.; Czech, J.; Czechtizky, W.; Evers, A.; Follmann, M.; Kohlmann, M.; Schiell, M.; Kurz, M.; Schreuder, H.A.; et al. Novel Small Molecule Inhibitors of Activated Thrombin Activatable Fibrinolysis Inhibitor (TAFIa) from Natural Product Anabaenopeptin. *J. Med. Chem.* **2015**, *58*, 4839–4844. [CrossRef]

38. Halland, N.; Czech, J.; Czechtizky, W.; Evers, A.; Follmann, M.; Kohlmann, M.; Schreuder, H.A.; Kallus, C. Sulfamide as Zinc Binding Motif in Small Molecule Inhibitors of Activated Thrombin Activatable Fibrinolysis Inhibitor (TAFIa). *J. Med. Chem.* **2016**, *59*, 9567–9573. [CrossRef] [PubMed]

39. Willemse, J.L.; Polla, M.; Hendriks, D.F. The intrinsic enzymatic activity of plasma procarboxypeptidase U (TAFI) can interfere with plasma carboxypeptidase N assays. *Anal. Biochem.* **2006**, *356*, 157–159. [CrossRef] [PubMed]

40. Foley, J.H.; Kim, P.; Nesheim, M.E. Thrombin-activable fibrinolysis inhibitor zymogen does not play a significant role in the attenuation of fibrinolysis. *J. Biol. Chem.* **2008**, *283*, 8863–8867. [CrossRef]

41. Bajzar, L.; Morser, J.; Nesheim, M. TAFI, or plasma procarboxypeptidase B, couples the coagulation and fibrinolytic cascades through the thrombin-thrombomodulin complex. *J. Biol. Chem.* **1996**, *271*, 16603–16608. [CrossRef] [PubMed]

42. Bajzar, L.; Nesheim, M.; Morser, J.; Tracy, P.B. Both cellular and soluble forms of thrombomodulin inhibit fibrinolysis by potentiating the activation of thrombin-activable fibrinolysis inhibitor. *J. Biol. Chem.* **1998**, *273*, 2792–2798. [CrossRef]

43. Bajzar, L.; Nesheim, M. The effect of activated protein C on fibrinolysis in cell-free plasma can be attributed specifically to attenuation of prothrombin activation. *J. Biol. Chem.* **1993**, *268*, 8608–8616. [CrossRef]

44. Mao, S.S.; Cooper, C.M.; Wood, T.; Shafer, J.A.; Gardell, S.J. Characterization of plasmin-mediated activation of plasma procarboxypeptidase B. Modulation by glycosaminoglycans. *J. Biol. Chem.* **1999**, *274*, 35046–35052. [CrossRef]

45. Binette, T.M.; Taylor, F.B., Jr.; Peer, G.; Bajzar, L. Thrombin-thrombomodulin connects coagulation and fibrinolysis: More than an in vitro phenomenon. *Blood* **2007**, *110*, 3168–3175. [CrossRef]

46. Leurs, J.; Wissing, B.M.; Nerme, V.; Schatteman, K.; Björquist, P.; Hendriks, D. Different mechanisms contribute to the biphasic pattern of carboxypeptidase U (TAFIa) generation during in vitro clot lysis in human plasma. *Thromb. Haemost.* **2003**, *89*, 264–271. [CrossRef] [PubMed]

47. Leenaerts, D.; Aernouts, J.; Van Der Veken, P.; Sim, Y.; Lambeir, A.M.; Hendriks, D. Plasma carboxypeptidase U (CPU, CPB2, TAFIa) generation during in vitro clot lysis and its interplay between coagulation and fibrinolysis. *Thromb. Haemost.* **2017**, *117*, 1498–1508. [CrossRef]

48. Mishra, N.; Vercauteren, E.; Develter, J.; Bammens, R.; Declerck, P.J.; Gils, A. Identification and characterisation of monoclonal antibodies that impair the activation of human thrombin activatable fibrinolysis inhibitor through different mechanisms. *Thromb. Haemost.* **2011**, *106*, 90–101. [CrossRef] [PubMed]

49. Vercauteren, E.; Emmerechts, J.; Peeters, M.; Hoylaerts, M.F.; Declerck, P.J.; Gils, A. Evaluation of the profibrinolytic properties of an anti-TAFI monoclonal antibody in a mouse thromboembolism model. *Blood* **2011**, *117*, 4615–4622. [CrossRef]

50. Fuentes-Prior, P.; Iwanaga, Y.; Huber, R.; Pagila, R.; Rumennik, G.; Seto, M.; Morser, J.; Light, D.R.; Bode, W. Structural basis for the anticoagulant activity of the thrombin-thrombomodulin complex. *Nature* **2000**, *404*, 518–525. [CrossRef]

51. Barbosa Pereira, P.J.; Segura-Martín, S.; Oliva, B.; Ferrer-Orta, C.; Avilés, F.X.; Coll, M.; Gomis-Rüth, F.X.; Vendrell, J. Human procarboxypeptidase B: Three-dimensional structure and implications for thrombin-activatable fibrinolysis inhibitor (TAFI). *J. Mol. Biol.* **2002**, *321*, 537–547. [CrossRef]

52. Wu, C.; Kim, P.Y.; Manuel, R.; Seto, M.; Whitlow, M.; Nagashima, M.; Morser, J.; Gils, A.; Declerck, P.; Nesheim, M.E. The roles of selected arginine and lysine residues of TAFI (Pro-CPU) in its activation to TAFIa by the thrombin-thrombomodulin complex. *J. Biol. Chem.* **2009**, *284*, 7059–7067. [CrossRef]

53. Wang, W.; Nagashima, M.; Schneider, M.; Morser, J.; Nesheim, M. Elements of the primary structure of thrombomodulin required for efficient thrombin-activable fibrinolysis inhibitor activation. *J. Biol. Chem.* **2000**, *275*, 22942–22947. [CrossRef]

54. Plug, T.; Kramer, G.; Meijers, J.C. A role for arginine-12 in thrombin-thrombomodulin-mediated activation of thrombin-activatable fibrinolysis inhibitor. *J. Thromb. Haemost.* **2014**, *12*, 1717–1725. [CrossRef] [PubMed]

55. Wu, C.; Stafford, A.R.; Fredenburgh, J.C.; Weitz, J.I.; Gils, A.; Declerck, P.J.; Kim, P.Y. Lys 42/43/44 and Arg 12 of thrombin-activable fibrinolysis inhibitor comprise a thrombomodulin exosite essential for its antifibrinolytic potential. *Thromb. Haemost.* **2017**, *117*, 1509–1517. [CrossRef] [PubMed]

56. Zhou, X.; Declerck, P.J. Generation of a stable thrombin-activatable fibrinolysis inhibitor deletion mutant exerting full carboxypeptidase activity without activation. *J. Thromb. Haemost.* **2015**, *13*, 1084–1089. [CrossRef]

57. Boffa, M.B.; Wang, W.; Bajzar, L.; Nesheim, M.E. Plasma and recombinant thrombin-activable fibrinolysis inhibitor (TAFI) and activated TAFI compared with respect to glycosylation, thrombin/thrombomodulin-dependent activation, thermal stability, and enzymatic properties. *J. Biol. Chem.* **1998**, *273*, 2127–2135. [CrossRef]

58. Boffa, M.B.; Bell, R.; Stevens, W.K.; Nesheim, M.E. Roles of thermal instability and proteolytic cleavage in regulation of activated thrombin-activable fibrinolysis inhibitor. *J. Biol. Chem.* **2000**, *275*, 12868–12878. [CrossRef]

59. Ceresa, E.; Van de Borne, K.; Peeters, M.; Lijnen, H.R.; Declerck, P.J.; Gils, A. Generation of a stable activated thrombin activable fibrinolysis inhibitor variant. *J. Biol. Chem.* **2006**, *281*, 15878–15883. [CrossRef]

60. Knecht, W.; Willemse, J.; Stenhamre, H.; Andersson, M.; Berntsson, P.; Furebring, C.; Harrysson, A.; Hager, A.C.; Wissing, B.M.; Hendriks, D.; et al. Limited mutagenesis increases the stability of human carboxypeptidase U (TAFIa) and demonstrates the importance of CPU stability over proCPU concentration in down-regulating fibrinolysis. *FEBS J.* **2006**, *273*, 778–792. [CrossRef]

61. Ceresa, E.; Peeters, M.; Declerck, P.J.; Gils, A. Announcing a TAFIa mutant with a 180-fold increased half-life and concomitantly a strongly increased antifibrinolytic potential. *J. Thromb. Haemost.* **2007**, *5*, 418–420. [CrossRef]

62. Marx, P.F.; Hackeng, T.M.; Dawson, P.E.; Griffin, J.H.; Meijers, J.C.; Bouma, B.N. Inactivation of active thrombin-activable fibrinolysis inhibitor takes place by a process that involves conformational instability rather than proteolytic cleavage. *J. Biol. Chem.* **2000**, *275*, 12410–12415. [CrossRef]

63. Marx, P.F.; Dawson, P.E.; Bouma, B.N.; Meijers, J.C. Plasmin-mediated activation and inactivation of thrombin-activatable fibrinolysis inhibitor. *Biochemistry* **2002**, *41*, 6688–6696. [CrossRef] [PubMed]

64. Santos, I.R.; Fernandes, A.P.; Carvalho, M.G.; Sousa, M.O.; Ferreira, C.N.; Gomes, K.B. Thrombin-activatable fibrinolysis inhibitor (TAFI) levels and its polymorphism rs3742264 are associated with dyslipidemia in a cohort of Brazilian subjects. *Clin. Chim. Acta* **2014**, *433*, 76–83. [CrossRef] [PubMed]

65. Juhan-Vague, I.; Morange, P.E.; Aubert, H.; Henry, M.; Aillaud, M.F.; Alessi, M.C.; Samnegård, A.; Hawe, E.; Yudkin, J.; Margaglione, M.; et al. Plasma thrombin-activatable fibrinolysis inhibitor antigen concentration and genotype in relation to myocardial infarction in the north and south of Europe. *Arterioscler. Thromb. Vasc. Biol.* **2002**, *22*, 867–873. [CrossRef] [PubMed]

66. Fawzy, M.S.; Mohammed, E.A.; Ahmed, A.S.; Fakhr-Eldeen, A. Thrombin-activatable fibrinolysis inhibitor Thr325Ile polymorphism and plasma level in breast cancer: A pilot study. *Meta Gene* **2015**, *4*, 73–84. [CrossRef]

67. Nagashima, M.; Yin, Z.F.; Zhao, L.; White, K.; Zhu, Y.; Lasky, N.; Halks-Miller, M.; Broze, G.J., Jr.; Fay, W.P.; Morser, J. Thrombin-activatable fibrinolysis inhibitor (TAFI) deficiency is compatible with murine life. *J. Clin. Investig.* **2002**, *109*, 101–110. [CrossRef]

68. Wang, X.; Smith, P.L.; Hsu, M.Y.; Tamasi, J.A.; Bird, E.; Schumacher, W.A. Deficiency in thrombin-activatable fibrinolysis inhibitor (TAFI) protected mice from ferric chloride-induced vena cava thrombosis. *J. Thromb. Thrombolysis* **2007**, *23*, 41–49. [CrossRef]

69. Orikaza, C.M.; Morelli, V.M.; Matos, M.F.; Lourenço, D.M. Haplotypes of TAFI gene and the risk of cerebral venous thrombosis–a case-control study. *Thromb. Res.* **2014**, *133*, 120–124. [CrossRef]

70. Verdú, J.; Marco, P.; Benlloch, S.; Lucas, J. Association between the Thr325Ile and Ala147Thr polymorphisms of the TAFI gene and the risk of venous thromboembolic disease. *Clin. Appl. Thromb. Hemost.* **2008**, *14*, 494–495. [CrossRef]

71. Meltzer, M.E.; Bol, L.; Rosendaal, F.R.; Lisman, T.; Cannegieter, S.C. Hypofibrinolysis as a risk factor for recurrent venous thrombosis; results of the LETS follow-up study. *J. Thromb. Haemost.* **2010**, *8*, 605–607. [CrossRef]

72. Isordia-Salas, I.; Martínez-Marino, M.; Alberti-Minutti, P.; Ricardo-Moreno, M.T.; Castro-Calvo, R.; Santiago-Germán, D.; Alvarado-Moreno, J.A.; Calleja-Carreño, C.; Hernández-Juárez, J.; Leaños-Miranda, A.; et al. Genetic Polymorphisms Associated with Thrombotic Disease Comparison of Two Territories: Myocardial Infarction and Ischemic Stroke. *Dis. Mark.* **2019**, *2019*, 3745735. [CrossRef] [PubMed]

73. Kamal, H.M.; Ahmed, A.S.; Fawzy, M.S.; Mohamed, F.A.; Elbaz, A.A. Plasma thrombin-activatable fibrinolysis inhibitor levels and Thr325Ile polymorphism as a risk marker of myocardial infarction in Egyptian patients. *Acta Cardiol.* **2011**, *66*, 483–488. [CrossRef]

74. Kozian, D.H.; Lorenz, M.; März, W.; Cousin, E.; Mace, S.; Deleuze, J.F. Association between the Thr325Ile polymorphism of the thrombin-activatable fibrinolysis inhibitor and stroke in the Ludwigshafen Risk and Cardiovascular Health Study. *Thromb. Haemost.* **2010**, *103*, 976–983. [CrossRef]

75. Rattanawan, C.; Komanasin, N.; Settasatian, N.; Settasatian, C.; Kukongviriyapan, U.; Intharapetch, P.; Senthong, V. Association of TAFI gene polymorphisms with severity of coronary stenosis in stable coronary artery disease. *Thromb. Res.* **2018**, *171*, 171–176. [CrossRef]

76. de Bruijne, E.L.; Gils, A.; Guimarães, A.H.; Dippel, D.W.; Deckers, J.W.; van den Meiracker, A.H.; Poldermans, D.; Rijken, D.C.; Declerck, P.J.; de Maat, M.P.; et al. The role of thrombin activatable fibrinolysis inhibitor in arterial thrombosis at a young age: The ATTAC study. *J. Thromb. Haemost.* **2009**, *7*, 919–927. [CrossRef]

77. Zee, R.Y.; Hegener, H.H.; Ridker, P.M. Carboxypeptidase B2 gene polymorphisms and the risk of venous thromboembolism. *J. Thromb. Haemost.* **2005**, *3*, 2819–2821. [CrossRef]

78. Martini, C.H.; Brandts, A.; de Bruijne, E.L.; van Hylckama Vlieg, A.; Leebeek, F.W.; Lisman, T.; Rosendaal, F.R. The effect of genetic variants in the thrombin activatable fibrinolysis inhibitor (TAFI) gene on TAFI-antigen levels, clot lysis time and the risk of venous thrombosis. *Br. J. Haematol* **2006**, *134*, 92–94. [CrossRef]

79. Tregouet, D.A.; Schnabel, R.; Alessi, M.C.; Godefroy, T.; Declerck, P.J.; Nicaud, V.; Munzel, T.; Bickel, C.; Rupprecht, H.J.; Lubos, E.; et al. Activated thrombin activatable fibrinolysis inhibitor levels are associated with the risk of cardiovascular death in patients with coronary artery disease: The AtheroGene study. *J. Thromb. Haemost.* **2009**, *7*, 49–57. [CrossRef]

80. Tokgoz, S.; Zamani, A.G.; Durakbasi-Dursun, H.G.; Yilmaz, O.; Ilhan, N.; Demirel, S.; Tavli, M.; Sinan, A. TAFI gene polymorphisms in patients with cerebral venous thrombosis. *Acta Neurol. Belg.* **2013**, *113*, 291–297. [CrossRef]

81. Arauz, A.; Argüelles, N.; Jara, A.; Guerrero, J.; Barboza, M.A. Thrombin-Activatable Fibrinolysis Inhibitor Polymorphisms and Cerebral Venous Thrombosis in Mexican Mestizo Patients. *Clin. Appl. Thromb. Hemost.* **2018**, *24*, 1291–1296. [CrossRef] [PubMed]

82. Biswas, A.; Tiwari, A.K.; Ranjan, R.; Meena, A.; Akhter, M.S.; Yadav, B.K.; Behari, M.; Saxena, R. Thrombin activatable fibrinolysis inhibitor gene polymorphisms are associated with antigenic levels in the Asian-Indian population but may not be a risk for stroke. *Br. J. Haematol.* **2008**, *143*, 581–588. [CrossRef]

83. Meltzer, M.E.; Lisman, T.; de Groot, P.G.; Meijers, J.C.; le Cessie, S.; Doggen, C.J.; Rosendaal, F.R. Venous thrombosis risk associated with plasma hypofibrinolysis is explained by elevated plasma levels of TAFI and PAI-1. *Blood* **2010**, *116*, 113–121. [CrossRef] [PubMed]

84. Eichinger, S.; Schönauer, V.; Weltermann, A.; Minar, E.; Bialonczyk, C.; Hirschl, M.; Schneider, B.; Quehenberger, P.; Kyrle, P.A. Thrombin-activatable fibrinolysis inhibitor and the risk for recurrent venous thromboembolism. *Blood* **2004**, *103*, 3773–3776. [CrossRef]

85. van Tilburg, N.H.; Rosendaal, F.R.; Bertina, R.M. Thrombin activatable fibrinolysis inhibitor and the risk for deep vein thrombosis. *Blood* **2000**, *95*, 2855–2859. [CrossRef] [PubMed]

86. Ladenvall, C.; Gils, A.; Jood, K.; Blomstrand, C.; Declerck, P.J.; Jern, C. Thrombin activatable fibrinolysis inhibitor activation peptide shows association with all major subtypes of ischemic stroke and with TAFI gene variation. *Arterioscler. Thromb. Vasc. Biol.* **2007**, *27*, 955–962. [CrossRef]

87. Leebeek, F.W.; Goor, M.P.; Guimaraes, A.H.; Brouwers, G.J.; Maat, M.P.; Dippel, D.W.; Rijken, D.C. High functional levels of thrombin-activatable fibrinolysis inhibitor are associated with an increased risk of first ischemic stroke. *J. Thromb. Haemost.* **2005**, *3*, 2211–2218. [CrossRef]

88. Schroeder, V.; Wilmer, M.; Buehler, B.; Kohler, H.P. TAFI activity in coronary artery disease: A contribution to the current discussion on TAFI assays. *Thromb. Haemost.* **2006**, *96*, 236–237. [CrossRef]

89. Morange, P.E.; Tregouet, D.A.; Frere, C.; Luc, G.; Arveiler, D.; Ferrieres, J.; Amouyel, P.; Evans, A.; Ducimetiere, P.; Cambien, F.; et al. TAFI gene haplotypes, TAFI plasma levels and future risk of coronary heart disease: The PRIME Study. *J. Thromb. Haemost.* **2005**, *3*, 1503–1510. [CrossRef]

90. Meltzer, M.E.; Doggen, C.J.; de Groot, P.G.; Meijers, J.C.; Rosendaal, F.R.; Lisman, T. Low thrombin activatable fibrinolysis inhibitor activity levels are associated with an increased risk of a first myocardial infarction in men. *Haematologica* **2009**, *94*, 811–818. [CrossRef]

91. Leung, L.L.; Nishimura, T.; Myles, T. Regulation of tissue inflammation by thrombin-activatable carboxypeptidase B (or TAFI). *Adv. Exp. Med. Biol.* **2008**, *632*, 61–69. [PubMed]

92. Koschinsky, M.L.; Boffa, M.B.; Nesheim, M.E.; Zinman, B.; Hanley, A.J.; Harris, S.B.; Cao, H.; Hegele, R.A. Association of a single nucleotide polymorphism in CPB2 encoding the thrombin-activable fibrinolysis inhibitor (TAF1) with blood pressure. *Clin. Genet.* **2001**, *60*, 345–349. [CrossRef]

93. Myles, T.; Nishimura, T.; Yun, T.H.; Nagashima, M.; Morser, J.; Patterson, A.J.; Pearl, R.G.; Leung, L.L. Thrombin activatable fibrinolysis inhibitor, a potential regulator of vascular inflammation. *J. Biol. Chem.* **2003**, *278*, 51059–51067. [CrossRef] [PubMed]

94. Fujimoto, H.; Gabazza, E.C.; Taguchi, O.; Nishii, Y.; Nakahara, H.; Bruno, N.E.; D'Alessandro-Gabazza, C.N.; Kasper, M.; Yano, Y.; Nagashima, M.; et al. Thrombin-activatable fibrinolysis inhibitor deficiency attenuates bleomycin-induced lung fibrosis. *Am. J. Pathol.* **2006**, *168*, 1086–1096. [CrossRef] [PubMed]

95. Herren, T.; Swaisgood, C.; Plow, E.F. Regulation of plasminogen receptors. *Front. Biosci.* **2003**, *8*, d1–d8. [CrossRef]

96. Te Velde, E.A.; Wagenaar, G.T.; Reijerkerk, A.; Roose-Girma, M.; Borel Rinkes, I.H.; Voest, E.E.; Bouma, B.N.; Gebbink, M.F.; Meijers, J.C. Impaired healing of cutaneous wounds and colonic anastomoses in mice lacking thrombin-activatable fibrinolysis inhibitor. *J. Thromb. Haemost.* **2003**, *1*, 2087–2096. [CrossRef]

97. Guimarães, A.H.; Laurens, N.; Weijers, E.M.; Koolwijk, P.; van Hinsbergh, V.W.; Rijken, D.C. TAFI and pancreatic carboxypeptidase B modulate in vitro capillary tube formation by human microvascular endothelial cells. *Arterioscler. Thromb. Vasc. Biol.* **2007**, *27*, 2157–2162. [CrossRef]

98. Swaisgood, C.M.; Schmitt, D.; Eaton, D.; Plow, E.F. In vivo regulation of plasminogen function by plasma carboxypeptidase B. *J. Clin. Investig.* **2002**, *110*, 1275–1282. [CrossRef] [PubMed]

99. Morser, J.; Gabazza, E.C.; Myles, T.; Leung, L.L. What has been learnt from the thrombin-activatable fibrinolysis inhibitor-deficient mouse? *J. Thromb. Haemost.* **2010**, *8*, 868–876. [CrossRef]

100. Margetic, S. Inflammation and haemostasis. *Biochem. Med.* **2012**, *22*, 49–62. [CrossRef]

101. Sattar, N.; McCarey, D.W.; Capell, H.; McInnes, I.B. Explaining how "high-grade" systemic inflammation accelerates vascular risk in rheumatoid arthritis. *Circulation* **2003**, *108*, 2957–2963. [CrossRef]

102. Biondi, R.B.; Salmazo, P.S.; Bazan, S.G.Z.; Hueb, J.C.; de Paiva, S.A.R.; Sassaki, L.Y. Cardiovascular Risk in Individuals with Inflammatory Bowel Disease. *Clin. Exp. Gastroenterol.* **2020**, *13*, 107–113. [CrossRef]

103. Colucci, M.; Semeraro, N. Thrombin activatable fibrinolysis inhibitor: At the nexus of fibrinolysis and inflammation. *Thromb. Res.* **2012**, *129*, 314–319. [CrossRef] [PubMed]

104. Hugenholtz, G.C.; Meijers, J.C.; Adelmeijer, J.; Porte, R.J.; Lisman, T. TAFI deficiency promotes liver damage in murine models of liver failure through defective down-regulation of hepatic inflammation. *Thromb. Haemost.* **2013**, *109*, 948–955. [CrossRef]

105. Naito, M.; Taguchi, O.; Kobayashi, T.; Takagi, T.; D'Alessandro-Gabazza, C.N.; Matsushima, Y.; Boveda-Ruiz, D.; Gil-Bernabe, P.; Matsumoto, T.; Chelakkot-Govindalayathil, A.L.; et al. Thrombin-activatable fibrinolysis inhibitor protects against acute lung injury by inhibiting the complement system. *Am. J. Respir. Cell. Mol. Biol.* **2013**, *49*, 646–653. [CrossRef]

106. Schneider, M.; Nesheim, M. Reversible inhibitors of TAFIa can both promote and inhibit fibrinolysis. *J. Thromb. Haemost.* **2003**, *1*, 147–154. [CrossRef] [PubMed]

107. Reverter, D.; Vendrell, J.; Canals, F.; Horstmann, J.; Avilés, F.X.; Fritz, H.; Sommerhoff, C.P. A carboxypeptidase inhibitor from the medical leech Hirudo medicinalis. Isolation, sequence analysis, cDNA cloning, recombinant expression, and characterization. *J. Biol. Chem.* **1998**, *273*, 32927–32933. [CrossRef] [PubMed]

108. Arolas, J.L.; Lorenzo, J.; Rovira, A.; Castellà, J.; Aviles, F.X.; Sommerhoff, C.P. A carboxypeptidase inhibitor from the tick Rhipicephalus bursa: Isolation, cDNA cloning, recombinant expression, and characterization. *J. Biol. Chem.* **2005**, *280*, 3441–3448. [CrossRef] [PubMed]

109. Walker, J.B.; Hughes, B.; James, I.; Haddock, P.; Kluft, C.; Bajzar, L. Stabilization versus inhibition of TAFIa by competitive inhibitors in vitro. *J. Biol. Chem.* **2003**, *278*, 8913–8921. [CrossRef]

110. Wang, W.; Hendriks, D.F.; Scharpé, S.S. Carboxypeptidase U, a plasma carboxypeptidase with high affinity for plasminogen. *J. Biol. Chem.* **1994**, *269*, 15937–15944. [CrossRef]

111. Mao, S.S.; Colussi, D.; Bailey, C.M.; Bosserman, M.; Burlein, C.; Gardell, S.J.; Carroll, S.S. Electrochemiluminescence assay for basic carboxypeptidases: Inhibition of basic carboxypeptidases and activation of thrombin-activatable fibrinolysis inhibitor. *Anal. Biochem.* **2003**, *319*, 159–170. [CrossRef]

112. Yoshimoto, N.; Sasaki, T.; Sugimoto, K.; Ishii, H.; Yamamoto, K. Design and characterization of a selenium-containing inhibitor of activated thrombin-activatable fibrinolysis inhibitor (TAFIa), a zinc-containing metalloprotease. *J. Med. Chem.* **2012**, *55*, 7696–7705. [CrossRef]

113. Hillmayer, K.; Vancraenenbroeck, R.; De Maeyer, M.; Compernolle, G.; Declerck, P.J.; Gils, A. Discovery of novel mechanisms and molecular targets for the inhibition of activated thrombin activatable fibrinolysis inhibitor. *J. Thromb. Haemost.* **2008**, *6*, 1892–1899. [CrossRef] [PubMed]

114. Gils, A.; Ceresa, E.; Macovei, A.M.; Marx, P.F.; Peeters, M.; Compernolle, G.; Declerck, P.J. Modulation of TAFI function through different pathways–implications for the development of TAFI inhibitors. *J. Thromb. Haemost.* **2005**, *3*, 2745–2753. [CrossRef] [PubMed]

115. Semeraro, F.; Ammollo, C.T.; Gils, A.; Declerck, P.J.; Colucci, M. Monoclonal antibodies targeting the antifibrinolytic activity of activated thrombin-activatable fibrinolysis inhibitor but not the anti-inflammatory activity on osteopontin and C5a. *J. Thromb. Haemost.* **2013**, *11*, 2137–2147. [CrossRef] [PubMed]

116. Hendrickx, M.L.; Zatloukalova, M.; Hassanzadeh-Ghassabeh, G.; Muyldermans, S.; Gils, A.; Declerck, P.J. Identification of a novel, nanobody-induced, mechanism of TAFI inactivation and its in vivo application. *J. Thromb. Haemost.* **2014**, *12*, 229–236. [CrossRef]

117. Buelens, K.; Hassanzadeh-Ghassabeh, G.; Muyldermans, S.; Gils, A.; Declerck, P.J. Generation and characterization of inhibitory nanobodies towards thrombin activatable fibrinolysis inhibitor. *J. Thromb. Haemost.* **2010**, *8*, 1302–1312. [CrossRef]

118. Wyseure, T.; Gils, A.; Declerck, P.J. Evaluation of the profibrinolytic properties of a bispecific antibody-based inhibitor against human and mouse thrombin-activatable fibrinolysis inhibitor and plasminogen activator inhibitor-1. *J. Thromb. Haemost.* **2013**, *11*, 2069–2071. [CrossRef]

119. Denorme, F.; Wyseure, T.; Peeters, M.; Vandeputte, N.; Gils, A.; Deckmyn, H.; Vanhoorelbeke, K.; Declerck Paul, J.; De Meyer Simon, F. Inhibition of thrombin-activatable fibrinolysis inhibitor and plasminogen activator inhibitor-1 reduces ischemic brain damage in mice. *Stroke* **2016**, *47*, 2419–2422. [CrossRef] [PubMed]

120. Wyseure, T.; Rubio, M.; Denorme, F.; Martinez de Lizarrondo, S.; Peeters, M.; Gils, A.; De Meyer, S.F.; Vivien, D.; Declerck, P.J. Innovative thrombolytic strategy using a heterodimer diabody against TAFI and PAI-1 in mouse models of thrombosis and stroke. *Blood* **2015**, *125*, 1325–1332. [CrossRef] [PubMed]

International Journal of
Molecular Sciences

MDPI

Review

Fibrinolysis: A Primordial System Linked to the Immune Response

Robert L. Medcalf * and Charithani B. Keragala

Molecular Neurotrauma and Haemostasis Laboratory, Australian Centre for Blood Diseases, Central Clinical School Melbourne, Monash University, Melbourne, VIC 3004, Australia; Charithani.Keragala@monash.edu
* Correspondence: robert.medcalf@monash.edu; Tel.: +61-3-9903-0133

Abstract: The fibrinolytic system provides an essential means to remove fibrin deposits and blood clots. The actual protease responsible for this is plasmin, formed from its precursor, plasminogen. Fibrin is heralded as it most renowned substrate but for many years plasmin has been known to cleave many other substrates, and to also activate other proteolytic systems. Recent clinical studies have shown that the promotion of plasmin can lead to an immunosuppressed phenotype, in part via its ability to modulate cytokine expression. Almost all immune cells harbor at least one of a dozen plasminogen receptors that allows plasmin formation on the cell surface that in turn modulates immune cell behavior. Similarly, a multitude of pathogens can also express their own plasminogen activators, or contain surface proteins that provide binding sites host plasminogen. Plasmin formed under these circumstances also empowers these pathogens to modulate host immune defense mechanisms. Phylogenetic studies have revealed that the plasminogen activating system predates the appearance of fibrin, indicating that plasmin did not evolve as a fibrinolytic protease but perhaps has its roots as an immune modifying protease. While its fibrin removing capacity became apparent in lower vertebrates these primitive under-appreciated immune modifying functions still remain and are now becoming more recognised.

Keywords: fibrinolysis; plasminogen activation; immune response; inflammation

Citation: Medcalf, R.L.; Keragala, C.B. Fibrinolysis: A Primordial System Linked to the Immune Response. *Int. J. Mol. Sci.* **2021**, 22, 3406. https://doi.org/10.3390/ijms22073406

Academic Editor: Amedeo Amedei

Received: 27 February 2021
Accepted: 24 March 2021
Published: 26 March 2021

Publisher's Note: MDPI stays neutral with regard to jurisdictional claims in published maps and institutional affiliations.

1. Introduction

The plasminogen activating ("fibrinolytic") system is one of the most important proteolytic cascades in all mammals and indeed in a variety of other species. While conventionally associated with blood clot removal via the generation of the key protease, plasmin, this system also performs a multitude of other important functions, some of which are beginning to impact on clinical medicine. Some of these developments, notably on the actions of plasmin on the immune response have been recently reviewed [1]. Direct evidence is now emerging, not only from animal models, but also in humans that the modulation of the plasminogen activating system does indeed impact on immune function. This review will discuss the link between the plasminogen activating system with immune function and also argue the case that the evolution of this system may have been initially directed at immune modulation but which subsequently became adapted for other functions, including fibrin removal.

2. Plasminogen Activation: A Universal System with a Broad Repertoire

For centuries, blood has been known to clot when outside the body. It was also observed that clotted blood could also spontaneously dissolve and this was initially put down to being a result of putrefaction (i.e., just simple protein decay, see [2]). However, in 1893 biochemical studies on fibrin in clotted human blood revealed that the longer blood rested on the bench, the lower the concentration of fibrin that was found [3]. Hence, fibrin was not simply rotting away, but was being removed via some apparent enzymatic process. This finding led to the first coining of the term "fibrinolysis" [3]. The biochemical underpinnings of this newly discovered process were of clear academic interest, but identifying the protease responsible for attacking the fibrin in the blood was not easy. In fact, it took another 40 years before scientists stumbled upon an exogenous source of fibrinolytic activity: the first in the saliva of blood feeding vampire bats by Otto Bier in 1932 [4] and in the same year, Aoi described a "Fibrolytic" activity in isolates of staphylococcus [5] that was subsequently identified as staphylokinase (see [6]). In 1933, a fibrinolytic activity was found in the culture broth of β-haemolytic streptococci [7]. These bacteria were also responsible for severe bleeding complications in patients and this newly identified fibrinolytic activity was the likely culprit. The nuts and bolts of the fibrinolytic pathway were largely revealed using these bacterial sources, mostly from streptococci. Initially the entity found in streptococci was referred to as "fibrinolysin" but was eventually designated as "streptokinase". Critically important experiments revealed streptokinase (and also staphylokinase) first needed to activate a plasma precursor as first reported by Milstone in 1941 [8]. This plasma "zymogen" was referred to as "pro-fibrinolysin" (later renamed as plasminogen) and the active fibrinolytic form as "fibrinolysin" (i.e., plasmin); see [9]. In the meantime, the elusive human-derived plasminogen activators were subsequently described, the first in 1947 [10] that was detected in human cells (initially referred to as tissue-cytofibrinolysin, and later as tissue-type plasminogen activator; t-PA) while a urinary source (urokinase-type plasminogen activator; u-PA) was described in 1952 [11]. While the fundamental studies using streptokinase and staphylokinase were ground breaking, mechanistically these bacterial entities were later found to function very differently to the renowned human plasminogen activators, t-PA and uPA. Indeed, neither streptokinase nor staphylokinase are plasminogen activators nor kinases. In fact, they are not even proteases but are rather plasminogen binding proteins: when complexed with plasminogen, the resulting complex itself becomes a plasminogen activator [12].

During this important period in fibrinolysis research, it did not take long to consider the possibility that harnessing this process might be of clinical use in patients with thromboembolic disease. Indeed, streptokinase was first used clinically in 1949 (in patients with pleural exudates [13]) and continues to be used today being listed as an essential medicine by the World Health Organisation (WHO). Staphylokinase was also pursued and in later years was shown to have distinct benefits over streptokinase; see [14]. Both tPA and uPA are also widely used agents and therapeutic thrombolysis is now mainstream in clinical medicine.

On the other side of the coin, it had become apparent early on that excessive fibrinolysis could promote the premature removal of blood clots and cause devastating bleeding. This led to the development of anti-fibrinolytic agents [15], notably tranexamic acid (TXA) in the early 1960′s, that is also listed today as an essential medicine by the WHO. Therefore, after the serendipitous finding that a fibrinolytic process existed in 1893 and the subsequent laboratory studies that followed into the mid 1940′s, mainstream medicine eventually exploited this enzymatic pathway for clinical benefit either to remove clots by forcibly generating plasmin, or by blocking plasmin generation or activity to preserve blood clots and stop bleeding, even to this day.

It would seem that this story could simply end here, and it probably could if not for the fact that the plasminogen activating process did not evolve for the sole purpose of removing fibrin. It also seems an anomaly that the renowned endogenous human plasminogen activators (t-PA and u-PA) are relatively specific for plasminogen, yet the formed plasmin is not at all solely specific for fibrin. Indeed, the promiscuous, trypsin-like substrate specificity of plasmin has empowered it with the capacity to cleave many targets. Hence, the target specificity of plasmin is derived from the plasminogen activators themselves that decide on where plasmin is produced. However, depending on the location and circumstances, plasmin has the potential to act upon numerous targets and in doing so influence a variety of physiological and even pathological processes.

Adding more complexity to this is that the list of plasminogen activators has expanded quite impressively. Even in humans, plasminogen can be activated by other serine proteases including kallikrein [16] and complement [17] while thrombin can have bidirectional effects on plasminogen activation [18]. However, there is an even longer list of proteases that activate human plasminogen from various pathogens as discussed below in Section 4.

Plasminogen itself is expressed at high concentration in human plasma (2 μM), but it is also present in essentially every other compartment at varying levels, ranging from seminal fluid [19] to the central nervous system [20] (and see [21]). The primary inhibitor of plasmin, α2-antiplasmin, is also expressed in many of these locations [22,23]. Nonetheless, with the range of plasminogen activators available from human and non-human sources (below), plasmin has the potential to be produced and to be largely controlled by antiplasmin (and by other molecules) almost anywhere, including locations where fibrin is not even present.

It is not surprising that the broad spectrum of plasmin substrates has implicated the plasminogen activating system in many processes ranging from wound repair to synaptic plasticity [24]. There is extensive literature that has summarized these non-canonical activities in the early 2000′s [25,26] and in more recent times [1,17,27–29]. However, one key function that may underpin the broad role of this system is its effect on the immune response. When reflecting on the early findings of the pioneers in this field, a link between fibrinolysis and the immune response was evident at the very outset but its profound effect at removing fibrin and the subsequent clinical development of fibrinolytic and anti-fibrinolytic agents overshadowed any other possible role. Additionally, off-target effects of fibrinolytic agents in clinical medicine were essentially only related to bleeding (which was predicted), while side-effects of the use of anti-fibrinolytic agents were generally negligible [30,31]. However, then again, any possible effects on immune function were not on the radar anyway. Additionally, off-target effects may not necessarily be deleterious anyway and would go undetected.

3. Plasminogen, Fibrinolysis and Immune Function

Plasmin(ogen) is now appreciated for its role as an immune modulator interacting directly with numerous cells of the immune response including monocytes, macrophages and dendritic cells [1]. Plasmin, formed on the surface of immune cells, is capable of activating several pro-inflammatory pathways resulting in cytokine production [32]. Even in healthy humans, simply administering an antifibrinolytic agent results in a significant increase in pro-inflammatory cytokines within 2h of treatment [33] suggesting that inherent plasmin formation provides a means to keep inflammation repressed, but this can quickly change as soon as plasmin is inhibited. This further implicates the entire plasminogen activation system as an essential part of an innate surveillance network geared to respond to immune challenges. Much of this immune/inflammatory signalling is mediated by at least 12 plasminogen receptors that are differentially expressed on the cell surface of almost all immune cells [34]. For example, cell surface-bound plasmin(ogen) is required for the efficient migration of macrophages to sites of inflammation [35]. Plasminogen has also been shown to promote macrophage phagocytosis in mice, with plasminogen-deficient mice exhibiting a 60% delay in clearing apoptotic thymocytes by spleen and an 85% reduction in uptake of immunoglobulin opsonised bodies by peritoneal macrophages [36]. Phagocytosis of antibody-mediated erythrocyte clearance by liver Kupffer cells was also reduced in plasminogen-deficient mice compared to Plg+/+ mice. Gene array studies of tissues from these Plg−/− mice revealed downregulation of several genes involved in phagocytosis, suggesting that plasminogen may be able to change the expression of certain genes contributing to phagocytosis [36]. Plasmin also engages other elements of innate immunity including the complement cascade, components of the extracellular matrix as wells as cells of the vasculature including endothelial and smooth muscle cells [1]. In stark contrast to its pro-inflammatory properties, plasmin also exhibits several anti-inflammatory and immunosuppressive responses. Plasmin(ogen) appears to regulate, via its receptor, annexin A1, key aspects of inflammation resolution, in particular, macrophage reprogramming, neutrophil apoptosis and efferocytosis [37]. Plasmin-treated dendritic cells for instance fail to undergo maturation following phagocytosis, exhibit reduced migration to lymph nodes and also stimulate the release of significant amounts of transforming growth factor-β (TGF-β) which has immunosuppressive properties. These dendritic cells also have reduced ability to induce allogeneic lymphocyte proliferation. These properties of plasmin are important in maintaining tissue homeostasis where it aids in initiating a response to tissue injury while preventing self-reactivity/autoimmunity [36,38].

Hence, it is apparent that pathogens, by harnessing the plasminogen activating pathway, might gain an additional advantage by counteracting immune defense processes that are initiated, at least in part, by cell-surface plasmin generation. While there is clearly an argument that these pathogens would survive host defenses by clearing fibrin, there is now looming evidence that also suggests that the hijacking of host plasminogen might aid pathogen survival by using plasmin to disengage some of the key immune pathways as an effective countermeasure.

4. Microorganisms, Plasmin Formation and Fibrinolysis

As mentioned above, streptokinase and staphylokinase were discovered as a key molecule released from some strains of β-haemolytic streptococci and staphylococcus, respectively. However, this is not unique to these particular strains as similar molecules are produced by many other bacteria (below). Even if a molecule is not produced endogenously, bacteria almost universally have the capacity to bind plasminogen and to use the host plasminogen activators as a means to generate localized plasmin. The formed plasmin is harnessed by these pathogens not only to remove the confines of fibrin, but also to suppress the host immune response and evade local immune attack [39]. Over 40 binding proteins have been reported in bacterial species which target plasmin(ogen) [40]. For example, Mycobacterium tuberculosis has 13 proteins with plasminogen binding potential and *Mycoplasma pneumoniae* exhibits 6 [41,42].

Plasminogen binding to these proteins activates a variety of mechanisms aimed at infiltrating host defences. These include for example the degradation of extracellular matrix proteins by *Leptospira*, where urokinase (u-PA) activates bound plasminogen to plasmin which then degrades fibronectin and laminin [43]. Remarkably, streptokinase of the non-human streptococci show evolutionary species-specificity for the plasminogen of the animal host they infect [44–46]. Staphylokinase, secreted by *Staphyloccocus aureus*, exhibits high affinity binding to plasminogen in plasma, thus forming a complex that can effectively activate plasminogen (akin to streptokinase) while also evading the inactivating capacity of α2-antiplasmin by binding to fibrin [14]. Interestingly, *Yersinia Pestis*, the causative pathogen of the plague, which killed a third of the European population in the 14th century, expresses a plasmid gene *pla* which encodes a surface plasminogen activator protease with unusual kinetic properties. The expression of this protease increases the virulence of *Yersinia Pestis* and is also likely to cleave and inactivate plasminogen activator inhibitor-1 (PAI-1), increasing the conversion of plasminogen to plasmin and promote virulence in the host [47–49].

Plasminogen-dependent extracellular matrix degeneration is utilised by the main pathogens causing bacterial meningitis, *H. influenzae, S. pneumonia* and *N. meningitides* [44,50]. Furthermore, plasmin's proteolytic role is harnessed by bacteria in degrading plasma proteins and peptides, including complement and immunoglobulins, which are critical in antigen presentation and processing within the host innate immune repertoire [51]. For example, *Leptospira* surface protein Lsa23 not only has the ability to block activation of both the alternative and classical complement pathways, but binds to and activates plasminogen to plasmin which in turn degrades complement proteins C3b and C4b, together improving the chances of evading host immunity [52].

Plasminogen receptors are also expressed on fungi including several Candida species, *Aspergillus, Cryptococcus neoformans* and *Pneumocystis carinii* [53,54]. Many of the receptors on Cryptococcus have the ability to activate the host PA system to allow the fungus to cross tissue barriers including the critical blood–brain barrier [55].

Plasminogen is also important in the invasiveness and pathogenesis of several parasites. *Trypanosoma cruzi, Leishmania* and the malarial parasites *Plasmodium falciparum* are known to engage enolase-plasminogen binding as well as uPA in aspects of their pathogenicity and replication [56,57]. More recently, the fibrinolytic system was reported to be essential for parasite migration across the dermis and liver [58]. Helminth parasites also exhibit multiple plasminogen binding proteins as they are in contact with fibrinolytic proteins within the intravascular space. Recruitment of plasminogen on the worm's surface appears to be one method of host immune evasion [40].

Most of the discussion above relates to the consequences of plasmin in modulating immune surveillance and how this can be intercepted by pathogens. There is also evidence that the plasminogen activators themselves, and independently of activating plasminogen, can also modulate immune function. Indeed, catalytically inactive t-PA has been reported to express inflammatory mediators by macrophages in vitro in a process dependent on LRP-1 [59]. Another report from the same group implicated a key role for NMDA-1 receptor signalling in this process and also reported that inactive tPA could block LPS toxicity in vivo in mice [60]. This same group just recently indicated that enzymatically inactive t-PA was also protective in a mouse model of inflammatory bowel disease [61].

A summary of the variety of effects of the fibrinolytic system on the immune and inflammatory responses is presented in Table 1.

Table 1. Properties of plasmin(ogen) in inflammation and immunity.

Target Effects	Properties of Plasmin(ogen)	References
Proinflammatory **Macrophage and Monocyte effects**	Interacts with macrophage migration and activation via plasminogen receptors. Promotes cytokine production in macrophages.	[32,35,62]
	Directly alters gene expression in macrophages by binding plasminogen receptors and enhancing phagocytic capacity, efferocytosis and foam cell formation.	[37,63,64]
	Promote macrophage phagocytosis in mice	[36]
	Potent chemoattractant of monocytes, induces actin polymerisation.	[65]
	Activates 5-lipoxygenase pathway in monocytes and macrophages resulting in the synthesis of proinflammatory leukotrienes.	[66]
	Stimulates JAK/STAT signalling in monocytes resulting in MCP-1 release, further promoting monocyte recruitment.	[67]
Dendritic Cell (DC) effects	Increases phagocytic activity of DCs without causing activation. This interaction is involved in the chemoattraction of dendritic cells, T- and B cells.	[38]
	Triggers chemotaxis of monocyte derived DCs and a T helper type-1 (Th1) phenotype in CD4+ T cells.	[68]
	Indirectly promotes neutrophil recruitment by binding to mast cells and stimulating release of leukotrienes	[69]
Other inflammatory effects	Induce expression of CCR6-activating chemokine CCL20 in dermis via induction of NF-kB.	[70]
	Stimulates NF-kB and AP-1, resulting in the production of tumor necrosis factor (TNF)-α, interleukin (IL)-1α, IL-1β, and tissue factor.	[71]
	Activates phospholipase A2 in endothelial cells, releasing arachidonic acid and subsequent production of prostacyclin, enhanced nitric oxide (NO)-mediated vasorelaxation and chemotactic monocyte chemotactic protein (MCP)-1 release.	[39,72,73]
	In pulmonary epithelial cells, plasmin induces cyclooxygenase (COX)-2, resulting in the release of antifibrotic prostaglandin E-2 (PGE-2).	[74]
	Promotes complement activation.	[17,75]
	Binds to platelets via PAR-4 and dose-dependently activate or inhibit platelet activation and aggregation. Interacts with extracellular matrix, endothelial cells, smooth muscle.	[76,77]
	Plasmin can activate the Matrix Metalloproteinases, transforming growth factor (TGF)-β, and neurotrophic factors.	[78–80]
	Inhibition of DC maturation following phagocytosis thereby inducing a tolerogenic phenotype. Reduced migration of DCs to lymph nodes and increase release of TGF-β. Reduction in DC ability to induce allogeneic lymphocyte proliferation.	[38]
Immuno-suppression **DC effects**	Suppression of proinflammatory cytokines in vivo, reversed by tranexamic acid. Inhibition of plasmin reduces post-surgical infection rates.	[33]

5. Phylogenetic Links with Plasminogen Activation

Phylogenetic studies have further provided compelling evidence to suggest that plasminogen and the plasminogen activators did not co-evolve but eventually became brothers in arms perhaps as a matter of convenience. Although plasmin, t-PA and u-PA are serine proteases, the codon usage used for the active serine in plasminogen (AGT) differs at two position to that used to encode the serine in t-PA (TCG) and u-PA (TCA) [81]. This strongly suggests that within the serine protease family, plasminogen evolved separately (and earlier, see below) from u-PA and t-PA

It has also been reported that the primordial ancestor of plasminogen first appeared in protochordates (e.g., amphioxus, sea squirt and related species) [82,83], animals that contain hemolymph that does not even clot but which cross-reacts with anti-human plasminogen antibodies and was localized to the hepatic diverticulum [84]. Plasminogen cDNA was cloned from Amphioxus *B. belcheri* and expressed in *E. coli*. Studies on this recombinant protein indicated that it contained two kringle domains in the N-terminus (not five as in humans) and a serine protease domain in the N-terminus. This molecule also lacked the PAN domain [85]. It also appeared that a lysine binding domain was conserved in one of these kringles [85]. Moreover, the amphioxus plasminogen harboured the putative t-PA/u-PA cleavage site (Arg-Val). The catalytic triad (His-Asp-Ser), critical for protease function was also present and located at positions corresponding to human plasminogen. Consistent with these finding the amphioxus plasminogen was shown to generate plasmin when incubated with human uPA [85]. It is not clear what endogenous proteases were used to activate plasminogen, since the classical plasminogen activators, t-PA and u-PA, only appeared around 20 million years later in cartilaginous fish, together with PAI-1 (see [83]). While protochordates cannot generate fibrin, they do contain a primitive yet full length fibrinogen molecule that does not harbour thrombin cleavage sites [86]. Hence, the primary function of this early plasminogen/plasmin molecule had nothing to do with fibrinolysis per se, yet it remains possible that the co-existing primitive fibrinogen itself could have been a substrate for this early plasmin. As cell clumping had been observed in sea squirts, it was speculated by Russell Doolittle that this fibrinogen molecule may have been involved in mediating cell–cell interactions and perhaps having some innate immune function [86]. It is possible that plasminogen had some regulatory function with fibrinogen that also may have had some relationship with ancestral complement proteins that were also present in these animals.

As biological complexity evolved with the appearance of vertebrates that included a sophisticated clotting system, existing molecules gained new or additional enzymatic functions (perhaps even re-purposed) to generate and to keep fibrin and other proteins in check. It is acknowledged that caution is needed in extrapolating primitive functions and the evolutionary time course of fibrin-targeted proteases using extant species. Nonetheless, the finding that a bonafide plasminogen exists in an extant species that does not make fibrin is indicative of another in vivo role of plasminogen, which may have been a primitive immune function.

6. Conclusions, Clinical Implications and Future Potential

There is no doubt about the importance of the fibrinolytic system in the removal of fibrin. Indeed, modulation of this process, either by increasing or decreasing plasmin levels has critically important effects in clinical medicine in relation to the removal of occlusive clots, or to reduce bleeding, respectively. It could be well argued that fibrin is the most relevant substrate for plasmin in the setting of thrombosis simply due to the mass of fibrin that accumulates and the risk that this poses. Under physiological conditions, where fibrin formation is at background levels, plasmin formation still occurs. Plasmin is a promiscuous protease, so while it might still be performing some degree of fibrinolysis under normal conditions, its broad substrate specificity empowers it to cleave other targets, some of which act in an immune surveillance capacity modulating inflammation. This is apparent from studies on volunteers administered antifibrinolytic drugs where baseline levels of proinflammatory cytokines rapidly increase [33]. There are now many other reports implicating plasmin, tPA and in fact the entire fibrinolytic system in the immune response. These immune/inflammatory functions may indeed have been the original role of this enzymatic system given that plasmin generation occurs in lower order species (protochordates) where fibrin itself is absent. Understanding the evolutionary origins of the plasminogen activating system and the realization that perhaps it is not as fibrin centric as initially thought reveals opportunities to apply and harness these properties for means not previously considered. Could thrombolysis in acute ischaemic stroke impair the host immune response in a plasmin-mediated manner while also running the risk of intracerebral haemorrhage? Could antifibrinolytic therapy in major surgery confer an immune advantage in recovering patients while also avoiding excessive bleeding? The answers to these questions are slowly emerging as is the potential for the wider implications of plasmin(ogen) activation beyond just haemostasis, influencing scientific research, clinical practice and ultimately patient outcomes. Further research aimed at harnessing the plasminogen activating system for its immune modulatory properties may potentially open doors to understanding this unique process further.

Author Contributions: R.L.M. and C.B.K. both contributed to the writing and editing of this review. Both authors have read and agreed to the published version of the manuscript.

Funding: This research was funded by the National Health and Medical Research Foundation (NHMRC) to RLM, grant numbers GNT1126636 and GNT1044152.

Conflicts of Interest: The authors declare no conflict of interest.

References

1. Draxler, D.F.; Sashindranath, M.; Medcalf, R.L. Plasmin: A Modulator of Immune Function. *Semin. Thromb. Hemost.* **2017**, *43*, 143–153. [CrossRef]
2. Green, J.R. Note on the Action of Sodium Chloride in dissolving Fibrin. *J. Physiol.* **1887**, *8*, 372–377. [CrossRef] [PubMed]
3. Dastre, A. Fibrinolyse dans le sang. *Arch. Physiol.* **1893**, *5*, 661.
4. Bier, O.E. Action anticoagulante et fibrinolytique de l'extract des glands salivaires d'une chauve-souris hematophage (*desmodus rufus*). *C.R. Soc Biol. (Paris)* **1932**, *110*, 129–131.
5. Aoi, F. On the fibrolysis of the staphylococcus. *Kitasato Arch. Exptl. Med.* **1932**, *9*, 171–201.
6. Lack, C.H. Staphylokinase: An activator of plasma Protease. *Nature* **1948**, 559–560. [CrossRef]
7. Tillett, W.S.; Garner, R.L. The Fibrinolytic Activity of Hemolytic Streptococci. *J. Exp. Med.* **1933**, *58*, 485–502. [CrossRef]
8. Milstone, H. A factor in normal human blood which particiaptes in streptococcal fibrinolysis. *J. Immunol.* **1941**, *42*, 109–116.
9. Macfarlane, R.G.; Pilling, J. Observations on fibrinolysis; plasminogen, plasmin, and antiplasmin content of human blood. *Lancet* **1946**, *2*, 562–565. [CrossRef]
10. Astrup, T.; Permin, P.M. Fibrinolysis in the animal organism. *Nature* **1947**, *159*, 681. [CrossRef] [PubMed]
11. Sobel, G.W.; Mohler, S.R.; Jones, N.W.; Dowdy, A.B.C.; Guest, M.M. Urokinase: An activator of plasma profibrinolysin extracted from urine. *Am. J. Physiol.* **1952**, *171*, 768–769.
12. Parry, M.A.; Zhang, X.C.; Bode, I. Molecular mechanisms of plasminogen activation: Bacterial cofactors provide clues. *Trends Biochem. Sci.* **2000**, *25*, 53–59. [CrossRef]
13. Tillett, W.S.; Sherry, S. The effect in patients of streptococcal fibrinolysin and streptococcal desoxyribonuclease on fibrinous, purulent, and sanguinous pleural exudations. *J. Clin. Investig.* **1949**, *28*, 173–190. [CrossRef] [PubMed]
14. Lijnen, H.R.; Van Hoef, B.; De Cock, F.; Okada, K.; Ueshima, S.; Matsuo, O.; Collen, D. On the mechanism of fibrin-specific plasminogen activation by staphylokinase. *J. Biol. Chem.* **1991**, *266*, 11826–11832. [CrossRef]
15. Okamoto, S.O.U. A new potent antifibrinolytic substance and its effects on blood of animals. *Keio J. Med.* **1962**, *11*, 105–115. [CrossRef]
16. Jorg, M.; Binder, B.R. Kinetic analysis of plasminogen activation by purified plasma kallikrein. *Thromb. Res.* **1985**, *39*, 323–331. [CrossRef]
17. Foley, J.H. Plasmin(ogen) at the Nexus of Fibrinolysis, Inflammation, and Complement. *Semin. Thromb. Hemost.* **2017**, *43*, 135–142. [CrossRef] [PubMed]
18. Tomczyk, M.; Suzuki, Y.; Sano, H.; Brzoska, T.; Tanaka, H.; Urano, T. Bidirectional functions of thrombin on fibrinolysis: Evidence of thrombin-dependent enhancement of fibrinolysis provided by spontaneous plasma clot lysis. *Thromb. Res.* **2016**, *143*, 28–33. [CrossRef] [PubMed]
19. Lwaleed, B.A.; Greenfield, R.; Stewart, A.; Birch, B.; Cooper, A.J. Seminal clotting and fibrinolytic balance: A possible physiological role in the male reproductive system. *Thromb. Haemost.* **2004**, *92*, 752–766. [CrossRef] [PubMed]

20. Basham, M.E.; Seeds, N.W. Plasminogen expression in the neonatal and adult mouse brain. *J. Neurochem.* **2001**, *77*, 318–325. [CrossRef] [PubMed]

21. Zhang, L.; Seiffert, D.; Fowler, B.J.; Jenkins, G.R.; Thinnes, T.C.; Loskutoff, D.J.; Parmer, R.J.; Miles, L.A. Plasminogen Has a Broad Extrahepatic Distribution. *Thromb. Haemost.* **2002**, *87*, 493–501. [PubMed]

22. Del Giudice, P.T.; da Silva, B.F.; Lo Turco, E.G.; Fraietta, R.; Spaine, D.M.; Santos, L.F.; Pilau, E.J.; Gozzo, F.C.; Cedenho, A.P.; Bertolla, R.P. Changes in the seminal plasma proteome of adolescents before and after varicocelectomy. *Fertil. Steril.* **2013**, *100*, 667–672. [CrossRef]

23. Kawashita, E.; Kanno, Y.; Asayama, H.; Okada, K.; Ueshima, S.; Matsuo, O.; Matsuno, H. Involvement of alpha2-antiplasmin in dendritic growth of hippocampal neurons. *J. Neurochem.* **2013**, *126*, 58–69. [CrossRef] [PubMed]

24. Medcalf, R.L. Fibrinolysis: From blood to the brain. *J. Thromb. Haemost.* **2017**, *15*, 2089–2098. [CrossRef]

25. Yepes, M.; Lawrence, D.A. New functions for an old enzyme: Nonhemostatic roles for tissue-type plasminogen activator in the central nervous system. *Exp. Biol. Med. (Maywood)* **2004**, *229*, 1097–1104. [CrossRef]

26. Samson, A.L.; Medcalf, R.L. Tissue-type plasminogen activator: A multifaceted modulator of neurotransmission and synaptic plasticity. *Neuron* **2006**, *50*, 673–678. [CrossRef]

27. Draxler, D.F.; Medcalf, R.L. The fibrinolytic system-more than fibrinolysis? *Transfus. Med. Rev.* **2015**, *29*, 102–109. [CrossRef] [PubMed]

28. Fredriksson, L.; Lawrence, D.A.; Medcalf, R.L. tPA Modulation of the Blood-Brain Barrier: A Unifying Explanation for the Pleiotropic Effects of tPA in the CNS. *Semin. Thromb. Hemost.* **2017**, *43*, 154–168. [CrossRef]

29. Medcalf, R.L.; Keragala, C.; Draxler, D.F. Fibrinolysis and the immune response in trauma. *Semin. Thromb. Haemost.* **2019**, in press.

30. Warnaar, N.; Mallett, S.V.; Klinck, J.R.; de Boer, M.T.; Rolando, N.; Burroughs, A.K.; Jamieson, N.V.; Rolles, K.; Porte, R.J. Aprotinin and the risk of thrombotic complications after liver transplantation: A retrospective analysis of 1492 patients. *Liver Transpl.* **2009**, *15*, 747–753. [CrossRef]

31. Henry, D.A.; Carless, P.A.; Moxey, A.J.; O'Connell, D.; Stokes, B.J.; Fergusson, D.A.; Ker, K. Anti-fibrinolytic use for minimising perioperative allogeneic blood transfusion. *Cochrane Database Syst. Rev.* **2011**, CD001886. [CrossRef]

32. Li, Q.; Laumonnier, Y.; Syrovets, T.; Simmet, T. Plasmin triggers cytokine induction in human monocyte-derived macrophages. *Arter. Thromb Vasc Biol* **2007**, *27*, 1383–1389. [CrossRef]

33. Draxler, D.F.; Yep, K.; Hanafi, G.; Winton, A.; Daglas, M.; Ho, H.; Sashindranath, M.; Wutzlhofer, L.M.; Forbes, A.; Goncalves, I.; et al. Tranexamic acid modulates the immune response and reduces postsurgical infection rates. *Blood Adv.* **2019**, *3*, 1598–1609. [CrossRef]

34. Miles, L.A.; Parmer, R.J. Plasminogen receptors: The first quarter century. *Semin Thromb Hemost* **2013**, *39*, 329–337. [CrossRef] [PubMed]

35. Silva, L.M.; Lum, A.G.; Tran, C.; Shaw, M.W.; Gao, Z.; Flick, M.J.; Moutsopoulos, N.M.; Bugge, T.H.; Mullins, E.S. Plasmin-mediated fibrinolysis enables macrophage migration in a murine model of inflammation. *Blood* **2019**, *134*, 291–303. [CrossRef]

36. Das, R.; Ganapathy, S.; Settle, M.; Plow, E.F. Plasminogen promotes macrophage phagocytosis in mice. *Blood* **2014**, *124*, 679–688. [CrossRef] [PubMed]

37. Sugimoto, M.A.; Ribeiro, A.L.C.; Costa, B.R.C.; Vago, J.P.; Lima, K.M.; Carneiro, F.S.; Ortiz, M.M.O.; Lima, G.L.N.; Carmo, A.A.F.; Rocha, R.M.; et al. Plasmin and plasminogen induce macrophage reprogramming and regulate key steps of inflammation resolution via annexin A1. *Blood* **2017**, *129*, 2896–2907. [CrossRef]

38. Borg, R.J.; Samson, A.L.; Au, A.E.; Scholzen, A.; Fuchsberger, M.; Kong, Y.Y.; Freeman, R.; Mifsud, N.A.; Plebanski, M.; Medcalf, R.L. Dendritic Cell-Mediated Phagocytosis but Not Immune Activation Is Enhanced by Plasmin. *PLoS ONE* **2015**, *10*, e0131216. [CrossRef] [PubMed]

39. Syrovets, T.; Lunov, O.; Simmet, T. Plasmin as a proinflammatory cell activator. *J. Leukoc. Biol.* **2012**, *92*, 509–519. [CrossRef] [PubMed]

40. Ayon-Nunez, D.A.; Fragoso, G.; Bobes, R.J.; Laclette, J.P. Plasminogen-binding proteins as an evasion mechanism of the host's innate immunity in infectious diseases. *Biosci. Rep.* **2018**, *38*. [CrossRef]

41. Grundel, A.; Friedrich, K.; Pfeiffer, M.; Jacobs, E.; Dumke, R. Subunits of the Pyruvate Dehydrogenase Cluster of Mycoplasma pneumoniae Are Surface-Displayed Proteins that Bind and Activate Human Plasminogen. *PLoS ONE* **2015**, *10*, e0126600. [CrossRef]

42. Xolalpa, W.; Vallecillo, A.J.; Lara, M.; Mendoza-Hernandez, G.; Comini, M.; Spallek, R.; Singh, M.; Espitia, C. Identification of novel bacterial plasminogen-binding proteins in the human pathogen Mycobacterium tuberculosis. *Proteomics* **2007**, *7*, 3332–3341. [CrossRef] [PubMed]

43. Vieira, M.L.; Atzingen, M.V.; Oliveira, R.; Mendes, R.S.; Domingos, R.F.; Vasconcellos, S.A.; Nascimento, A.L. Plasminogen binding proteins and plasmin generation on the surface of Leptospira spp.: The contribution to the bacteria-host interactions. *J. Biomed. Biotechnol.* **2012**, *2012*, 758513. [CrossRef] [PubMed]

44. Peetermans, M.; Vanassche, T.; Liesenborghs, L.; Lijnen, R.H.; Verhamme, P. Bacterial pathogens activate plasminogen to breach tissue barriers and escape from innate immunity. *Crit. Rev. Microbiol.* **2016**, *42*, 866–882. [CrossRef]

45. Sun, H.; Ringdahl, U.; Homeister, J.W.; Fay, W.P.; Engleberg, N.C.; Yang, A.Y.; Rozek, L.S.; Wang, X.; Sjobring, U.; Ginsburg, D. Plasminogen is a critical host pathogenicity factor for group A streptococcal infection. *Science* **2004**, *305*, 1283–1286. [CrossRef] [PubMed]

46. Gladysheva, I.P.; Turner, R.B.; Sazonova, I.Y.; Liu, L.; Reed, G.L. Coevolutionary patterns in plasminogen activation. *Proc. Natl. Acad. Sci. USA* **2003**, *100*, 9168–9172. [CrossRef] [PubMed]

47. Eddy, J.L.; Schroeder, J.A.; Zimbler, D.L.; Caulfield, A.J.; Lathem, W.W. Proteolysis of plasminogen activator inhibitor-1 by Yersinia pestis remodulates the host environment to promote virulence. *J. Thromb. Haemost.* **2016**, *14*, 1833–1843. [CrossRef]

48. Sodeinde, O.A.; Subrahmanyam, Y.V.; Stark, K.; Quan, T.; Bao, Y.; Goguen, J.D. A surface protease and the invasive character of plague. *Science* **1992**, *258*, 1004–1007. [CrossRef] [PubMed]

49. Banerjee, S.K.; Crane, S.D.; Pechous, R.D. A Dual Role for the Plasminogen Activator Protease During the Preinflammatory Phase of Primary Pneumonic Plague. *J. Infect. Dis.* **2020**, *222*, 407–416. [CrossRef] [PubMed]

50. Knaust, A.; Weber, M.V.; Hammerschmidt, S.; Bergmann, S.; Frosch, M.; Kurzai, O. Cytosolic proteins contribute to surface plasminogen recruitment of Neisseria meningitidis. *J. Bacteriol.* **2007**, *189*, 3246–3255. [CrossRef]

51. Crane, D.D.; Warner, S.L.; Bosio, C.M. A novel role for plasmin-mediated degradation of opsonizing antibody in the evasion of host immunity by virulent, but not attenuated, Francisella tularensis. *J. Immunol.* **2009**, *183*, 4593–4600. [CrossRef] [PubMed]

52. Siqueira, G.H.; Atzingen, M.V.; de Souza, G.O.; Vasconcellos, S.A.; Nascimento, A. Leptospira interrogans Lsa23 protein recruits plasminogen, factor H and C4BP from normal human serum and mediates C3b and C4b degradation. *Microbiology (Reading)* **2016**, *162*, 295–308. [CrossRef]

53. Crowe, J.D.; Sievwright, I.K.; Auld, G.C.; Moore, N.R.; Gow, N.A.; Booth, N.A. Candida albicans binds human plasminogen: Identification of eight plasminogen-binding proteins. *Mol. Microbiol.* **2003**, *47*, 1637–1651. [CrossRef]

54. Funk, J.; Schaarschmidt, B.; Slesiona, S.; Hallstrom, T.; Horn, U.; Brock, M. The glycolytic enzyme enolase represents a plasminogen-binding protein on the surface of a wide variety of medically important fungal species. *Int. J. Med. Microbiol.* **2016**, *306*, 59–68. [CrossRef] [PubMed]

55. Stie, J.; Fox, D. Blood-brain barrier invasion by Cryptococcus neoformans is enhanced by functional interactions with plasmin. *Microbiology (Reading)* **2012**, *158*, 240–258. [CrossRef]

56. Vanegas, G.; Quinones, W.; Carrasco-Lopez, C.; Concepcion, J.L.; Albericio, F.; Avilan, L. Enolase as a plasminogen binding protein in Leishmania mexicana. *Parasitol. Res.* **2007**, *101*, 1511–1516. [CrossRef]

57. Roggwiller, E.; Fricaud, A.C.; Blisnick, T.; Braun-Breton, C. Host urokinase-type plasminogen activator participates in the release of malaria merozoites from infected erythrocytes. *Mol. Biochem. Parasitol.* **1997**, *86*, 49–59. [CrossRef]

58. Alves, E.S.T.L.; Radtke, A.; Balaban, A.; Pascini, T.V.; Pala, Z.R.; Roth, A.; Alvarenga, P.H.; Jeong, Y.J.; Olivas, J.; Ghosh, A.K.; et al. The fibrinolytic system enables the onset of Plasmodium infection in the mosquito vector and the mammalian host. *Sci. Adv.* **2021**, *7*. [CrossRef]

59. Mantuano, E.; Brifault, C.; Lam, M.S.; Azmoon, P.; Gilder, A.S.; Gonias, S.L. LDL receptor-related protein-1 regulates NFkappaB and microRNA-155 in macrophages to control the inflammatory response. *Proc. Natl. Acad. Sci. USA* **2016**, *113*, 1369–1374. [CrossRef]

60. Mantuano, E.; Azmoon, P.; Brifault, C.; Banki, M.A.; Gilder, A.S.; Campana, W.M.; Gonias, S.L. Tissue-type plasminogen activator regulates macrophage activation and innate immunity. *Blood* **2017**, *130*, 1364–1374. [CrossRef] [PubMed]

61. Das, L.; Banki, M.A.; Azmoon, P.; Pizzo, D.; Gonias, S.L. Enzymatically Inactive Tissue-Type Plasminogen Activator Reverses Disease Progression in the Dextran Sulfate Sodium Mouse Model of Inflammatory Bowel Disease. *Am. J. Pathol.* **2021**. [CrossRef]

62. Thaler, B.; Baik, N.; Hohensinner, P.J.; Baumgartner, J.; Panzenbock, A.; Stojkovic, S.; Demyanets, S.; Huk, I.; Rega-Kaun, G.; Kaun, C.; et al. Differential expression of Plg-RKT and its effects on migration of proinflammatory monocyte and macrophage subsets. *Blood* **2019**, *134*, 561–567. [CrossRef] [PubMed]

63. Das, R.; Ganapathy, S.; Mahabeleshwar, G.H.; Drumm, C.; Febbraio, M.; Jain, M.K.; Plow, E.F. Macrophage gene expression and foam cell formation are regulated by plasminogen. *Circulation* **2013**, *127*, 1209–1218, e1201–e1216. [CrossRef]

64. Vago, J.P.; Sugimoto, M.A.; Lima, K.M.; Negreiros-Lima, G.L.; Baik, N.; Teixeira, M.M.; Perretti, M.; Parmer, R.J.; Miles, L.A.; Sousa, L.P. Plasminogen and the Plasminogen Receptor, Plg-RKT, Regulate Macrophage Phenotypic, and Functional Changes. *Front. Immunol.* **2019**, *10*, 1458. [CrossRef] [PubMed]

65. Syrovets, T.; Tippler, B.; Rieks, M.; Simmet, T. Plasmin is a potent and specific chemoattractant for human peripheral monocytes acting via a cyclic guanosine monophosphate-dependent pathway. *Blood* **1997**, *89*, 4574–4583. [CrossRef] [PubMed]

66. Weide, I.; Tippler, B.; Syrovets, T.; Simmet, T. Plasmin is a specific stimulus of the 5-lipoxygenase pathway of human peripheral monocytes. *Thromb. Haemost.* **1996**, *76*, 561–568. [CrossRef]

67. Burysek, L.; Syrovets, T.; Simmet, T. The serine protease plasmin triggers expression of MCP-1 and CD40 in human primary monocytes via activation of p38 MAPK and janus kinase (JAK)/STAT signaling pathways. *J. Biol. Chem.* **2002**, *277*, 33509–33517. [CrossRef]

68. Li, X.; Syrovets, T.; Genze, F.; Pitterle, K.; Oberhuber, A.; Orend, K.H.; Simmet, T. Plasmin triggers chemotaxis of monocyte-derived dendritic cells through an Akt2-dependent pathway and promotes a T-helper type-1 response. *Arterioscler. Thromb. Vasc. Biol.* **2010**, *30*, 582–590. [CrossRef] [PubMed]

69. Uhl, B.; Zuchtriegel, G.; Puhr-Westerheide, D.; Praetner, M.; Rehberg, M.; Fabritius, M.; Hessenauer, M.; Holzer, M.; Khandoga, A.; Furst, R.; et al. Tissue plasminogen activator promotes postischemic neutrophil recruitment via its proteolytic and nonproteolytic properties. *Arter. Thromb. Vasc. Biol.* **2014**, *34*, 1495–1504. [CrossRef]

70. Li, Q.; Ke, F.; Zhang, W.; Shen, X.; Xu, Q.; Wang, H.; Yu, X.Z.; Leng, Q.; Wang, H. Plasmin plays an essential role in amplification of psoriasiform skin inflammation in mice. *PLoS ONE* **2011**, *6*, e16483. [CrossRef]

71. Syrovets, T.; Jendrach, M.; Rohwedder, A.; Schule, A.; Simmet, T. Plasmin-induced expression of cytokines and tissue factor in human monocytes involves AP-1 and IKKbeta-mediated NF-kappaB activation. *Blood* **2001**, *97*, 3941–3950. [CrossRef]
72. Chang, W.C.; Shi, G.Y.; Chow, Y.H.; Chang, L.C.; Hau, J.S.; Lin, M.T.; Jen, C.J.; Wing, L.Y.; Wu, H.L. Human plasmin induces a receptor-mediated arachidonate release coupled with G proteins in endothelial cells. *Am. J. Physiol.* **1993**, *264*, C271–C281. [CrossRef]
73. Fujiyoshi, T.; Hirano, K.; Hirano, M.; Nishimura, J.; Takahashi, S.; Kanaide, H. Plasmin induces endothelium-dependent nitric oxide-mediated relaxation in the porcine coronary artery. *Arterioscler. Thromb. Vasc. Biol.* **2007**, *27*, 949–954. [CrossRef]
74. Bauman, K.A.; Wettlaufer, S.H.; Okunishi, K.; Vannella, K.M.; Stoolman, J.S.; Huang, S.K.; Courey, A.J.; White, E.S.; Hogaboam, C.M.; Simon, R.H.; et al. The antifibrotic effects of plasminogen activation occur via prostaglandin E2 synthesis in humans and mice. *J. Clin. Invest.* **2010**, *120*, 1950–1960. [CrossRef]
75. Zhao, X.J.; Larkin, T.M.; Lauver, M.A.; Ahmad, S.; Ducruet, A.F. Tissue plasminogen activator mediates deleterious complement cascade activation in stroke. *PLoS ONE* **2017**, *12*, e0180822. [CrossRef]
76. Quinton, T.M.; Kim, S.; Derian, C.K.; Jin, J.; Kunapuli, S.P. Plasmin-mediated activation of platelets occurs by cleavage of protease-activated receptor 4. *J. Biol. Chem.* **2004**, *279*, 18434–18439. [CrossRef]
77. Syrovets, T.; Simmet, T. Novel aspects and new roles for the serine protease plasmin. *Cell Mol. Life Sci.* **2004**, *61*, 873–885. [CrossRef]
78. Lijnen, H.R. Matrix metalloproteinases and cellular fibrinolytic activity. *Biochemistry (Mosc.)* **2002**, *67*, 92–98. [CrossRef]
79. Pang, P.T.; Teng, H.K.; Zaitsev, E.; Woo, N.T.; Sakata, K.; Zhen, S.; Teng, K.K.; Yung, W.H.; Hempstead, B.L.; Lu, B. Cleavage of proBDNF by tPA/plasmin is essential for long-term hippocampal plasticity. *Science* **2004**, *306*, 487–491. [CrossRef] [PubMed]
80. Yee, J.A.; Yan, L.; Dominguez, J.C.; Allan, E.H.; Martin, T.J. Plasminogen-dependent activation of latent transforming growth factor beta (TGF beta) by growing cultures of osteoblast-like cells. *J. Cell Physiol.* **1993**, *157*, 528–534. [CrossRef] [PubMed]
81. Brenner, S. The molecular evolution of genes and proteins: A tale of two serines. *Nature* **1988**, *334*, 528–530. [CrossRef] [PubMed]
82. Doolittle, R.F. Step-by-step evolution of vertebrate blood coagulation. *Cold Spring Harb. Symp. Quant. Biol.* **2009**, *74*, 35–40. [CrossRef] [PubMed]
83. Chana-Munoz, A.; Jendroszek, A.; Sonnichsen, M.; Wang, T.; Ploug, M.; Jensen, J.K.; Andreasen, P.A.; Bendixen, C.; Panitz, F. Origin and diversification of the plasminogen activation system among chordates. *BMC Evol. Biol.* **2019**, *19*, 27. [CrossRef] [PubMed]
84. Liang, Y.J.; Zhang, S.C. Demonstration of plasminogen-like protein in amphioxus with implications for the origin of vertebrate liver. *Acta Zool.* **2006**, *87*, 141–145. [CrossRef]
85. Liu, M.; Zhang, S. A kringle-containing protease with plasminogen-like activity in the basal chordate Branchiostoma belcheri. *Biosci. Rep.* **2009**, *29*, 385–395. [CrossRef]
86. Doolittle, R.F. The protochordate Ciona intestinalis has a protein like full-length vertebrate fibrinogen. *J. Innate Immun.* **2012**, *4*, 219–222. [CrossRef]

International Journal of *Molecular Sciences*

Review

Coagulation and Fibrinolysis in Obstructive Sleep Apnoea

Andras Bikov [1,2,*], Martina Meszaros [3,4] and Esther Irene Schwarz [4,5]

[1] North West Lung Centre, Manchester University NHS Foundation Trust, Manchester M23 9LT, UK
[2] Division of Infection, Immunity and Respiratory Medicine, University of Manchester, Manchester M13 9MT, UK
[3] Department of Pulmonology, Semmelweis University, 1083 Budapest, Hungary; martina.meszaros@usz.ch
[4] Department of Pulmonology and Sleep Disorders Centre, University Hospital Zurich, 8006 Zurich, Switzerland; estherirene.schwarz@usz.ch
[5] Centre of Competence Sleep & Health Zurich, University of Zurich, 8091 Zurich, Switzerland
* Correspondence: andras.bikov@gmail.com; Tel.: +44-161-291-2493; Fax: +44-161-291-5730

Abstract: Obstructive sleep apnoea (OSA) is a common disease which is characterised by repetitive collapse of the upper airways during sleep resulting in chronic intermittent hypoxaemia and frequent microarousals, consequently leading to sympathetic overflow, enhanced oxidative stress, systemic inflammation, and metabolic disturbances. OSA is associated with increased risk for cardiovascular morbidity and mortality, and accelerated coagulation, platelet activation, and impaired fibrinolysis serve the link between OSA and cardiovascular disease. In this article we briefly describe physiological coagulation and fibrinolysis focusing on processes which could be altered in OSA. Then, we discuss how OSA-associated disturbances, such as hypoxaemia, sympathetic system activation, and systemic inflammation, affect these processes. Finally, we critically review the literature on OSA-related changes in markers of coagulation and fibrinolysis, discuss potential reasons for discrepancies, and comment on the clinical implications and future research needs.

Keywords: obstructive sleep apnoea; OSA; coagulation; fibrinolysis; platelets

Citation: Bikov, A.; Meszaros, M.; Schwarz, E.I. Coagulation and Fibrinolysis in Obstructive Sleep Apnoea. *Int. J. Mol. Sci.* **2021**, *22*, 2834. https://doi.org/10.3390/ijms22062834

Academic Editor: Hau C. Kwaan

Received: 21 February 2021
Accepted: 8 March 2021
Published: 11 March 2021

Publisher's Note: MDPI stays neutral with regard to jurisdictional claims in published maps and institutional affiliations.

1. Introduction

Obstructive sleep apnoea (OSA) is a common sleep-related breathing disorder that has been associated with an increased incidence of thromboembolic, cardio- and cerebrovascular events. The main direct pathophysiological consequences of the repetitive collapse of the upper airway during sleep resulting in apnoeas and hypopnoeas are intermittent hypoxaemia, intrathoracic pressure swings, and arousals. Sympathetic overdrive, hypertension, oxidative stress, shear stress, and metabolic derangements promote endothelial dysfunction in OSA [1–4]. In addition, there is growing awareness that there are different phenotypes of OSA based on pathophysiology, symptoms, comorbidities, and long-term cardiovascular consequences [5,6]. The factors promoting endothelial dysfunction might differ between phenotypes, and findings of mechanistic studies might not be applicable to all patients with OSA.

OSA has also been shown to result in a hypercoagulable state, and hypercoagulability has been proposed as one of the contributing mechanisms for the observed increased risk of vascular events in OSA [7,8]. Several studies have reported that fibrinogen and other prothrombotic factors are increased and that fibrinolytic capacity is reduced in OSA. Endothelial dysfunction, an increase in prothrombotic factors [9], a decrease in fibrinolytic activity, platelet activation, and changed rheology and viscosity (e.g., increase in haematocrit as a consequence of nocturnal hypoxaemia) are potential mechanisms explaining a prothrombotic state in OSA [10]. Although there are several convincing theories for how the pathophysiological consequences of OSA might result in a procoagulant state, either via endothelial dysfunction or through interfering with the coagulation cascade or fibrinolysis, there is only limited quality evidence on a direct causative relationship between OSA and a procoagulant state.

To identify relevant articles in this field, the search engine of PubMed was used. The search strategy was conducted using a combination of keywords in the following order: "coagulation AND sleep apnoea" OR "coagulation AND OSA" OR "fibrinolysis AND OSA" OR "coagulation AND hypoxia" OR "coagulation AND inflammation" OR "fibrinolysis AND hypoxia" OR "fibrinolysis AND inflammation OR "platelet AND sleep apnoea", OR "platelet and OSA" OR "platelet AND hypoxia" OR "platelet AND inflammation". Moreover, we used all molecule names that were found in the review, for example "factor XII AND sleep apnoea" OR "factor XII AND OSA" OR "TF AND sleep apnoea" OR "TF AND OSA" OR "fibrinogen AND sleep apnoea" OR "fibrinogen AND OSA" OR "PAI-1 AND sleep apnoea" OR "PAI-1 AND OSA". Animal and human studies were included. We selected case control, randomised controlled, and interventional studies and also thematic reviews and systematic reviews/meta-analyses. We included articles written in English published prior to 15.01.2021.

Hypotheses and current evidence on how OSA might promote a disturbance of haemostasis and coagulation–fibrinolysis balance which results in a prothrombotic state, data on the effects of OSA treatment on coagulation activity, and the fibrinolytic system as well as knowledge gaps and future perspectives are discussed.

2. Overview of the Coagulation System and Fibrinolysis and the Role of Platelets

The coagulation system works by adhesion and aggregation of activated platelets (known as "primary haemostasis") and formation of a fibrin network via the coagulation cascade (known as "secondary haemostasis") [11]. This review focuses mainly on the secondary homeostasis and its role in the pathogenesis of OSA.

The coagulation cascade includes a series of proteolytic events in which serine proteases activate proenzymes on the surface of activated platelets. The cascade has traditionally been divided into "intrinsic" and "extrinsic" pathways. Current literature divides the process of coagulation into "initiation", "amplification", "propagation", and "stabilization" phases [11] (Figure 1).

The activation of the tissue factor (TF, also known as factor III) in combination with factor VII has been considered as the first step of the coagulation cascade. TF is a membrane-bound glycoprotein which is constitutively presented in the subendothelium [12]. In case of endothelial injury, TF is exposed to plasma procoagulants and binds factor VII which is activated (factor VIIa). Moreover, inducible TF can be expressed by inflammatory cells, for example in monocytes in response to endotoxin [13] and by vascular endothelial cells following induction by cytokines such as tumour necrosis factor α (TNFα) [14]. Via calcium signalling, TF–factor VIIa complex binds factor X and catalyses its conversion to factor Xa [15]. Factor Xa can be generated as well by the complex of factor IXa and factor VIIIa after the activation of factor XII–factor XI complex (kallikrein–kinin system, reviewed in detail [16]). Factor Xa binds and activates factor V, and together they form the prothrombinase complex (factor Xa–factor Va–factor II). The prothrombinase complex generates thrombin (factor IIa) from prothrombin (factor II) resulting in the conversion of circulating fibrinogen (factor I) to insoluble fibrin (factor Ia). This complex is the most efficient in the presence of calcium and phospholipid surface of the activated platelets [11]. Finally, fibrin is covalently cross-linked by factor XIII resulting in fibrin polymers which are the major constituents of the clot [17]. Factor XIII binds other anti-fibrinolytic proteins to fibrin such as α2-antiplasmin (A2AP), thrombin activated fibrinolysis inhibitor (TAFI), and complement C3. C3 deposits are associated with thinner fibrin fibres, while they do not affect plasmin formation [18]. Another essential mechanism for clot stability is the presence of activated platelets [19]. Platelets compress and reduce the volume of the thrombus, and they interact with fibrin fibres via the GP IIb–IIIa complex.

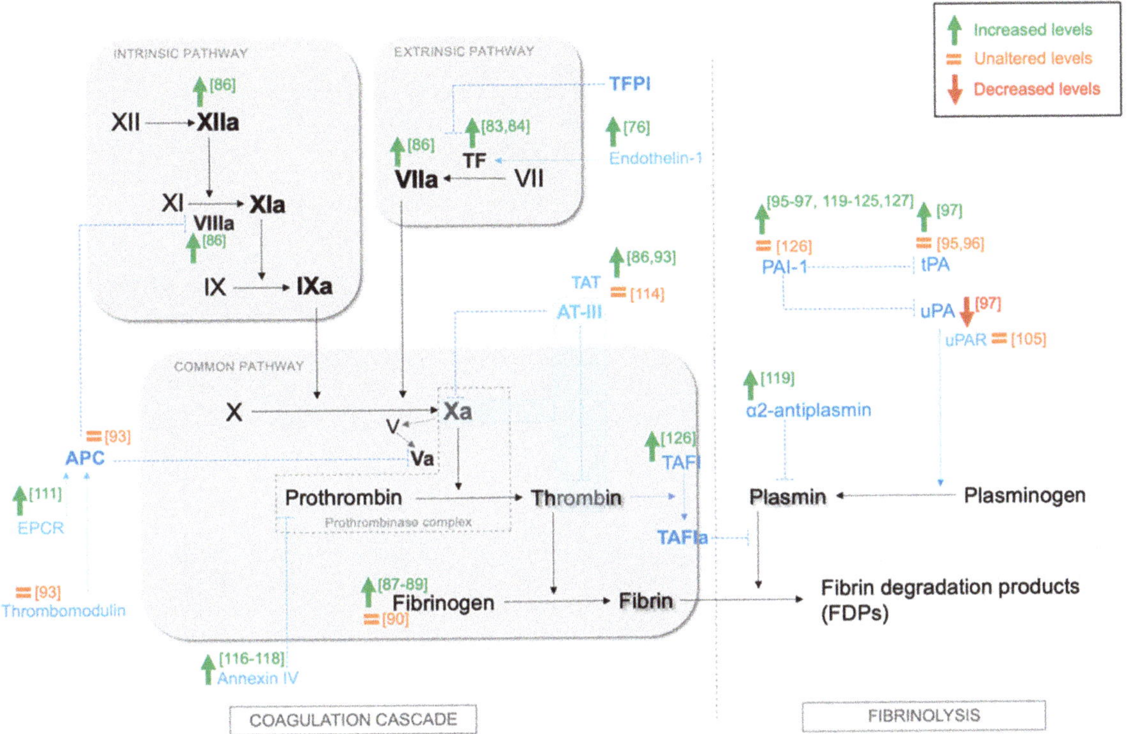

Figure 1. The coagulation cascade and fibrinolysis and how obstructive sleep apnoea (OSA) interacts with them. APC—activated protein C; AT-III—antithrombin III; EPCR—endothelial protein C receptor; IX—factor IX; IXa—activated factor IX; PAI-1—plasminogen activator inhibitor-1; TAFI—thrombin activated fibrinolysis inhibitor; TAFIa—activated thrombin activated fibrinolysis inhibitor; TAT—thrombin-antithrombin complex; TF—tissue factor; TFPI—tissue factor pathway inhibitor; tPA—tissue plasminogen activator; uPA—urokinase plasminogen activator; uPAR—urokinase type plasminogen activator receptor; Va—activated factor V; VIIIa—activated factor VIII; X—factor X; Xa—activated factor X; XI—factor XI; Xia—activated factor XI; XII—Factor XII; XIIa—activated factor XII.

The thrombus is lysed by plasmin following activation of plasminogen, which is a zymogen produced by the liver. The activation can occur on the surface of the thrombus or the endothelial cell mainly through the tissue plasminogen activator (tPA) or urokinase (uPA) [20]. The production of both enzymes is induced by thrombin [21]. In plasma the main activator is the tPA as it has higher affinity to plasminogen than uPA, whilst uPA is more important in the extravascular activation of plasmin and is involved in cell migration and wound healing [22]. The tPA is predominantly released by the endothelial cells and requires fibrin as a cofactor, while uPA is produced by monocytes and the urinary epithelium and does not need fibrin for its action. Apart from tPA and uPA, members of the contact pathway, such as activated factor XII, activated factor XI, and kallikrein can also activate plasminogen [23].

Fibrinolysis occurs both on the thrombus and the surface of cells; however, the former process is more effective. The fibrin-bound tPa has significantly higher catalytic activity to activate plasminogen compared to the fluid phase [24]. In addition, the effect of A2AP is inactivated if the plasmin is bound to fibrin [25]. Moreover, both fibrinogen and fibrin facilitate plasmin conversion [26]. Cell-surface-related fibrinolysis is achieved by two main mechanisms. Annexin II may form a complex with S100A10 that binds plasminogen and tPA independently from fibrin [27]. The other mechanism is mediated by the urokinase type plasminogen activator receptor (uPAR) which binds urokinase with plasminogen [28]. The expression and cleavage of uPAR is upregulated by TNFα, interleukin 1β (IL-1β), and IL-6 [28]. The cleavage is also facilitated by uPA and plasmin leading to soluble uPAR (suPAR) [28] which is a proinflammatory molecule but can also act as a scavenger receptor for uPA, inhibiting its function [29].

Once fibrin polymers are degraded by plasmin, fibrin degradation products (FDPs) are formed. Some of these have immunoregulatory and thrombosis modulatory roles. The most commonly used FDP in clinical practice is the d-dimer, which reflects thrombus formation and fibrinolysis [20].

The physiological coagulation cascade is regulated by three main processes. First, antithrombin forms an inhibiting complex with thrombin and factor Xa, called the thrombin–antithrombin (TAT) complex [30]. Second, tissue factor pathway inhibitor (TFPI), which is presented on endothelial cells, inhibits the action of the TF–factor VIIa complex. Third, thrombin forms a complex with endothelial membrane-anchored thrombomodulin and activates protein C. Activated protein C (APC) with its cofactor protein S degrades factors Va and VIIIa resulting in downregulation of the coagulation system [11].

Plasminogen activators are inhibited by plasminogen activator inhibitor-1 (PAI-1), PAI-2, A2AP, α2-macroglobulin, C1-esterase inhibitor, and protease nexin-1. PAI-1 is released by the endothelial cells and platelets and is the most important inhibitor of tPA and uPA [31]. It is upregulated by thrombin; various proinflammatory cytokines, such as TNFα, IL-6, C-reactive protein (CRP), and transforming growth factor beta (TGF-β); as well as hormones, such as insulin and cortisol [32]. α2-antiplasmin is produced by the liver and is a potent inhibitor of plasmin. It regulates fibrinolysis in three ways: by inhibiting adsorption of plasminogen to fibrin, forming complexes with plasmin, and making fibrin more resistant to plasmin through cross-linking with factor XIIIa [33]. It seems that factor XIII is essential to its mechanism [34] as it is inactive if the plasmin is bound to fibrin without XIII [25]. A further mechanism contributing to the regulation of fibrinolysis involves TAFI. This molecule is synthesised in the liver and decreases the number of available plasminogen binding sites, slowing down the fibrinolysis [20].

Activated platelets have a pivotal role in coagulation by providing an activated membrane (such as surface phospholipids) to the coagulation factors and aggregating in the haemostatic plug. Following endothelial injury, platelets are exposed to the highly thrombogenic subendothelium. Subendothelial proteins, such as von Willebrand factor (vWF), collagen, thrombospondin, and vitronectin, bind to several surface glycoprotein (GP) receptors of the platelets [11]. vWF multimers bind the platelet GP–Ib–IX–V complex which reduces platelet velocity. Thus, collagen fibres are able to bind platelet GP–VI and GP–Ia–IIa complexes, anchoring the platelets to the subendothelium [35]. Notably, vWF is also secreted by the endothelial cells and protects circulating factor VIII from the proteolytical degradation [36]. Adhesion leads to cytoskeleton rearrangement and a change in platelet shape resulting in platelet activation [37]. Furthermore, platelets can also be activated directly by thrombin [38], fibrinogen [39], or proinflammatory cytokines such as platelet-activating factor [40]. Activated platelets release a wide range of mediators to facilitate further activation and aggregation of other platelets. P-selectin is translocated from alpha-granules to the platelet surface and binds its ligand P-selectin glycoprotein ligand 1 (PSGL-1) on leukocytes and endothelial cells [41,42]. This interaction promotes rolling of leukocytes and platelets on the activated endothelium. Thus, leukocytes can form a scaffold to fibrin formation. Moreover, P-selectin itself induces fibrin deposition [43]. Adenosine diphosphate (ADP) and thromboxane-A_2 (TxA$_2$) released from dense granules induce vasoconstriction and further platelet activation with increased platelet GP IIb–IIIa expression [37]. Activation of platelet GP IIb–IIIa leads to plug formation by binding vWF and causing fibrinogen deposition on the platelet surface. Activated platelets release coagulation factors, such as factor V and factor VIII, resulting in further fibrin formation [44,45] (Figure 2).

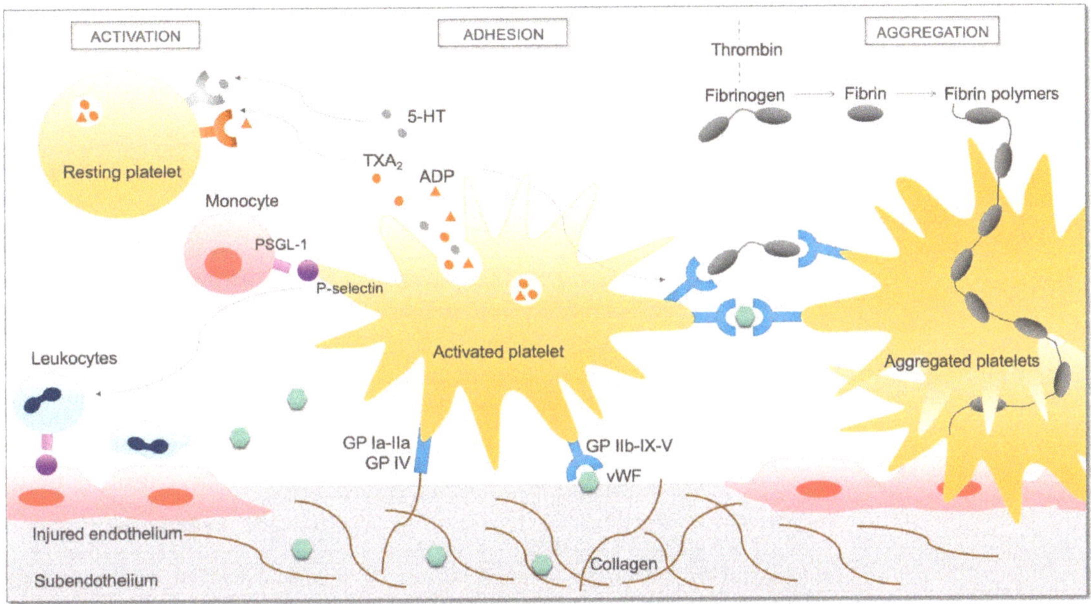

Figure 2. The mechanism of platelet adhesion and aggregation. 5-HT—serotonin; ADP—adenosine diphosphate; GP Ia-IIa—glycoprotein Ia-IIa; GP IIb-IX-V—glycoprotein IIb-IX-V; GP IV—glycoprotein IV; PSGL-1—P-selectin glycoprotein ligand 1; TXA$_2$—thromboxane-A2; vWF—von Willebrand factor.

The two main inhibitors of platelet activation and aggregation are the vasoactive nitric oxide (NO) and prostacyclin (PGI$_2$) released by the endothelium which work synergistically in platelets [46,47].

In conclusion, physiological haemostasis is regulated by complex interactions between coagulation factors and their regulators, platelets, adhesion molecules, and immune cells and endothelial cells.

3. Current Knowledge on the Effects of OSA on Coagulation, Fibrinolysis, and Platelet Activation

In theory, OSA can affect all pathways of the Virchow triad and result in endothelial damage, stasis, and hypercoagulability. Described alterations in the coagulation system induced by OSA are outlined here.

Intermittent hypoxia is one of the primary proposed mechanisms of haemostatic alterations in OSA. After 4 weeks of exposure to intermittent hypoxia, the levels of fibrinogen, factor VIII, and vWF were elevated in an animal model [48]. Intermittent hypoxia modifies the hepatic protein synthesis and aggravates inflammation in the liver which is the major source of coagulant and anticoagulant factors [49]. Increased expression of hypoxia-inducible factor-1 (HIF-1) and transcription factor nuclear-kB (NF-kB) is also mediated by intermittent hypoxia in OSA [50–54] leading to upregulated expression of procoagulant factors, such as TF and factor VIII [55–57]. Moreover, PAI-I, VEGF, and NOS genes are also targeted genes of HIF-1 and regulate haemostatic processes [58]. Furthermore, intermittent hypoxia itself increases the production of TF by suppressing the protein C anticoagulant pathway in endothelial cells [59]. Enhanced platelet activity and aggregation were also documented under hypoxic conditions [60,61], and the degree of hypoxia was a significant predictor of platelet activation [62]. In contrast, a recent study demonstrated reduced activation of platelet GP IIb–IIIa under hypoxia resulting in an impaired platelet adhesive function [63].

Increased sympathetic activity caused by intermittent hypoxia and sleep fragmentation has also emerged as an important factor in OSA-associated hypercoagulability [64]. Catecholamines increase the levels of circulating factor V and vWF and directly activate the platelets [65,66]. In the study of Eisensehr et al., elevated epinephrine levels in the morning correlated with increased haemostasis in patients with OSA [67].

Finally, it has previously been described that chronic inflammation itself leads to abnormal haemostasis [68]. Intermittent hypoxia and accompanying oxidative stress may induce the production of proinflammatory cytokines in OSA [69]. Cytokines and chemokines directly and indirectly activate platelets; thus, they release stored proinflammatory substances [70,71]. Cytokines, such as TNFα and IL-1β, also increase the expression of TF [14,72]. The extrinsic pathway can be enhanced in parallel by endothelial dysfunction which is consequently caused by hypoxic and inflammatory processes in OSA [73,74].

The endothelium is the most important factor which regulates coagulation by ensuring adequate blood flow, serving a barrier to subepithelial prothrombotic extracellular matrix components and releasing vasoactive regulatory molecules. It is known that intermittent hypoxaemia leads to endothelial injury contributing to impaired regulation of the coagulation [75]. Endothelin-1 is overexpressed by the endothelial cells in OSA [76] and leads to increased expression of vWF and TF [77,78]. The glycocalyx is a layer covering the endothelium which is composed of glycosaminoglycans, proteoglycans, and plasma proteins [79]. The main regulator of the coagulation cascade, antithrombin III, is bound to heparan sulphate proteoglycans of the glycocalyx [80]. Systemic inflammatory stimuli, hyperglycaemia, oxidised low-density lipoprotein cholesterol (LDL-C), and oxidative stress could damage the glycocalyx, leading to impaired regulation of coagulation [79]. In line with this, increased turnover of hyaluronic acid, an important component of the glycocalyx, has recently been reported in OSA [81]. In addition, endothelial dysfunction also promotes platelet activation and aggregation by increased release of vWF and plasminogen activator inhibitor-1 (PAI-1) and decreased production of NO and PGI$_2$ from the endothelial cells [82] (Figure 3).

Clinical studies in patients with OSA supported the findings from basic science and model studies. The levels of TF were elevated in OSA [83,84]. The concentration of TF was directly related to the percentage of time spent with an oxygen saturation < 90% [83] and oxygen desaturation index (ODI) [84] highlighting the role of intermittent hypoxia. Notably, higher plasma levels of TF were also associated with polysomnographic indices of sleep fragmentation in participants without a history of OSA, indicating that sleep disruption itself alters the coagulation system [85]. However, another study did not show a correlation between TF levels and the arousal index in patients with OSA [83]. In the study by Robinson et al., the plasma levels of factor VIIa and factor XIIa were significantly higher in the OSA group compared to controls. However, there was no correlation between these coagulation factors and the severity of OSA [86]. Plasma fibrinogen levels were elevated in patients with OSA in most [87–89] but not all studies [90]. In a recent meta-analysis on prothrombotic markers including 2190 participants from 15 studies, patients with OSA had significantly higher plasma fibrinogen levels compared with controls [9]. Wessendorf et al. demonstrated that elevated circulating fibrinogen was associated with the severity of OSA in patients with concomitant history of stroke suggesting a possible link between OSA-associated hypercoagulation and cerebrovascular complications [89] (Supplementary Table S1).

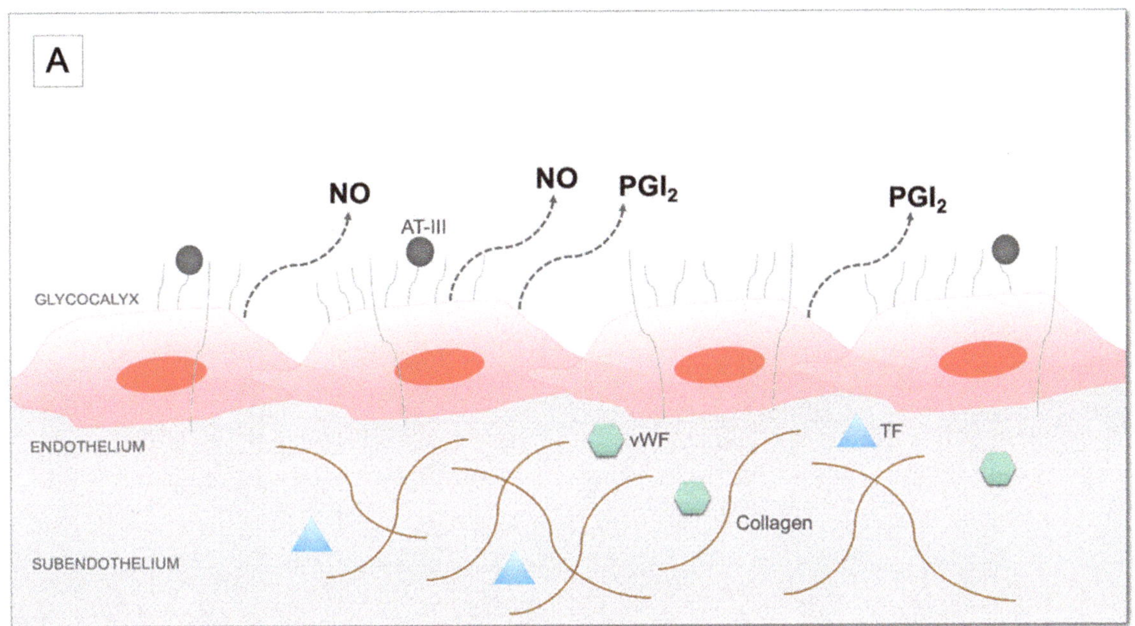

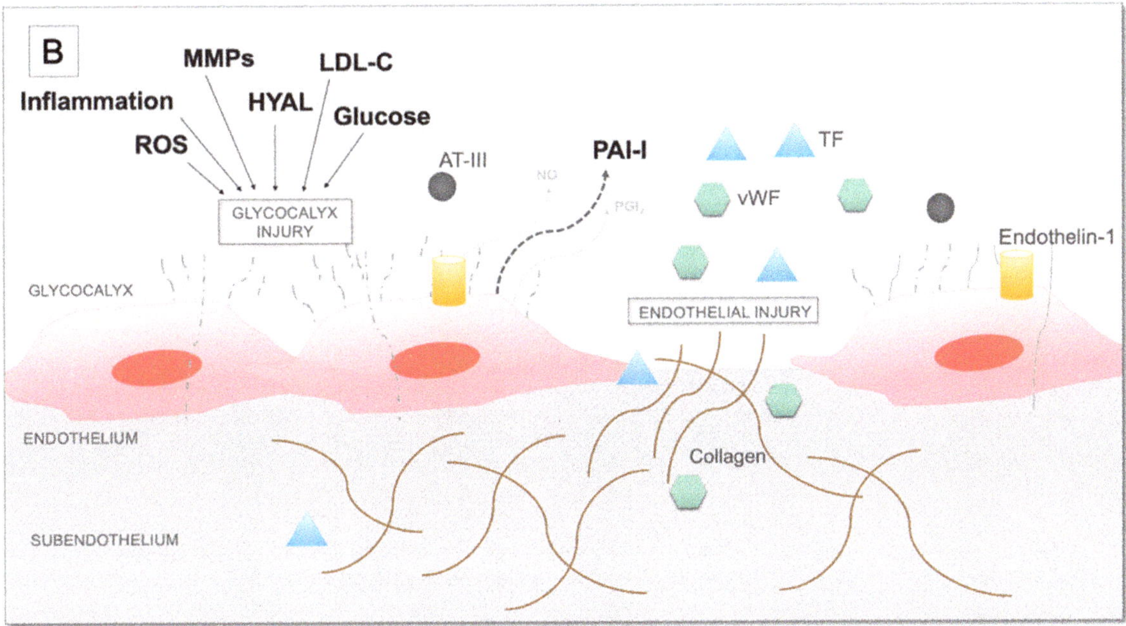

Figure 3. The role of endothelium in regulation of haemostasis (**A**) and endothelial injury (**B**) in OSA. AT-III—antithrombin III; HYAL—hyaluronidase; LDL-C—low-density lipoprotein cholesterol; MMPs—matrix metalloproteinases; NO—nitric oxide; PAI-1—plasminogen activator inhibitor-1; PGI₂—prostacyclin; ROS—reactive oxygen species; TF—tissue factor; vWF—von Willebrand factor.

Tissue plasminogen activator is released by thrombin, proinflammatory cytokines, and vascular endothelial growth factor (VEGF) from the storage granules in endothelial cells [91]. Markers of systemic inflammation [92], thrombin [86,93], and VEGF levels [94] are increased in OSA, theoretically contributing to high tPA levels. However, the data on the levels of tPA are contradictory, as no difference in tPA levels [95] and activity [96], higher tPA levels [97], and lower tPA activity [95] were consistently reported in OSA. The circadian variation in tPA levels and activity could be a possible explanation for the contradictory results [95]. tPA together with uPA are rapidly cleared by the liver following forming complexes with LDL-receptor like protein [91]; however, it is not clear if this mechanism is altered in OSA. Several drugs, such as steroids, statins, and valproic acid, may induce tPA release. Differences in medication usage of the studied populations could also lead to discrepancies. Only one study investigated uPA levels in OSA and reported lower concentrations [97]. The expression of uPA is induced by oestradiol [98] and survivin [99], the levels of which are decreased in OSA [100,101]. Urokinase is activated by plasmin and kallikrein [102]. The latter was found to show decreased expression in OSA [103]. The soluble levels of uPAR were found unaltered in OSA [104,105]. The expression of uPAR is upregulated by proinflammatory cytokines [28,106–108] and TGF-β [109,110]. While inflammation is accelerated [92], reduced levels of TGF-β were reported in OSA [97]. Although suPAR levels in general reflect uPAR expression, it is noteworthy that the cleavage of uPAR may be reduced in OSA due to the low levels of uPA and activated plasmin. Factor XII is a weak activator of plasminogen. Its levels were reported to be increased in OSA [86]. The urinary levels of kallikrein, another weak plasminogen activator were found to be reduced in children with OSA [103] (Supplementary Table S2).

The regulator pathways of the coagulation cascade have not been extensively investigated. The levels of APC were comparable between the OSA and non-OSA groups in the study of Takagi et al. [93]. Endothelial protein C receptor (EPCR) and thrombomodulin promote the activation of protein C. Whilst the blood and urinary levels of EPCR were increased in OSA [111], there was no difference in thrombomodulin [93]. Apolipoprotein H inhibits the activation of protein C [112], and their levels were increased in OSA [113]. Whilst antithrombin itself has not been measured in OSA before, the levels of TAT complex were higher in patients with OSA compared to the control group, and there was an association between TAT levels and OSA severity measured by ODI or AHI in some [86,93] but not all studies [114]. Annexin V also has anticoagulant properties as it competes with prothrombin for phosphatidylserine binding sites [115]. Increased frequency of Annexin V+ endothelial cells [116] and Annexin V+ microparticles [117,118] were reported in OSA suggesting that this molecule may serve in a negative feedback mechanism of OSA-related coagulation (Supplementary Table S1).

Most studies reported elevated PAI-1 levels in blood samples of patients with adult [95,96,119–124] and paediatric [125] OSA. Although the study by Nizankowska-Jedrzejczyk et al. did not find a difference in PAI-1 levels [126], the total number of patients and controls was relatively low ($n = 38$). The levels of PAI-1 are directly related to disease severity [95,119,122,127], emphasising the role of OSA in increased PAI-1 expression. The increased PAI-1 levels in OSA are not surprising, as PAI-1 expression is increased by hypoxaemia [128], systemic inflammation [129], oxidative stress [130], cortisol, and angiotensin II [129]. Another potential mechanism explaining increased PAI-1 levels could be the decreased expression of the klotho in OSA [131]. Klotho is an anti-inflammatory, anti-aging protein, and increased PAI-1 levels were found in klotho deficient mice [132]. However, PAI-1 levels need to be interpreted carefully, as a significant circadian variation of PAI-1 has been described previously [95,124]. This variation is due to both direct control of PAI-1 expression by the circadian genes and diurnal variation of hormones [129]. Interestingly, the variation is larger in patients with OSA than in non-OSA controls [95]. Significantly higher levels of A2AP levels were reported in OSA, and they were related to disease severity [119]. The reason for these changes was not investigated in detail but could be due to the increased IL-6 levels in OSA which induce A2AP formation [133]. The levels of TAFI were reported to be higher in OSA [126]. This molecule is activated by thrombin, plasmin, trypsin, and neutrophil elastase [134]. However, the potential of thrombin to active TAFI is multiplied by thrombomodulin which is unaltered in OSA [93] suggesting that TAFI activation is weak in OSA. The fibrin is stabilised by the complement C3 which has been previously reported to be higher in OSA [135] (Supplementary Table S2).

Increased platelet activation and aggregation were described in OSA in several [136–139] but not all studies [96,140]. Platelet aggregation was higher in severe OSA compared to mild disease [67] and correlated with AHI [136]. GP–Ib is a marker of platelet activity and it is downregulated and internalised during platelet activation [141]. GP–Ib receptor density in platelets was downregulated in OSA indicating increased platelet activation and platelet reactivity; however, GP IIb–IIIa expression did not differ between the OSA and control group [62]. Another marker of platelet activation, P-selectin was measured in higher concentrations in patients with OSA compared to controls in some [86,126,142,143] but not all studies [62,117,144]. The results of studies on the role of vWF in OSA are inconsistent. In some studies, vWF levels were significantly higher in the OSA group compared to controls [83,104]. However, Zamarrón-Sanz et al. did not detect any difference in vWF levels between OSA and controls [145]. Platelet-derived microparticles (PMPs) are generated during platelet activation. PMPs are also suggested to provide an activated surface for the coagulation cascade with 50–100× higher procoagulant activity compared to the activated platelets [146,147]. The levels of PMPs were higher in OSA [148,149] and correlated with disease severity [149] in most but not all studies [139]. Significantly higher platelet counts were detected in OSA [150] and children with OSA [151,152] compared to healthy individuals. Moreover, there was an association between platelet count and disease severity in patients with OSA and manifest cardiovascular disease [153] (Supplementary Table S3).

Elevated blood coagulability was confirmed by clinical coagulation tests in OSA. Prothrombin time (PT) is used in clinical practice to evaluate the function of the extrinsic and common coagulation pathways, and activated partial thromboplastin time (aPTT) reflects the abnormalities of the intrinsic pathway [154]. A recent study measured a significantly shorter PT and unchanged aPTT especially in patients with moderate to severe OSA compared to controls, suggesting an activated extrinsic pathway in OSA [150].

OSA is associated with a procoagulant state due to increased levels of coagulation factors, enhanced platelet activation and aggregation, and endothelial dysfunction induced by intermittent hypoxia and inflammatory processes. However, the relationship between OSA and some individual coagulation factors or regulator molecules is controversial. Future randomised controlled studies are warranted to gain a more precise understanding of haemostasis in OSA.

4. The Effect of OSA Treatment on Coagulation, Fibrinolysis, and Platelet Activation

Limited data are available on the effects of OSA therapies on haemostatic alterations, and the findings are inconsistent. One month of continuous positive airway pressure (CPAP) therapy failed to decrease the levels of activated factor VIIa, factor XIIa, and factor VIIIa [86]. In contrast, another study demonstrated a significant post-CPAP decrease in 24 h concentrations of factor V, factor VIII, and vWF especially in the nocturnal and morning periods. However, factor VII levels remained unchanged after 2 months of treatment with CPAP [155], yet another study reported a significant decrease in factor VII levels after 6 months of therapy [156]. Whilst some authors found no difference in plasma fibrinogen levels [86,140,155,157], another group detected decreased fibrinogen concentrations in response to CPAP therapy; however, the sample size was small (*n* = 11) in this study [158] (Supplementary Table S1).

Neither uPA nor tPA concentrations changed following CPAP treatment [97]. Two studies reported that PAI-1 levels decreased following two weeks [159] and one month [97] of CPAP treatment, respectively. However, another well-designed study did not find any change in PAI-1 levels following two months of CPAP usage [155]. This may suggest that the short-term beneficial effect of CPAP may be reversed by homeostatic factors in the long term. More concordant is the lack of effect of CPAP on the diurnal variability of PAI-1 levels in OSA [121,155]. Treatment with a mandibular advancement device (MAD) did not change PAI-1 concentrations [126]. In contrast, PAI-1 levels significantly decreased following adenotonsillectomy in children with OSA [160]. Finally, PAI-1 concentrations were decreased following sleeve gastrectomy [161] compatible with the fact that PAI-1 partly originates from adipose tissue [129]. TAFI levels significantly decreased following treatment with MAD [126] (Supplementary Table S2).

Several studies found a decrease in platelet activation and aggregation after one night or one to three months of CPAP therapy in OSA [136,138,162,163]. However, whilst platelet aggregation was reduced following 90 days of CPAP therapy, there was no difference at 30 days [164]. P-selectin was not influenced either by CPAP [86] or by MAD [126]; however, in another study CPAP resulted in a decrease in P-selectin levels [138]. Most of the studies did not report significant changes in the levels of vWF [83,86,121,159], with one exception [155] which had a cross over design with a 1 month washout period. Significantly decreased levels of vWF were detected 6 months after upper airway surgery [165]. Two weeks of CPAP withdrawal resulted in an elevation in the levels of PMPs [166]; however, the same workgroup also reported conflicting results [167]. Higher body mass index and disease severity of participants in the former study may explain the different results. PT and aPTT were increased following 30 days of CPAP therapy [163] and upper airway surgery [168] (Supplementary Table S3).

The effect of OSA therapies on haemostasis is inconclusive. Several studies demonstrated that short- and long-term CPAP therapy had beneficial effects on coagulation system and platelet function in OSA. However, CPAP failed to improve the procoagulant state in OSA in other reports. Further adequately powered randomised controlled studies with higher treatment efficacy and adherence are required to determine the effects of CPAP therapy on the haemostatic alterations.

5. Discussion of Major Findings

As outlined above, several pathophysiological consequences of OSA, particularly intermittent hypoxia, sympathetic activity, systemic inflammation, and consequential endothelial dysfunction, may result in a hypercoagulable state, platelet activation, and impaired fibrinolytic capacity. Some data from in-vitro studies and animal models have been confirmed in case control studies in patients with OSA. However, there are also controversial findings, particularly on the effect of OSA treatment on the finding of changes in coagulation or fibrinolysis. Many studies had methodological limitations or were not designed to primarily assess coagulation, and the few randomised controlled trials (RCTs) had a limited sample size. In addition, the severity of OSA and the associated hypoxic burden differed between studies. However, whether the discrepancies between studies are due to the differences in OSA severity and the phenotype or limitations in study design is speculative.

Some findings on hypercoagulability and a disturbed coagulation–fibrinolysis balance have been consistently shown and support the role of hypercoagulability as one of the mechanisms explaining the observed incidence of vascular events in OSA. For instance, elevated levels of PAI-1 are known to increase the risk for myocardial infarction [169,170].

How hypercoagulability and impaired fibrinolysis are affected by comorbidities has not been systematically studied in large sample sizes. Coexistent comorbidities, such as obesity, may lead to systemic inflammation and liver damage that also leads to abnormalities in coagulation [10,171]. This could be an explanation for the lack of change in coagulation factors following CPAP therapy [86]. Of note, most studies assessed short-term effects of CPAP therapy on the coagulation system. It is also unclear whether effective treatment of OSA with CPAP could reduce the cardiovascular risk that is attributable to disturbances in the coagulation system or fibrinolysis.

Limitations in study design that do not allow for causative associations to be established between OSA and changes in coagulation or that do not adequately control for comorbidities or OSA severity or phenotypes are a common problem. However, there are some data from RCTs and meta-analyses that strengthen the level of evidence, e.g., on increased levels of procoagulant and platelet-derived microvesicles or fibrinogen in OSA [166,172].

6. Clinical Implications

A procoagulant state in OSA may make OSA patients more susceptible to both venous thromboembolism and thrombus formation on arterial plaques resulting in pulmonary embolism, deep vein thrombosis, acute coronary syndrome, and stroke [1,2,173]. The findings on hypercoagulability and impaired fibrinolysis lead to the hypothesis that OSA both promotes occurrence of vascular events in patients with atherosclerosis due to hypercoagulability and results in more severe end organ damage in case of an ischemic vascular event as a consequence of the impaired clot lysis.

The current knowledge on OSA as risk factor of a procoagulant state should be implemented in treatment recommendations together with comorbidities and other cardiovascular risk factors. In addition, measures of coagulation and fibrinolysis could be used for phenotyping patients and risk assessment.

7. Implications for Research

Due to the limitations in study design of previous studies on coagulation and fibrinolysis in OSA and due to relevant confounding factors, such as obesity and comorbidities, conclusions on the causative relationship on OSA and hypercoagulability are somewhat limited, and the quality of evidence needs to be improved. Potential causative associations between OSA and hypercoagulability and the role of comorbidities is a research topic that needs to be further investigated in well-designed studies using different models and designs.

8. Summary

In summary, there is evidence that OSA results in elevated levels of fibrinogen and increased platelet activity, promotes platelet adhesion and aggregation, and results in an impaired fibrinolytic capacity. A hypercoagulable state and impaired fibrinolysis promote thrombotic events, and this might be one of the several underlying mechanisms linking OSA with an adverse cardio- and cerebrovascular outcome. Several other pathophysiological consequences of OSA that differ between phenotypes of OSA, clustering of cardiovascular risk factors, and comorbidities might define the vascular risk that OSA induces in an individual. However, hypercoagulability, fibrinolysis, and haemostasis are potential therapy targets that can be influenced either via the coagulation system or the endothelium.

Supplementary Materials: The following are available online at https://www.mdpi.com/article/10.3390/ijms22062834/s1.

Author Contributions: The manuscript was drafted and was critically reviewed and approved by all authors. The figures and tables were created by M.M. All authors have read and agreed to the published version of the manuscript.

Funding: This research received no external funding.

Institutional Review Board Statement: Not applicable.

Informed Consent Statement: Not applicable.

Data Availability Statement: Not applicable.

Acknowledgments: Andras Bikov is supported by the NIHR Manchester BRC.

Conflicts of Interest: The authors declare no conflict of interest.

References

1. Kohler, M.; Stoewhas, A.C.; Ayers, L.; Senn, O.; Bloch, K.E.; Russi, E.W.; Stradling, J.R. Effects of continuous positive airway pressure therapy withdrawal in patients with obstructive sleep apnea: A randomized controlled trial. *Am. J. Respir. Crit. Care Med.* **2011**, *184*, 1192–1199. [CrossRef]
2. Schwarz, E.I.; Puhan, M.A.; Schlatzer, C.; Stradling, J.R.; Kohler, M. Effect of CPAP therapy on endothelial function in obstructive sleep apnoea: A systematic review and meta-analysis. *Respirology* **2015**, *20*, 889–895. [CrossRef] [PubMed]
3. Dunet, V.; Rey-Bataillard, V.; Allenbach, G.; Beysard, N.; Lovis, A.; Prior, J.O.; Heinzer, R. Effects of continuous positive airway pressure treatment on coronary vasoreactivity measured by (82)Rb cardiac PET/CT in obstructive sleep apnea patients. *Sleep Breath.* **2016**, *20*, 673–679. [CrossRef]
4. Kohler, M.; Stradling, J.R. Mechanisms of vascular damage in obstructive sleep apnea. *Nat. Rev. Cardiol.* **2010**, *7*, 677–685. [CrossRef]
5. Zinchuk, A.V.; Gentry, M.J.; Concato, J.; Yaggi, H.K. Phenotypes in obstructive sleep apnea: A definition, examples and evolution of approaches. *Sleep Med. Rev.* **2017**, *35*, 113–123. [CrossRef]
6. Gabryelska, A.; Białasiewicz, P. Association between excessive daytime sleepiness, REM phenotype and severity of obstructive sleep apnea. *Sci. Rep.* **2020**, *10*, 34. [CrossRef]
7. Sanchez-de-la-Torre, M.; Campos-Rodriguez, F.; Barbe, F. Obstructive sleep apnoea and cardiovascular disease. *Lancet Respir. Med.* **2013**, *1*, 61–72. [CrossRef]
8. von Kanel, R.; Dimsdale, J.E. Hemostatic alterations in patients with obstructive sleep apnea and the implications for cardiovascular disease. *Chest* **2003**, *124*, 1956–1967. [CrossRef] [PubMed]
9. Qiu, Y.; Li, X.; Zhang, X.; Wang, W.; Chen, J.; Liu, Y.; Fang, X.; Ni, X.; Zhang, J.; Wang, S.; et al. Prothrombotic Factors in Obstructive Sleep Apnea: A Systematic Review with Meta-Analysis. *Ear Nose Throat J.* **2020**, 145561320965208. [CrossRef] [PubMed]
10. Liak, C.; Fitzpatrick, M. Coagulability in obstructive sleep apnea. *Can. Respir. J.* **2011**, *18*, 338–348. [CrossRef]
11. Palta, S.; Saroa, R.; Palta, A. Overview of the coagulation system. *Indian J. Anaesth.* **2014**, *58*, 515–523. [CrossRef] [PubMed]
12. Mann, K.G.; van't Veer, C.; Cawthern, K.; Butenas, S. The role of the tissue factor pathway in initiation of coagulation. *Blood Coagul. Fibrinolysis* **1998**, *9* (Suppl. S1), S3–S7.
13. Osterud, B. Tissue factor expression by monocytes: Regulation and pathophysiological roles. *Blood Coagul. Fibrinolysis* **1998**, *9* (Suppl. S1), S9–S14. [PubMed]
14. Edgington, T.S.; Mackman, N.; Fan, S.T.; Ruf, W. Cellular immune and cytokine pathways resulting in tissue factor expression and relevance to septic shock. *Nouv. Rev. Fr. Hematol.* **1992**, *34*, S15–S27. [PubMed]
15. ten Cate, H.; Bauer, K.A.; Levi, M.; Edgington, T.S.; Sublett, R.D.; Barzegar, S.; Kass, B.L.; Rosenberg, R.D. The activation of factor X and prothrombin by recombinant factor VIIa in vivo is mediated by tissue factor. *J. Clin. Investig.* **1993**, *92*, 1207–1212. [CrossRef]
16. Bryant, J.W.; Shariat-Madar, Z. Human plasma kallikrein-kinin system: Physiological and biochemical parameters. *Cardiovasc. Hematol. Agents Med. Chem.* **2009**, *7*, 234–250. [CrossRef]
17. Doolittle, R.F. Fibrinogen and fibrin. *Annu. Rev. Biochem.* **1984**, *53*, 195–229. [CrossRef] [PubMed]
18. Hess, K.; Alzahrani, S.H.; Mathai, M.; Schroeder, V.; Carter, A.M.; Howell, G.; Koko, T.; Strachan, M.W.; Price, J.F.; Smith, K.A.; et al. A novel mechanism for hypofibrinolysis in diabetes: The role of complement C3. *Diabetologia* **2012**, *55*, 1103–1113. [CrossRef]
19. Whyte, C.S.; Mitchell, J.L.; Mutch, N.J. Platelet-Mediated Modulation of Fibrinolysis. *Semin. Thromb. Hemost.* **2017**, *43*, 115–128. [CrossRef] [PubMed]
20. Chapin, J.C.; Hajjar, K.A. Fibrinolysis and the control of blood coagulation. *Blood Rev.* **2015**, *29*, 17–24. [CrossRef]
21. Foley, J.H. Plasmin(ogen) at the Nexus of Fibrinolysis, Inflammation, and Complement. *Semin. Thromb. Hemost.* **2017**, *43*, 135–142. [CrossRef] [PubMed]
22. Draxler, D.F.; Sashindranath, M.; Medcalf, R.L. Plasmin: A Modulator of Immune Function. *Semin. Thromb. Hemost.* **2017**, *43*, 143–153. [CrossRef]
23. Miles, L.A.; Greengard, J.S.; Griffin, J.H. A comparison of the abilities of plasma kallikrein, beta-Factor XIIa, Factor XIa and urokinase to activate plasminogen. *Thromb. Res.* **1983**, *29*, 407–417. [CrossRef]
24. Hoylaerts, M.; Rijken, D.C.; Lijnen, H.R.; Collen, D. Kinetics of the activation of plasminogen by human tissue plasminogen activator. Role of fibrin. *J. Biol. Chem.* **1982**, *257*, 2912–2919. [CrossRef]
25. Schneider, M.; Nesheim, M. A study of the protection of plasmin from antiplasmin inhibition within an intact fibrin clot during the course of clot lysis. *J. Biol. Chem.* **2004**, *279*, 13333–13339. [CrossRef]
26. Thorsen, S. The mechanism of plasminogen activation and the variability of the fibrin effector during tissue-type plasminogen activator-mediated fibrinolysis. *Ann. N. Y. Acad. Sci.* **1992**, *667*, 52–63. [CrossRef]
27. Flood, E.C.; Hajjar, K.A. The annexin A2 system and vascular homeostasis. *Vasc. Pharmacol.* **2011**, *54*, 59–67. [CrossRef] [PubMed]
28. Enocsson, H.; Sjowall, C.; Wettero, J. Soluble urokinase plasminogen activator receptor–a valuable biomarker in systemic lupus erythematosus? *Clin. Chim. Acta Int. J. Clin. Chem.* **2015**, *444*, 234–241. [CrossRef] [PubMed]
29. Masucci, M.T.; Pedersen, N.; Blasi, F. A soluble, ligand binding mutant of the human urokinase plasminogen activator receptor. *J. Biol. Chem.* **1991**, *266*, 8655–8658. [CrossRef]

30. Takano, K.; Yamaguchi, T.; Uchida, K. Markers of a hypercoagulable state following acute ischemic stroke. *Stroke* **1992**, *23*, 194–198. [CrossRef] [PubMed]

31. Sprengers, E.D.; Kluft, C. Plasminogen activator inhibitors. *Blood* **1987**, *69*, 381–387. [CrossRef]

32. Rahman, F.A.; Krause, M.P. PAI-1, the Plasminogen System, and Skeletal Muscle. *Int. J. Mol. Sci.* **2020**, *21*. [CrossRef]

33. Carpenter, S.L.; Mathew, P. Alpha2-antiplasmin and its deficiency: Fibrinolysis out of balance. *Haemoph. Off. J. World Fed. Hemoph.* **2008**, *14*, 1250–1254. [CrossRef]

34. Rijken, D.C.; Uitte de Willige, S. Inhibition of Fibrinolysis by Coagulation Factor XIII. *BioMed Res. Int.* **2017**, *2017*, 1209676. [CrossRef]

35. Ruggeri, Z.M. Platelets in atherothrombosis. *Nat. Med.* **2002**, *8*, 1227–1234. [CrossRef] [PubMed]

36. Weiss, H.J.; Sussman, I.I.; Hoyer, L. W. Stabilization of factor VIII in plasma by the von Willebrand factor. Studies on post-transfusion and dissociated factor VIII and in patients with von Willebrand's disease. *J. Clin. Investig.* **1977**, *60*, 390–404. [CrossRef]

37. Periayah, M.H.; Halim, A.S.; Mat Saad, A.Z. Mechanism Action of Platelets and Crucial Blood Coagulation Pathways in Hemostasis. *Int. J. Hematol. Oncol. Stem Cell Res.* **2017**, *11*, 319–327. [PubMed]

38. Lisman, T.; Weeterings, C.; de Groot, P.G. Platelet aggregation: Involvement of thrombin and fibrin(ogen). *Front. Biosci.* **2005**, *10*, 2504–2517. [CrossRef] [PubMed]

39. Shattil, S.J.; Hoxie, J.A.; Cunningham, M.; Brass, L.F. Changes in the platelet membrane glycoprotein IIb.IIIa complex during platelet activation. *J. Biol. Chem.* **1985**, *260*, 11107–11114. [CrossRef]

40. Zimmerman, G.A.; McIntyre, T.M.; Prescott, S.M.; Stafforini, D.M. The platelet-activating factor signaling system and its regulators in syndromes of inflammation and thrombosis. *Crit. Care Med.* **2002**, *30*, S294–S301. [CrossRef] [PubMed]

41. Yago, T.; Shao, B.; Miner, J.J.; Yao, L.; Klopocki, A.G.; Maeda, K.; Coggeshall, K.M.; McEver, R.P. E-selectin engages PSGL-1 and CD44 through a common signaling pathway to induce integrin alphaLbeta2-mediated slow leukocyte rolling. *Blood* **2010**, *116*, 485–494. [CrossRef] [PubMed]

42. Carlow, D.A.; Gossens, K.; Naus, S.; Veerman, K.M.; Seo, W.; Ziltener, H.J. PSGL-1 function in immunity and steady state homeostasis. *Immunol. Rev.* **2009**, *230*, 75–96. [CrossRef] [PubMed]

43. André, P.; Hartwell, D.; Hrachovinová, I.; Saffaripour, S.; Wagner, D.D. Pro-coagulant state resulting from high levels of soluble P-selectin in blood. *Proc. Natl. Acad. Sci. USA* **2000**, *97*, 13835–13840. [CrossRef] [PubMed]

44. Monković, D.D.; Tracy, P.B. Functional characterization of human platelet-released factor V and its activation by factor Xa and thrombin. *J. Biol. Chem.* **1990**, *265*, 17132–17140. [CrossRef]

45. Heemskerk, J.W.; Bevers, E.M.; Lindhout, T. Platelet activation and blood coagulation. *Thromb. Haemost.* **2002**, *88*, 186–193. [PubMed]

46. Michelson, A.D.; Benoit, S.E.; Furman, M.I.; Breckwoldt, W.L.; Rohrer, M.J.; Barnard, M.R.; Loscalzo, J. Effects of nitric oxide/EDRF on platelet surface glycoproteins. *Am. J. Physiol.* **1996**, *270*, H1640–H1648. [CrossRef] [PubMed]

47. McAdam, B.F.; Catella-Lawson, F.; Mardini, I.A.; Kapoor, S.; Lawson, J.A.; FitzGerald, G.A. Systemic biosynthesis of prostacyclin by cyclooxygenase (COX)-2: The human pharmacology of a selective inhibitor of COX-2. *Proc. Natl. Acad. Sci. USA* **1999**, *96*, 272–277. [CrossRef]

48. Li, C.; Yang, X.; Feng, J.; Lei, P.; Wang, Y. Proinflammatory and prothrombotic status in emphysematous rats exposed to intermittent hypoxia. *Int. J. Clin. Exp. Pathol.* **2015**, *8*, 374–383. [PubMed]

49. Amitrano, L.; Guardascione, M.A.; Brancaccio, V.; Balzano, A. Coagulation disorders in liver disease. *Semin. Liver Dis.* **2002**, *22*, 83–96. [CrossRef] [PubMed]

50. Gabryelska, A.; Szmyd, B.; Szemraj, J.; Stawski, R.; Sochal, M.; Białasiewicz, P. Patients with obstructive sleep apnea present with chronic upregulation of serum HIF-1α protein. *J. Clin. Sleep Med.* **2020**, *16*, 1761–1768. [CrossRef]

51. Gabryelska, A.; Szmyd, B.; Panek, M.; Szemraj, J.; Kuna, P.; Białasiewicz, P. Serum hypoxia-inducible factor-1α protein level as a diagnostic marker of obstructive sleep apnea. *Pol. Arch. Intern. Med.* **2020**, *130*, 158–160. [CrossRef] [PubMed]

52. Lu, D.; Li, N.; Yao, X.; Zhou, L. Potential inflammatory markers in obstructive sleep apnea-hypopnea syndrome. *Bosn. J. Basic Med. Sci.* **2017**, *17*, 47–53. [CrossRef] [PubMed]

53. Gabryelska, A.; Stawski, R.; Sochal, M.; Szmyd, B.; Białasiewicz, P. Influence of one-night CPAP therapy on the changes of HIF-1α protein in OSA patients: A pilot study. *J. Sleep Res.* **2020**, *29*, e12995. [CrossRef]

54. Gabryelska, A.; Sochal, M.; Turkiewicz, S.; Białasiewicz, P. Relationship between HIF-1 and Circadian Clock Proteins in Obstructive Sleep Apnea Patients-Preliminary Study. *J. Clin. Med.* **2020**, *9*. [CrossRef]

55. Taylor, C.T.; Cummins, E.P. The role of NF-kappaB in hypoxia-induced gene expression. *Ann. N. Y. Acad. Sci.* **2009**, *1177*, 178–184. [CrossRef]

56. Mackman, N.; Brand, K.; Edgington, T.S. Lipopolysaccharide-mediated transcriptional activation of the human tissue factor gene in THP-1 monocytic cells requires both activator protein 1 and nuclear factor kappa B binding sites. *J. Exp. Med.* **1991**, *174*, 1517–1526. [CrossRef]

57. Begbie, M.; Notley, C.; Tinlin, S.; Sawyer, L.; Lillicrap, D. The Factor VIII acute phase response requires the participation of NFkappaB and C/EBP. *Thromb. Haemost.* **2000**, *84*, 216–222.

58. Liu, W.; Shen, S.-M.; Zhao, X.-Y.; Chen, G.-Q. Targeted genes and interacting proteins of hypoxia inducible factor-1. *Int. J. Biochem. Mol. Biol.* **2012**, *3*, 165–178. [PubMed]

59. Ten, V.S.; Pinsky, D.J. Endothelial response to hypoxia: Physiologic adaptation and pathologic dysfunction. *Curr. Opin. Crit. Care* **2002**, *8*, 242–250. [CrossRef]

60. Delaney, C.; Davizon-Castillo, P.; Allawzi, A.; Posey, J.; Gandjeva, A.; Neeves, K.; Tuder, R.M.; Paola, J.D.; Stenmark, K.R.; Nozik, E.S. Platelet activation contributes to hypoxia-induced inflammation. *Am. J. Physiol. Lung Cell. Mol. Physiol.* **2021**, *320*, L413–L421. [CrossRef]

61. Wedzicha, J.A.; Syndercombe-Court, D.; Tan, K.C. Increased platelet aggregate formation in patients with chronic airflow obstruction and hypoxaemia. *Thorax* **1991**, *46*, 504–507. [CrossRef]

62. Rahangdale, S.; Yeh, S.Y.; Novack, V.; Stevenson, K.; Barnard, M.R.; Furman, M.I.; Frelinger, A.L.; Michelson, A.D.; Malhotra, A. The influence of intermittent hypoxemia on platelet activation in obese patients with obstructive sleep apnea. *J. Clin. Sleep Med.* **2011**, *7*, 172–178. [CrossRef] [PubMed]

63. Kiouptsi, K.; Gambaryan, S.; Walter, E.; Walter, U.; Jurk, K.; Reinhardt, C. Hypoxia impairs agonist-induced integrin αIIbβ3 activation and platelet aggregation. *Sci. Rep.* **2017**, *7*, 7621. [CrossRef]

64. Ferreira, C.B.; Schoorlemmer, G.H.; Rocha, A.A.; Cravo, S.L. Increased sympathetic responses induced by chronic obstructive sleep apnea are caused by sleep fragmentation. *J. Appl. Physiol. (1985)* **2020**, *129*, 163–172. [CrossRef]

65. Olson, L.J.; Olson, E.J.; Somers, V.K. Obstructive Sleep Apnea and Platelet Activation: Another Potential Link Between Sleep-Disordered Breathing and Cardiovascular Disease. *Chest* **2004**, *126*, 339–341. [CrossRef]

66. von Känel, R.; Dimsdale, J.E. Effects of sympathetic activation by adrenergic infusions on hemostasis in vivo. *Eur. J. Haematol.* **2000**, *65*, 357–369. [CrossRef]

67. Eisensehr, I.; Ehrenberg, B.L.; Noachtar, S.; Korbett, K.; Byrne, A.; McAuley, A.; Palabrica, T. Platelet activation, epinephrine, and blood pressure in obstructive sleep apnea syndrome. *Neurology* **1998**, *51*, 188–195. [CrossRef] [PubMed]

68. Levi, M.; ten Cate, H.; Bauer, K.A.; van der Poll, T.; Edgington, T.S.; Büller, H.R.; van Deventer, S.J.; Hack, C.E.; ten Cate, J.W.; Rosenberg, R.D. Inhibition of endotoxin-induced activation of coagulation and fibrinolysis by pentoxifylline or by a monoclonal anti-tissue factor antibody in chimpanzees. *J. Clin. Investig.* **1994**, *93*, 114–120. [CrossRef] [PubMed]

69. Ryan, S.; Taylor, C.T.; McNicholas, W.T. Selective activation of inflammatory pathways by intermittent hypoxia in obstructive sleep apnea syndrome. *Circulation* **2005**, *112*, 2660–2667. [CrossRef] [PubMed]

70. Gabryelska, A.; Łukasik, Z.M.; Makowska, J.S.; Białasiewicz, P. Obstructive Sleep Apnea: From Intermittent Hypoxia to Cardiovascular Complications via Blood Platelets. *Front. Neurol.* **2018**, *9*, 635. [CrossRef] [PubMed]

71. Tyagi, T.; Ahmad, S.; Gupta, N.; Sahu, A.; Ahmad, Y.; Nair, V.; Chatterjee, T.; Bajaj, N.; Sengupta, S.; Ganju, L.; et al. Altered expression of platelet proteins and calpain activity mediate hypoxia-induced prothrombotic phenotype. *Blood* **2014**, *123*, 1250–1260. [CrossRef] [PubMed]

72. Taylor, F.B., Jr. The inflammatory-coagulant axis in the host response to gram-negative sepsis: Regulatory roles of proteins and inhibitors of tissue factor. *New Horiz.* **1994**, *2*, 555–565. [PubMed]

73. Hoyos, C.M.; Melehan, K.L.; Liu, P.Y.; Grunstein, R.R.; Phillips, C.L. Does obstructive sleep apnea cause endothelial dysfunction? A critical review of the literature. *Sleep Med. Rev.* **2015**, *20*, 15–26. [CrossRef]

74. Napoleone, E.; Di Santo, A.; Lorenzet, R. Monocytes upregulate endothelial cell expression of tissue factor: A role for cell-cell contact and cross-talk. *Blood* **1997**, *89*, 541–549. [CrossRef] [PubMed]

75. Budhiraja, R.; Parthasarathy, S.; Quan, S.F. Endothelial dysfunction in obstructive sleep apnea. *J. Clin. Sleep Med.* **2007**, *3*, 409–415. [CrossRef]

76. Phillips, B.G.; Narkiewicz, K.; Pesek, C.A.; Haynes, W.G.; Dyken, M.E.; Somers, V.K. Effects of obstructive sleep apnea on endothelin-1 and blood pressure. *J. Hypertens.* **1999**, *17*, 61–66. [CrossRef]

77. Halim, A.; Kanayama, N.; el Maradny, E.; Maehara, K.; Masahiko, H.; Terao, T. Endothelin-1 increased immunoreactive von Willebrand factor in endothelial cells and induced micro thrombosis in rats. *Thromb. Res.* **1994**, *76*, 71–78. [CrossRef]

78. Kambas, K.; Chrysanthopoulou, A.; Kourtzelis, I.; Skordala, M.; Mitroulis, I.; Rafail, S.; Vradelis, S.; Sigalas, I.; Wu, Y.Q.; Speletas, M.; et al. Endothelin-1 signaling promotes fibrosis in vitro in a bronchopulmonary dysplasia model by activating the extrinsic coagulation cascade. *J. Immunol.* **2011**, *186*, 6568–6575. [CrossRef]

79. Masola, V.; Zaza, G.; Onisto, M.; Lupo, A.; Gambaro, G. Glycosaminoglycans, proteoglycans and sulodexide and the endothelium: Biological roles and pharmacological effects. *Int. Angiol. J. Int. Union Angiol.* **2014**, *33*, 243–254.

80. Rosenberg, R.D.; Shworak, N.W.; Liu, J.; Schwartz, J.J.; Zhang, L. Heparan sulfate proteoglycans of the cardiovascular system. Specific structures emerge but how is synthesis regulated? *J. Clin. Investig.* **1997**, *99*, 2062–2070. [CrossRef]

81. Meszaros, M.; Kis, A.; Kunos, L.; Tarnoki, A.D.; Tarnoki, D.L.; Lazar, Z.; Bikov, A. The role of hyaluronic acid and hyaluronidase-1 in obstructive sleep apnoea. *Sci. Rep.* **2020**, *10*, 19484. [CrossRef]

82. Félétou, M.; Vanhoutte, P.M. Endothelial dysfunction: A multifaceted disorder (The Wiggers Award Lecture). *Am. J. Physiol. Heart Circ. Physiol.* **2006**, *291*, H985–H1002. [CrossRef]

83. El Solh, A.A.; Akinnusi, M.E.; Berim, I.G.; Peter, A.M.; Paasch, L.L.; Szarpa, K.R. Hemostatic implications of endothelial cell apoptosis in obstructive sleep apnea. *Sleep Breath.* **2008**, *12*, 331–337. [CrossRef]

84. Hayashi, M.; Fujimoto, K.; Urushibata, K.; Takamizawa, A.; Kinoshita, O.; Kubo, K. Hypoxia-sensitive molecules may modulate the development of atherosclerosis in sleep apnoea syndrome. *Respirology* **2006**, *11*, 24–31. [CrossRef]

85. von Känel, R.; Loredo, J.S.; Ancoli-Israel, S.; Mills, P.J.; Natarajan, L.; Dimsdale, J.E. Association between polysomnographic measures of disrupted sleep and prothrombotic factors. *Chest* **2007**, *131*, 733–739. [CrossRef]

86. Robinson, G.V.; Pepperell, J.C.; Segal, H.C.; Davies, R.J.; Stradling, J.R. Circulating cardiovascular risk factors in obstructive sleep apnoea: Data from randomised controlled trials. *Thorax* **2004**, *59*, 777–782. [CrossRef] [PubMed]

87. Mehra, R.; Xu, F.; Babineau, D.C.; Tracy, R.P.; Jenny, N.S.; Patel, S.R.; Redline, S. Sleep-disordered breathing and prothrombotic biomarkers: Cross-sectional results of the Cleveland Family Study. *Am. J. Respir. Crit. Care Med.* **2010**, *182*, 826–833. [CrossRef] [PubMed]

88. Peled, N.; Kassirer, M.; Kramer, M.R.; Rogowski, O.; Shlomi, D.; Fox, B.; Berliner, A.S.; Shitrit, D. Increased erythrocyte adhesiveness and aggregation in obstructive sleep apnea syndrome. *Thromb. Res.* **2008**, *121*, 631–636. [CrossRef] [PubMed]

89. Wessendorf, T.E.; Thilmann, A.F.; Wang, Y.M.; Schreiber, A.; Konietzko, N.; Teschler, H. Fibrinogen levels and obstructive sleep apnea in ischemic stroke. *Am. J. Respir. Crit. Care Med.* **2000**, *162*, 2039–2042. [CrossRef]

90. Akyüz, A.; Akkoyun, D.Ç.; Oran, M.; Değirmenci, H.; Alp, R. Mean platelet volume in patients with obstructive sleep apnea and its relationship with simpler heart rate derivatives. *Cardiol. Res. Pract.* **2014**, *2014*, 454701. [CrossRef] [PubMed]

91. Kruithof, E.K.; Dunoyer-Geindre, S. Human tissue-type plasminogen activator. *Thromb. Haemost.* **2014**, *112*, 243–254. [CrossRef]

92. Kent, B.D.; Ryan, S.; McNicholas, W.T. Obstructive sleep apnea and inflammation: Relationship to cardiovascular co-morbidity. *Respir. Physiol. Neurobiol.* **2011**, *178*, 475–481. [CrossRef]

93. Takagi, T.; Morser, J.; Gabazza, E.C.; Qin, L.; Fujiwara, A.; Naito, M.; Yamaguchi, A.; Kobayashi, T.; D'Alessandro-Gabazza, C.N.; Boveda Ruiz, D.; et al. The coagulation and protein C pathways in patients with sleep apnea. *Lung* **2009**, *187*, 209–213. [CrossRef]

94. Zhang, X.B.; Jiang, X.T.; Cai, F.R.; Zeng, H.Q.; Du, Y.P. Vascular endothelial growth factor levels in patients with obstructive sleep apnea: A meta-analysis. *Eur. Arch. Oto-Rhino-Laryngol.* **2017**, *274*, 661–670. [CrossRef]

95. Bagai, K.; Muldowney, J.A., 3rd; Song, Y.; Wang, L.; Bagai, J.; Artibee, K.J.; Vaughan, D.E.; Malow, B.A. Circadian variability of fibrinolytic markers and endothelial function in patients with obstructive sleep apnea. *Sleep* **2014**, *37*, 359–367. [CrossRef]

96. Rångemark, C.; Hedner, J.A.; Carlson, J.T.; Gleerup, G.; Winther, K. Platelet function and fibrinolytic activity in hypertensive and normotensive sleep apnea patients. *Sleep* **1995**, *18*, 188–194. [CrossRef]

97. Steffanina, A.; Proietti, L.; Antonaglia, C.; Palange, P.; Angelici, E.; Canipari, R. The Plasminogen System and Transforming Growth Factor-beta in Subjects with Obstructive Sleep Apnea Syndrome: Effects of CPAP Treatment. *Respir. Care* **2015**, *60*, 1643–1651. [CrossRef] [PubMed]

98. Lyngbaek, S.; Sehestedt, T.; Marott, J.L.; Hansen, T.W.; Olsen, M.H.; Andersen, O.; Linneberg, A.; Madsbad, S.; Haugaard, S.B.; Eugen-Olsen, J.; et al. CRP and suPAR are differently related to anthropometry and subclinical organ damage. *Int. J. Cardiol.* **2013**, *167*, 781–785. [CrossRef] [PubMed]

99. Baran, M.; Mollers, L.N.; Andersson, S.; Jonsson, I.M.; Ekwall, A.K.; Bjersing, J.; Tarkowski, A.; Bokarewa, M. Survivin is an essential mediator of arthritis interacting with urokinase signalling. *J. Cell. Mol. Med.* **2009**, *13*, 3797–3808. [CrossRef] [PubMed]

100. Kunos, L.; Horvath, P.; Kis, A.; Tarnoki, D.L.; Tarnoki, A.D.; Lazar, Z.; Bikov, A. Circulating Survivin Levels in Obstructive Sleep Apnoea. *Lung* **2018**, *196*, 417–424. [CrossRef] [PubMed]

101. Stavaras, C.; Pastaka, C.; Papala, M.; Gravas, S.; Tzortzis, V.; Melekos, M.; Seitanidis, G.; Gourgoulianis, K.I. Sexual function in pre- and post-menopausal women with obstructive sleep apnea syndrome. *Int. J. Impot. Res.* **2012**, *24*, 228–233. [CrossRef]

102. Kadir, R.R.A.; Bayraktutan, U. Urokinase Plasminogen Activator: A Potential Thrombolytic Agent for Ischaemic Stroke. *Cell. Mol. Neurobiol.* **2020**, *40*, 347–355. [CrossRef]

103. Gozal, D.; Jortani, S.; Snow, A.B.; Kheirandish-Gozal, L.; Bhattacharjee, R.; Kim, J.; Capdevila, O.S. Two-dimensional differential in-gel electrophoresis proteomic approaches reveal urine candidate biomarkers in pediatric obstructive sleep apnea. *Am. J. Respir. Crit. Care Med.* **2009**, *180*, 1253–1261. [CrossRef] [PubMed]

104. von Känel, R.; Malan, N.T.; Hamer, M.; Lambert, G.W.; Schlaich, M.; Reimann, M.; Malan, L. Three-year changes of prothrombotic factors in a cohort of South Africans with a high clinical suspicion of obstructive sleep apnea. *Thromb. Haemost.* **2016**, *115*, 63–72. [CrossRef] [PubMed]

105. Bocskei, R.M.; Meszaros, M.; Tarnoki, A.D.; Tarnoki, D.L.; Kunos, L.; Lazar, Z.; Bikov, A. Circulating Soluble Urokinase-Type Plasminogen Activator Receptor in Obstructive Sleep Apnoea. *Medicina (Kaunas, Lithuania)* **2020**, *56*. [CrossRef] [PubMed]

106. Nykjaer, A.; Moller, B.; Todd, R.F., 3rd; Christensen, T.; Andreasen, P.A.; Gliemann, J.; Petersen, C.M. Urokinase receptor. An activation antigen in human T lymphocytes. *J. Immunol.* **1994**, *152*, 505–516.

107. Chavakis, T.; Willuweit, A.K.; Lupu, F.; Preissner, K.T.; Kanse, S.M. Release of soluble urokinase receptor from vascular cells. *Thromb. Haemost.* **2001**, *86*, 686–693. [CrossRef]

108. Thuno, M.; Macho, B.; Eugen-Olsen, J. suPAR: The molecular crystal ball. *Disease Markers* **2009**, *27*, 157–172. [CrossRef]

109. Lund, L.R.; Ellis, V.; Ronne, E.; Pyke, C.; Dano, K. Transcriptional and post-transcriptional regulation of the receptor for urokinase-type plasminogen activator by cytokines and tumour promoters in the human lung carcinoma cell line A549. *Biochem. J.* **1995**, *310*, 345–352. [CrossRef]

110. Shetty, S.; Idell, S. A urokinase receptor mRNA binding protein from rabbit lung fibroblasts and mesothelial cells. *Am. J. Physiol.* **1998**, *274*, L871–L882. [CrossRef]

111. Kohli, M.; Sharma, S.K.; Upadhyay, V.; Varshney, S.; Sengupta, S.; Basak, T.; Sreenivas, V. Urinary EPCR and dermcidin as potential novel biomarkers for severe adult OSA patients. *Sleep Med.* **2019**, *64*, 92–100. [CrossRef]

112. Keeling, D.M.; Wilson, A.J.; Mackie, I.J.; Isenberg, D.A.; Machin, S.J. Role of beta 2-glycoprotein I and anti-phospholipid antibodies in activation of protein C in vitro. *J. Clin. Pathol.* **1993**, *46*, 908–911. [CrossRef] [PubMed]

113. Peng, Y.; Zhou, L.; Cao, Y.; Chen, P.; Chen, Y.; Zong, D.; Ouyang, R. Relation between serum leptin levels, lipid profiles and neurocognitive deficits in Chinese OSAHS patients. *Int. J. Neurosci.* **2017**, *127*, 981–987. [CrossRef]

114. García Suquia, A.; Alonso-Fernández, A.; de la Peña, M.; Romero, D.; Piérola, J.; Carrera, M.; Barceló, A.; Soriano, J.B.; Arque, M.; Fernández-Capitán, C.; et al. High D-dimer levels after stopping anticoagulants in pulmonary embolism with sleep apnoea. *Eur. Respir. J.* **2015**, *46*, 1691–1700. [CrossRef] [PubMed]

115. Russo-Marie, F. Annexin V and phospholipid metabolism. *Clin. Chem. Lab. Med.* **1999**, *37*, 287–291. [CrossRef]

116. El Solh, A.A.; Akinnusi, M.E.; Baddoura, F.H.; Mankowski, C.R. Endothelial cell apoptosis in obstructive sleep apnea: A link to endothelial dysfunction. *Am. J. Respir. Crit. Care Med.* **2007**, *175*, 1186–1191. [CrossRef] [PubMed]

117. Bikov, A.; Kunos, L.; Pállinger, É.; Lázár, Z.; Kis, A.; Horváth, G.; Losonczy, G.; Komlósi, Z.I. Diurnal variation of circulating microvesicles is associated with the severity of obstructive sleep apnoea. *Sleep Breath.* **2017**, *21*, 595–600. [CrossRef] [PubMed]

118. Yun, C.H.; Jung, K.H.; Chu, K.; Kim, S.H.; Ji, K.H.; Park, H.K.; Kim, H.C.; Lee, S.T.; Lee, S.K.; Roh, J.K. Increased circulating endothelial microparticles and carotid atherosclerosis in obstructive sleep apnea. *J. Clin. Neurol. (Seoul, Korea)* **2010**, *6*, 89–98. [CrossRef]

119. Zakrzewski, M.; Zakrzewska, E.; Kicinski, P.; Przybylska-Kuc, S.; Dybala, A.; Myslinski, W.; Pastryk, J.; Tomaszewski, T.; Mosiewicz, J. Evaluation of Fibrinolytic Inhibitors: Alpha-2-Antiplasmin and Plasminogen Activator Inhibitor 1 in Patients with Obstructive Sleep Apnoea. *PLoS ONE* **2016**, *11*, e0166725. [CrossRef]

120. Martin, R.A.; Strosnider, C.; Giersch, G.; Womack, C.J.; Hargens, T.A. The effect of acute aerobic exercise on hemostasis in obstructive sleep apnea. *Sleep Breath.* **2017**, *21*, 623–629. [CrossRef]

121. von Känel, R.; Natarajan, L.; Ancoli-Israel, S.; Mills, P.J.; Wolfson, T.; Gamst, A.C.; Loredo, J.S.; Dimsdale, J.E. Effect of continuous positive airway pressure on day/night rhythm of prothrombotic markers in obstructive sleep apnea. *Sleep Med.* **2013**, *14*, 58–65. [CrossRef] [PubMed]

122. Reggiani, M.; Karttunen, V.; Wartiovaara-Kautto, U.; Riutta, A.; Uchiyama, S.; Hillbom, M. Fibrinolytic activity and platelet function in subjects with obstructive sleep apnoea and a patent foramen ovale: Is there an option for prevention of ischaemic stroke? *Stroke Res. Treat.* **2012**, *2012*, 945849. [CrossRef]

123. Ifergane, G.; Ovanyan, A.; Toledano, R.; Goldbart, A.; Abu-Salame, I.; Tal, A.; Stavsky, M.; Novack, V. Obstructive Sleep Apnea in Acute Stroke: A Role for Systemic Inflammation. *Stroke* **2016**, *47*, 1207–1212. [CrossRef] [PubMed]

124. von Känel, R.; Natarajan, L.; Ancoli-Israel, S.; Mills, P.J.; Loredo, J.S.; Dimsdale, J.E. Day/Night rhythm of hemostatic factors in obstructive sleep apnea. *Sleep* **2010**, *33*, 371–377. [CrossRef]

125. Gileles-Hillel, A.; Alonso-Álvarez, M.L.; Kheirandish-Gozal, L.; Peris, E.; Cordero-Guevara, J.A.; Terán-Santos, J.; Martinez, M.G.; Jurado-Luque, M.J.; Corral-Peñafiel, J.; Duran-Cantolla, J.; et al. Inflammatory Markers and Obstructive Sleep Apnea in Obese Children: The NANOS Study. *Mediat. Inflamm.* **2014**, *2014*, 605280. [CrossRef]

126. Niżankowska-Jędrzejczyk, A.; Almeida, F.R.; Lowe, A.A.; Kania, A.; Nastałek, P.; Mejza, F.; Foley, J.H.; Niżankowska-Mogilnicka, E.; Undas, A. Modulation of inflammatory and hemostatic markers in obstructive sleep apnea patients treated with mandibular advancement splints: A parallel, controlled trial. *J. Clin. Sleep Med.* **2014**, *10*, 255–262. [CrossRef]

127. von Känel, R.; Loredo, J.S.; Ancoli-Israel, S.; Mills, P.J.; Dimsdale, J.E. Elevated plasminogen activator inhibitor 1 in sleep apnea and its relation to the metabolic syndrome: An investigation in 2 different study samples. *Metabolism* **2007**, *56*, 969–976. [CrossRef]

128. Liao, H.; Hyman, M.C.; Lawrence, D.A.; Pinsky, D.J. Molecular regulation of the PAI-1 gene by hypoxia: Contributions of Egr-1, HIF-1alpha, and C/EBPalpha. *FASEB J. Off. Publ. Fed. Am. Soc. Exp. Biol.* **2007**, *21*, 935–949. [CrossRef] [PubMed]

129. Rabieian, R.; Boshtam, M.; Zareei, M.; Kouhpayeh, S.; Masoudifar, A.; Mirzaei, H. Plasminogen Activator Inhibitor Type-1 as a Regulator of Fibrosis. *J. Cell. Biochem.* **2018**, *119*, 17–27. [CrossRef] [PubMed]

130. Vaughan, D.E.; Rai, R.; Khan, S.S.; Eren, M.; Ghosh, A.K. Plasminogen Activator Inhibitor-1 Is a Marker and a Mediator of Senescence. *Arterioscler. Thromb. Vasc. Biol.* **2017**, *37*, 1446–1452. [CrossRef]

131. Pákó, J.; Kunos, L.; Mészáros, M.; Tárnoki, D.L.; Tárnoki, Á.D.; Horváth, I.; Bikov, A. Decreased Levels of Anti-Aging Klotho in Obstructive Sleep Apnea. *Rejuvenation Res.* **2020**, *23*, 256–261. [CrossRef]

132. Takeshita, K.; Yamamoto, K.; Ito, M.; Kondo, T.; Matsushita, T.; Hirai, M.; Kojima, T.; Nishimura, M.; Nabeshima, Y.; Loskutoff, D.J.; et al. Increased expression of plasminogen activator inhibitor-1 with fibrin deposition in a murine model of aging, "Klotho" mouse. *Semin. Thromb. Hemost.* **2002**, *28*, 545–554. [CrossRef]

133. Inoue, K.; Takano, H.; Yanagisawa, R.; Sakurai, M.; Shimada, A.; Sato, M.; Yoshino, S.; Yoshikawa, T. Role of interleukin-6 in fibrinolytic changes induced by lipopolysaccharide in mice. *Blood Coagul. Fibrinolysis* **2006**, *17*, 307–309. [CrossRef]

134. Miljić, P.; Heylen, E.; Willemse, J.; Djordjević, V.; Radojković, D.; Colović, M.; Elezović, I.; Hendriks, D. Thrombin activatable fibrinolysis inhibitor (TAFI): A molecular link between coagulation and fibrinolysis. *Srp. Arh. Celok. Lek.* **2010**, *138* (Suppl. S1), 74–78. [CrossRef] [PubMed]

135. Horvath, P.; Tarnoki, D.L.; Tarnoki, A.D.; Karlinger, K.; Lazar, Z.; Losonczy, G.; Kunos, L.; Bikov, A. Complement system activation in obstructive sleep apnea. *J. Sleep Res.* **2018**, *27*, e12674. [CrossRef]

136. Hui, D.S.; Ko, F.W.; Fok, J.P.; Chan, M.C.; Li, T.S.; Tomlinson, B.; Cheng, G. The effects of nasal continuous positive airway pressure on platelet activation in obstructive sleep apnea syndrome. *Chest* **2004**, *125*, 1768–1775. [CrossRef] [PubMed]

137. Akinnusi, M.E.; Paasch, L.L.; Szarpa, K.R.; Wallace, P.K.; El Solh, A.A. Impact of Nasal Continuous Positive Airway Pressure Therapy on Markers of Platelet Activation in Patients with Obstructive Sleep Apnea. *Respiration* **2009**, *77*, 25–31. [CrossRef] [PubMed]

138. Bokinsky, G.; Miller, M.; Ault, K.; Husband, P.; Mitchell, J. Spontaneous platelet activation and aggregation during obstructive sleep apnea and its response to therapy with nasal continuous positive airway pressure. A preliminary investigation. *Chest* **1995**, *108*, 625–630. [CrossRef] [PubMed]

139. Geiser, T.; Buck, F.; Meyer, B.J.; Bassetti, C.; Haeberli, A.; Gugger, M. In vivo platelet activation is increased during sleep in patients with obstructive sleep apnea syndrome. *Respiration* **2002**, *69*, 229–234. [CrossRef] [PubMed]

140. Reinhart, W.H.; Oswald, J.; Walter, R.; Kuhn, M. Blood viscosity and platelet function in patients with obstructive sleep apnea syndrome treated with nasal continuous positive airway pressure. *Clin. Hemorheol. Microcirc.* **2002**, *27*, 201–207. [PubMed]

141. Michelson, A.D.; Ellis, P.A.; Barnard, M.R.; Matic, G.B.; Viles, A.F.; Kestin, A.S. Downregulation of the platelet surface glycoprotein Ib-IX complex in whole blood stimulated by thrombin, adenosine diphosphate, or an in vivo wound. *Blood* **1991**, *77*, 770–779. [CrossRef]

142. Horváth, P.; Lázár, Z.; Gálffy, G.; Puskás, R.; Kunos, L.; Losonczy, G.; Mészáros, M.; Tárnoki, Á.D.; Tárnoki, D.L.; Bikov, A. Circulating P-Selectin Glycoprotein Ligand 1 and P-Selectin Levels in Obstructive Sleep Apnea Patients. *Lung* **2020**, *198*, 173–179. [CrossRef] [PubMed]

143. Winiarska, H.M.; Cofta, S.; Bielawska, L.; Płóciniczak, A.; Piorunek, T.; Wysocka, E. Circulating P-Selectin and Its Glycoprotein Ligand in Nondiabetic Obstructive Sleep Apnea Patients. *Adv. Exp. Med. Biol.* **2020**, *1279*, 61–69. [CrossRef]

144. Bravo Mde, L.; Serpero, L.D.; Barceló, A.; Barbé, F.; Agustí, A.; Gozal, D. Inflammatory proteins in patients with obstructive sleep apnea with and without daytime sleepiness. *Sleep Breath.* **2007**, *11*, 177–185. [CrossRef]

145. Zamarrón-Sanz, C.; Ricoy-Galbaldon, J.; Gude-Sampedro, F.; Riveiro-Riveiro, A. Plasma levels of vascular endothelial markers in obstructive sleep apnea. *Arch. Med. Res.* **2006**, *37*, 552–555. [CrossRef] [PubMed]

146. Owens, A.P., 3rd; Mackman, N. Microparticles in hemostasis and thrombosis. *Circ. Res.* **2011**, *108*, 1284–1297. [CrossRef] [PubMed]

147. Sinauridze, E.I.; Kireev, D.A.; Popenko, N.Y.; Pichugin, A.V.; Panteleev, M.A.; Krymskaya, O.V.; Ataullakhanov, F.I. Platelet microparticle membranes have 50- to 100-fold higher specific procoagulant activity than activated platelets. *Thromb. Haemost.* **2007**, *97*, 425–434.

148. Ayers, L.; Ferry, B.; Craig, S.; Nicoll, D.; Stradling, J.R.; Kohler, M. Circulating cell-derived microparticles in patients with minimally symptomatic obstructive sleep apnoea. *Eur. Respir. J.* **2009**, *33*, 574–580. [CrossRef]

149. Maruyama, K.; Morishita, E.; Sekiya, A.; Omote, M.; Kadono, T.; Asakura, H.; Hashimoto, M.; Kobayashi, M.; Nakatsumi, Y.; Takada, S.; et al. Plasma levels of platelet-derived microparticles in patients with obstructive sleep apnea syndrome. *J. Atheroscler. Thromb.* **2012**, *19*, 98–104. [CrossRef] [PubMed]

150. Hong, S.N.; Yun, H.C.; Yoo, J.H.; Lee, S.H. Association Between Hypercoagulability and Severe Obstructive Sleep Apnea. *JAMA Otolaryngol. Head Neck Surg.* **2017**, *143*, 996–1002. [CrossRef]

151. Barceló, A.; Morell-Garcia, D.; Sanchís, P.; Peña-Zarza, J.A.; Bauça, J.M.; Piérola, J.; Peña, M.; Toledo-Pons, N.; Giménez, P.; Ribot, C.; et al. Prothrombotic state in children with obstructive sleep apnea. *Sleep Med.* **2019**, *53*, 101–105. [CrossRef]

152. Zicari, A.M.; Occasi, F.; Di Mauro, F.; Lollobrigida, V.; Di Fraia, M.; Savastano, V.; Loffredo, L.; Nicita, F.; Spalice, A.; Duse, M. Mean Platelet Volume, Vitamin D and C Reactive Protein Levels in Normal Weight Children with Primary Snoring and Obstructive Sleep Apnea Syndrome. *PLoS ONE* **2016**, *11*, e0152497. [CrossRef] [PubMed]

153. Saygin, M.; Ozturk, O.; Ozguner, M.F.; Akkaya, A.; Varol, E. Hematological Parameters as Predictors of Cardiovascular Disease in Obstructive Sleep Apnea Syndrome Patients. *Angiology* **2016**, *67*, 461–470. [CrossRef] [PubMed]

154. Dahlbäck, B. Blood coagulation. *Lancet* **2000**, *355*, 1627–1632. [CrossRef]

155. Phillips, C.L.; McEwen, B.J.; Morel-Kopp, M.C.; Yee, B.J.; Sullivan, D.R.; Ward, C.M.; Tofler, G.H.; Grunstein, R.R. Effects of continuous positive airway pressure on coagulability in obstructive sleep apnoea: A randomised, placebo-controlled crossover study. *Thorax* **2012**, *67*, 639–644. [CrossRef] [PubMed]

156. Chin, K.; Kita, H.; Noguchi, T.; Otsuka, N.; Tsuboi, T.; Nakamura, T.; Shimizu, K.; Mishima, M.; Ohi, M. Improvement of factor VII clotting activity following long-term NCPAP treatment in obstructive sleep apnoea syndrome. *QJM* **1998**, *91*, 627–633. [CrossRef]

157. Dorkova, Z.; Petrasova, D.; Molcanyiova, A.; Popovnakova, M.; Tkacova, R. Effects of Continuous Positive Airway Pressure on Cardiovascular Risk Profile in Patients with Severe Obstructive Sleep Apnea and Metabolic Syndrome. *Chest* **2008**, *134*, 686–692. [CrossRef] [PubMed]

158. Chin, K.; Ohi, M.; Kita, H.; Noguchi, T.; Otsuka, N.; Tsuboi, T.; Mishima, M.; Kuno, K. Effects of NCPAP therapy on fibrinogen levels in obstructive sleep apnea syndrome. *Am. J. Respir. Crit. Care Med.* **1996**, *153*, 1972–1976. [CrossRef] [PubMed]

159. von Kanel, R.; Loredo, J.S.; Ancoli-Israel, S.; Dimsdale, J.E. Association between sleep apnea severity and blood coagulability: Treatment effects of nasal continuous positive airway pressure. *Sleep Breath.* **2006**, *10*, 139–146. [CrossRef]

160. Kheirandish-Gozal, L.; Gileles-Hillel, A.; Alonso-Alvarez, M.L.; Peris, E.; Bhattacharjee, R.; Teran-Santos, J.; Duran-Cantolla, J.; Gozal, D. Effects of adenotonsillectomy on plasma inflammatory biomarkers in obese children with obstructive sleep apnea: A community-based study. *Int. J. Obes. (Lond)* **2015**, *39*, 1094–1100. [CrossRef] [PubMed]

161. Gumbau, V.; Bruna, M.; Canelles, E.; Guaita, M.; Mulas, C.; Bases, C.; Celma, I.; Puche, J.; Marcaida, G.; Oviedo, M.; et al. A prospective study on inflammatory parameters in obese patients after sleeve gastrectomy. *Obes. Surg.* **2014**, *24*, 903–908. [CrossRef] [PubMed]

162. Shimizu, M.; Kamio, K.; Haida, M.; Ono, Y.; Miyachi, H.; Yamamoto, M.; Shinohara, Y.; Ando, Y. Platelet activation in patients with obstructive sleep apnea syndrome and effects of nasal-continuous positive airway pressure. *Tokai J. Exp. Clin. Med.* **2002**, *27*, 107–112. [PubMed]

163. Zhang, X.; Yin, K.; Wang, H.; Su, M.; Yang, Y. Effect of continuous positive airway pressure treatment on elderly Chinese patients with obstructive sleep apnea in the prethrombotic state. *Chin. Med. J. (Engl.)* **2003**, *116*, 1426–1428.

164. Oga, T.; Chin, K.; Tabuchi, A.; Kawato, M.; Morimoto, T.; Takahashi, K.; Handa, T.; Takahashi, K.; Taniguchi, R.; Kondo, H.; et al. Effects of Obstructive Sleep Apnea with Intermittent Hypoxia on Platelet Aggregability. *J. Atheroscler. Thromb.* **2009**, *16*, 862–869. [CrossRef]

165. Wang, F.; Liu, Y.; Xu, H.; Qian, Y.; Zou, J.; Yi, H.; Guan, J.; Yin, S. Association between Upper-airway Surgery and Ameliorative Risk Markers of Endothelial Function in Obstructive Sleep Apnea. *Sci. Rep.* **2019**, *9*, 20157. [CrossRef] [PubMed]

166. Ayers, L.; Turnbull, C.; Petousi, N.; Ferry, B.; Kohler, M.; Stradling, J. Withdrawal of Continuous Positive Airway Pressure Therapy for 2 Weeks in Obstructive Sleep Apnoea Patients Results in Increased Circulating Platelet and Leucocyte-Derived Microvesicles. *Respiration* **2016**, *91*, 412–413. [CrossRef] [PubMed]

167. Ayers, L.; Stoewhas, A.C.; Ferry, B.; Stradling, J.; Kohler, M. Elevated levels of endothelial cell-derived microparticles following short-term withdrawal of continuous positive airway pressure in patients with obstructive sleep apnea: Data from a randomized controlled trial. *Respiration* **2013**, *85*, 478–485. [CrossRef] [PubMed]

168. Zhan, X.; Li, L.; Wang, N.; Ge, X.; Pinto, J.M.; Wu, X.; Wei, Y. Can upper airway surgery for OSA protect against cardiovascular sequelae via effects on coagulation? *Acta Otolaryngol.* **2016**, *136*, 293–297. [CrossRef] [PubMed]

169. Kohler, H.P.; Grant, P.J. Plasminogen-activator inhibitor type 1 and coronary artery disease. *N. Engl. J. Med.* **2000**, *342*, 1792–1801. [CrossRef] [PubMed]

170. Meade, T.W.; Ruddock, V.; Stirling, Y.; Chakrabarti, R.; Miller, G.J. Fibrinolytic activity, clotting factors, and long-term incidence of ischaemic heart disease in the Northwick Park Heart Study. *Lancet* **1993**, *342*, 1076–1079. [CrossRef]

171. Lévy, P.; Kohler, M.; McNicholas, W.T.; Barbé, F.; McEvoy, R.D.; Somers, V.K.; Lavie, L.; Pépin, J.L. Obstructive sleep apnoea syndrome. *Nat. Rev. Dis. Primers* **2015**, *1*, 15015. [CrossRef] [PubMed]

172. Lu, F.; Jiang, T.; Wang, W.; Hu, S.; Shi, Y.; Lin, Y. Circulating fibrinogen levels are elevated in patients with obstructive sleep apnea: A systemic review and meta-analysis. *Sleep Med.* **2020**, *68*, 115–123. [CrossRef] [PubMed]

173. Alonso-Fernández, A.; Toledo-Pons, N.; García-Río, F. Obstructive sleep apnea and venous thromboembolism: Overview of an emerging relationship. *Sleep Med. Rev.* **2020**, *50*, 101233. [CrossRef] [PubMed]

International Journal of
Molecular Sciences

Review

A Narrative Review on Plasminogen Activator Inhibitor-1 and Its (Patho)Physiological Role: To Target or Not to Target?

Machteld Sillen and Paul J. Declerck *

Laboratory for Therapeutic and Diagnostic Antibodies, Department of Pharmaceutical and Pharmacological Sciences, KU Leuven, B-3000 Leuven, Belgium; machteld.sillen@kuleuven.be
* Correspondence: paul.declerck@kuleuven.be

Abstract: Plasminogen activator inhibitor-1 (PAI-1) is the main physiological inhibitor of plasminogen activators (PAs) and is therefore an important inhibitor of the plasminogen/plasmin system. Being the fast-acting inhibitor of tissue-type PA (tPA), PAI-1 primarily attenuates fibrinolysis. Through inhibition of urokinase-type PA (uPA) and interaction with biological ligands such as vitronectin and cell-surface receptors, the function of PAI-1 extends to pericellular proteolysis, tissue remodeling and other processes including cell migration. This review aims at providing a general overview of the properties of PAI-1 and the role it plays in many biological processes and touches upon the possible use of PAI-1 inhibitors as therapeutics.

Keywords: plasminogen activator inhibitor-1; PAI-1; fibrinolysis; cardiovascular disease; cancer; inflammation; fibrosis; aging

check for
updates

Citation: Sillen, M.; Declerck, P.J. A Narrative Review on Plasminogen Activator Inhibitor-1 and Its (Patho)Physiological Role: To Target or Not to Target?. *Int. J. Mol. Sci.* **2021**, 22, 2721. https://doi.org/10.3390/ijms22052721

Academic Editor: Hau C. Kwaan

Received: 12 February 2021
Accepted: 3 March 2021
Published: 8 March 2021

Publisher's Note: MDPI stays neutral with regard to jurisdictional claims in published maps and institutional affiliations.

1. Introduction

Plasminogen activator inhibitor-1 (PAI-1) belongs to the family of serine protease inhibitors (serpins) and is an important regulator of the plasminogen/plasmin system (Figure 1) [1]. This system revolves around the conversion of the zymogen plasminogen into the active enzyme plasmin through proteolytic cleavage that is mediated by plasminogen activators (PAs). When mediated by tissue-type PA (tPA), plasmin is primarily involved in fibrinolysis as it degrades the insoluble fibrin meshwork that constitutes blood clots. Through urokinase-type PA (uPA)-mediated plasminogen activation, the function of the plasminogen/plasmin system extends to pericellular proteolysis associated with processes including tissue remodeling and cell migration. Since its discovery, the (patho)physiological role of PAI-1 has been extensively studied in humans as well as in diverse disease models in animals. A link has been demonstrated between PAI-1 and various diseases including cardiovascular disease (CVD), metabolic disturbances, aging, cancer, tissue fibrosis, inflammation, and neurodegenerative disease. As a consequence, several PAI-1 inhibitors have been developed to further study the role of PAI-1 in disease models and to explore their potential applications in a therapeutic setting. This narrative review aims at providing a general overview of the properties of PAI-1 and the role it plays in many biological processes and touches upon the possible use of PAI-1 inhibitors as therapeutics.

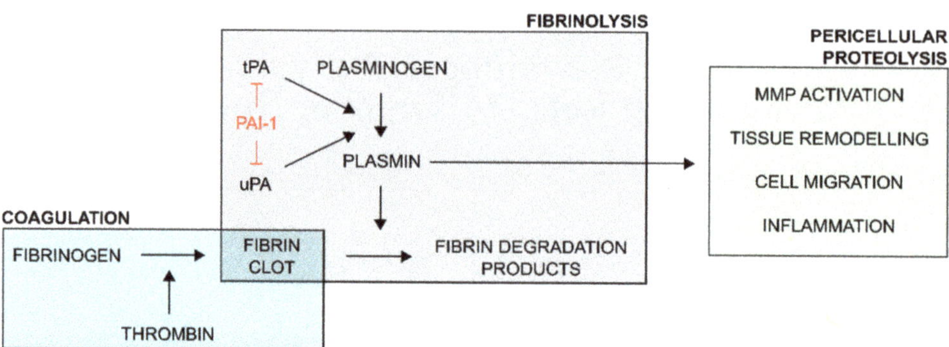

Figure 1. Schematic overview of the regulatory role of plasminogen activator inhibitor-1 (PAI-1) in the plasminogen activator/plasmin system. Upon vascular injury, the coagulation process ultimately generates thrombin which acts upon fibrinogen to form an insoluble fibrin clot. This coagulatory response is balanced by the fibrinolytic system. Plasminogen activators (PAs) tissue-type PA (tPA) and urokinase-type PA (uPA) convert plasminogen to proteolytically active plasmin. Plasmin, the key enzyme of the fibrinolytic system, degrades the fibrin clot into soluble fibrin degradation products. Through uPA-mediated plasminogen activation, the function of plasmin extends to pericellular proteolytic processes, involving activation of matrix metalloproteinases (MMPs), tissue remodeling, cell migration and adhesion, and inflammation. PAI-1 is an important regulator of the plasminogen activator/plasmin system as it interferes with plasminogen activation by directly inhibiting tPA and uPA.

2. PAI-1 Synthesis and Distribution

PAI-1 is the primary physiological inhibitor of plasminogen activators tPA and uPA. It was first detected almost four decades ago as an inhibitor of the fibrinolytic system produced by cultured bovine endothelial cells [2], but can be expressed by several other cell types in various tissues [3–5]. The expression and release of PAI-1 is strongly regulated by various factors, including growth factors, inflammatory cytokines, hormones, glucose, and endotoxins [6,7]. PAI-1 is synthesized as a 45-kDa single-chain glycoprotein comprising 379 or 381 amino acids depending on alternative cleavage sites for signal peptidases [8]. Three potential glycosylation sites have been identified based on the amino-acid sequence, of which Asn209 and Asn265 display a tissue-type specific glycosylation pattern [9]. In the blood, PAI-1 circulates in two distinct pools, free in plasma or retained in platelets [10]. Even though plasma PAI-1 circulates at relatively low levels (5–50 ng/mL), it mainly adopts the active conformation. In contrast, the main blood pool of PAI-1 is retained in platelets (up to approximately 300 ng/mL) and was found to be only 2–5% functionally active upon lysis of platelets [11,12]. However, more recent studies demonstrated that upon platelet activation platelet-derived PAI-1 may be present in the active conformation at significantly higher levels [13,14], presumably due to de novo PA-1 synthesis through translationally active PAI-1 messenger RNA of which the synthesis rate is importantly upregulated by platelet activation [13]. Furthermore, platelet activation results in the release of PAI-1 followed by partial retention of PAI-1 on the platelet membrane, thereby contributing to thrombolysis resistance of the clot [14–16].

3. PAI-1 Structure and Function

3.1. PAI-1 Is an Inhibitory Serpin

As a member of the serpin superfamily, PAI-1 displays their common highly ordered structure that is characterized by three β-sheets (A, B, and C) and nine α-helices (hA through hI) [17,18]. PAI-1 is synthesized in a metastable active conformation, exposing a flexible reactive center loop (RCL) at the top of the molecule. The RCL comprises 26 residues (designated P16 to P10′), including a bait peptide bond that mimics the normal substrate of the PAs (designated P1-P1′).

Upon interaction, the PA binds to PAI-1 through the P1-P1′ reactive center and several exosite regions adjacent to the RCL to form a noncovalent Michaelis complex. Subsequently, the PA active site serine (tPA-Ser478 or uPA-Ser195) attacks the P1-P1′ bond to form a tetrahedral intermediate with PAI-1. Successful cleavage of this bond yields the acyl-enzyme intermediate in which the PA is covalently linked to the P1 residue in PAI-1. The PAI-1/PA reaction then follows a branched pathway mechanism in which PAI-1 either acts as an inhibitor or a substrate towards the PA.

In the inhibitory pathway, the acyl-enzyme intermediate is converted to an irreversible inhibitory complex by full insertion of the N-terminal part of the RCL (P16-P1) as strand 4 into the central PAI-1 β-sheet A. Simultaneously, the PA is translocated to the opposite side of the PAI-1 molecule where a large part of the PA is deformed by compression against the body of PAI-1 [19,20]. Deformation of the PA and especially its active site prevents hydrolysis of the acyl-enzyme intermediate and traps the PA as a stable PAI-1/PA complex.

In contrast, the substrate pathway is characterized by hydrolysis of the acyl-enzyme intermediate prior to PA distortion, resulting in the release of regenerated PA from cleaved PAI-1 [21,22]. Substrate behavior of PAI-1 has been associated with either the existence of a conformational distinct substrate-like subset of PAI-1 [23,24], or can be induced by changing the kinetic parameters that underlie the branched pathway mechanism in favor of the substrate pathway [25,26].

3.2. PAI-1 Stability

PAI-1 is unique among serpins as active PAI-1 spontaneously converts into a thermodynamically stable latent form by slowly self-inserting the N-terminal part of the RCL into the core of the PAI-1 molecule. As a result, the P1-P1′ reactive center becomes inaccessible for PAs [27]. Whereas this transition occurs with a functional half-life of approximately two hours at 37 °C in vitro, the active form of PAI-1 is stabilized at least two-fold in vivo by the association with vitronectin in plasma and the extracellular matrix [28,29]. Furthermore, external conditions (ionic strength [29,30], pH [29], metal ions [31], arginine [32,33]), binding to other proteins (α_1-acid glycoprotein [34], antibodies [35]), and mutagenesis [36] have been shown to also affect the stability of PAI-1 (reviewed elsewhere [37]).

3.3. Interactions with Non-Proteinase Ligands

Apart from binding and inactivating PAs, PAI-1 can also interact with non-proteinase ligands including vitronectin and members of the low-density lipoprotein receptor (LDLR) family. Through these interactions, the functions of PAI-1 extend to various (patho)physiological processes not relying on its anti-protease activity.

As mentioned, active PAI-1 is importantly stabilized through allosteric modulation by its high-affinity interaction with the aminoterminal somatomedin B (SMB) domain of the glycoprotein vitronectin. Vitronectin is abundantly present in plasma (~300 µg/mL) and the extracellular matrix and plays a pivotal role in tissue remodeling, cell differentiation and migration, and inflammation. These effects are mediated by binding of the SMB domain or the neighboring Arg-Gly-Asp (RGD) motif [38,39] of vitronectin to cell surface-associated proteins including integrins and transmembrane receptors such as the uPA receptor (uPAR), which initiates intracellular signaling events. Due to the proximity of the binding sites for PAI-1, integrins, and uPAR, PAI-1 can interfere with vitronectin function as it competes with integrins and uPAR for binding to vitronectin.

Adjacent to the vitronectin binding site, PAI-1 also displays a high-affinity binding site for receptors of the LDLR family, such as LDLR-related protein 1 (LRP1) and the very-low-density lipoprotein receptor [40]. Whereas cleaved, latent, and native PAI-1 depend on LRP1 for cellular clearance, clearance of the PAI-1/uPA complex is greatly enhanced by the involvement of uPAR. By interacting with uPA, which is in turn associated with the cell surface uPAR, PAI-1 is localized to the cell-surface, thereby facilitating the interaction between the PAI-1 moiety within the PAI-1/uPA/uPAR complex and LRP1. Subsequently, internalization of the complex through endocytosis results in the degradation of the PAI-1/uPA complex and recycling of free uPAR to the cell surface. Since both the uPAR and LDLR receptors are important for intracellular signaling as well, binding of PAI-1 or the PAI-1/uPA complex can either indirectly affect signaling activity by regulating receptor levels on the cell surface, or directly by inducing receptor activity through a direct binding interaction.

4. Role of PAI-1 in Diverse Pathologies

The role of PAI-1 in diverse (patho)physiological processes has been extensively studied in humans, in mice being genetically or functionally deficient for PAI-1, or in transgenic mice overexpressing PAI-1. A link has been demonstrated between PAI-1 and various diseases including cardiovascular disease (CVD), metabolic disturbances, aging, cancer, tissue fibrosis, inflammation, and neurodegenerative disease.

4.1. PAI-1 in Cardiovascular Disease

Hemostasis is an important physiological process for maintaining vascular integrity and securing a sufficient blood flow throughout the circulatory system. Therefore, it requires a dynamic interplay between the vascular system, blood platelets, the coagulatory system, and the fibrinolytic system. As PAI-1 is a major inhibitor of the fibrinolytic system, elevated PAI-1 levels create a hypofibrinolytic or prothrombotic state that may contribute to the development of CVD. The recent 2018 report issued by the World Health Organization [41] once more underscores the high mortality rate for cardiovascular diseases, accounting for an estimated 17.8 million deaths worldwide in 2016. An important fraction, i.e., 87% of the deaths caused by CVD, can be contributed to ischemic heart disease (IHD) and stroke. Both IHD and stroke occur when the blood supply to either the heart or the brain is insufficient due to blockage of the blood vessels supplying the organ. This blockage is often caused by the formation of a blood clot (thrombosis) or a buildup of plaque (atherosclerosis), two processes that often coincide with increased plasma PAI-1 levels. Transgenic mice overexpressing wild-type or a stabilized active mutant of human PAI-1 were shown to develop either transient venous thrombosis or age-dependent coronary arterial thrombosis and myocardial infarction, respectively [42,43]. Myocardial infarction is often caused by occlusive thrombus formation that is triggered by the exposure of a procoagulatory surface following disruption of an atherosclerotic plaque in the coronary arteries [44]. Several cell types associated with atherosclerotic plaques in human coronary arteries have been shown to overexpress PAI-1, with the highest levels found in the vulnerable part of the plaque [45–48]. Through local inhibition of plasmin generation, PAI-1 can reduce tissue remodeling and potentially stabilize the fibrin matrix of the developing plaque. Furthermore, studies on the effects of pharmacological PAI-1 inhibition on atherogenesis in mice with obesity and metabolic syndrome also suggested a role for PAI-1 in adipose tissue inflammation, macrophage accumulation, and inducing senescence of smooth muscle cells through its interaction with LRP1 [49]. Whereas several studies demonstrated substantial evidence for PAI-1 as an independent risk factor for CVD including myocardial infarction and stroke [50–52], coronary heart disease [53], venous thrombosis [54], and atherosclerosis [45,55], other studies could not confirm these associations or the significance for the link was lost after adjusting for other risk factors, such as age, sex, and metabolic abnormities [52,56–58].

4.2. PAI-1 in Metabolic Disturbances

Several epidemiological studies have demonstrated that elevated circulating PAI-1 levels and PAI-1 activity are an important feature of or even a marker for the development of metabolic disturbances including obesity, type 2 diabetes, and metabolic syndrome [59–61]. Metabolic syndrome is a multifactorial disease characterized by a cluster of co-occurring metabolic abnormalities that include central obesity, impaired glucose tolerance, hyperinsulinemia, dyslipidemia, and hypertension. Furthermore, these symptoms are all well-documented risk factors for cardiovascular disease and diabetes [62] and it has been shown that individuals with the metabolic syndrome are at higher risk for developing these comorbidities [63]. Importantly, elevated levels of glucose [64] and insulin [65,66] and its precursors [67], and free fatty acids [66,68] have been shown to induce PAI-1 expression or to reduce the rate of PAI-1 mRNA degradation [69]. Moreover, it was demonstrated in obese mice [70] and later confirmed in human adipose tissue [71] that adipocytes are an important source of PAI-1, underscoring the contribution of enlarged adipose tissue to circulating PAI-1 levels. On the other hand, obesity, type 2 diabetes, and metabolic syndrome are often associated with a chronic state of inflammation that is characterized by overexpression of inflammatory adipokines, such as interleukin-6 (IL-6) and tumor necrosis factor-α (TNF-α) [72], which induce PAI-1 expression in adipose tissue [73,74]. These increased PAI-1 levels further contribute to the development of inflammation in adipose tissue by increasing the number of inflammatory macrophages that infiltrate in the tissue [75]. Apart from the positive correlation between inflammatory markers and PAI-1 levels, a link has also been observed between PAI-1 and lipid metabolism in obesity [76,77]. In this respect, higher PAI-1 levels coincide with a decreased mean low-density lipoprotein (LDL) size and higher amounts of small-dense LDL lipoprotein fraction, which importantly contribute to the atherogenic lipid profile and increased cardiovascular risk in obesity [76]. Of note, several studies have shown that PAI-1 deficiency [78–80], pharmacological inhibition of PAI-1 [81], and a reduction in plasma PAI-1 levels through dietary restrictions [82] are protective against the development of obesity and metabolic disorders.

4.3. PAI-1 in Inflammation and Infectious Disease

Acute phase proteins, such as PAI-1, play an important role in inflammatory and immune responses following infectious and/or noninfectious injuries. Several studies have shown that PAI-1 is a critical mediator of the early host defense response that is necessary for the eradication of various pathogens [83–85]. In contrast, uncontrolled inflammation is an essential component of several diseases including respiratory diseases, such as acute respiratory distress syndrome (ARDS) caused by bacterial or viral infections or sepsis [86]. Two characteristic features of ARDS, namely the formation of intravascular micro-thrombi and fibrin deposits in the alveolar space, are often the combined result of tissue factor generated by inflammatory cells (procoagulatory) and PAI-1 produced by endothelial cells (antifibrinolytic) [87]. Recently, it was shown that, in patients with severe coronavirus disease 2019 (COVID-19), plasma PAI-1 levels were as highly elevated as compared to patients with bacterial sepsis or ARDS [88]. As mentioned, expression of PAI-1 can be induced by a wide range of pro-inflammatory mediators including IL-6 and TNF-α, which are known as components of the cytokine release syndrome that is observed in many patients with severe COVID-19 [89]. Another study revealed that elevated levels of the PAI-1/tPA complex were related with disease severity and could be considered an independent risk factor for death in patients with COVID-19 [90]. By affecting multiple organs, including the lung as the predominant target, the formation of microthrombi in these organs may lead to multiorgan failure [91]. In this respect, the efficacy, i.e., an improved clinical outcome, a lower degree of organ failure, decreased PAI-1 levels, and ventilator free days, as well as the safety of a small molecule PAI-1 inhibitor, TM5614, are currently investigated in a phase 1/2 clinical trial for high-risk patients hospitalized with severe COVID-19 (ClinicalTrials.gov: NCT04634799).

4.4. PAI-1 in Cancer

Since motile cells focalize uPA on the cell surface through association with uPAR, which is highly expressed on tumor cells, uPA has been considered as the critical trigger for plasmin formation during tumor cell invasion and metastasis. Importantly, by having binding sites other than for uPA, uPAR interacts with vitronectin, integrins, and transmembrane receptors to facilitate intracellular signaling by effector molecules that are involved in cell migration [92]. By binding to vitronectin, PAI-1 prevents the interaction between vitronectin and two cell surface-associated proteins, namely uPAR and $\alpha_v\beta_3$

integrin, and as a result represses cell migration on vitronectin in the extracellular matrix (ECM). Apart from directly inhibiting uPA-mediated plasmin formation, PAI-1 also inhibits the activity of the uPA/uPAR complex by promoting its endocytosis via LRP1, followed by the degradation of uPA and recycling of uPAR. Furthermore, it causes the cell to detach from the ECM. PAI-1 was thus expected to have an anti-tumor effect. Interestingly, ample evidence has been provided for a paradoxical pro-tumorigenic function of PAI-1, being both pro-angiogenic [93] and anti-apoptotic [94], documented to be dependent on the stage of cancer progression, the cell type, the source (i.e., host or tumor) and on the relative concentration of PAI-1 [93,95–98]. By blocking $\alpha_v\beta_3$-mediated endothelial cell migration on vitronectin in the extracellular matrix, PAI-1 was shown to promote angiogenesis by stimulating integrin $\alpha_5\beta_1$-mediated endothelial cell migration toward fibronectin inside tumor tissue [99]. Apart from its interaction with vitronectin, PAI-1 can modulate cell migration by binding to surface receptor LRP1 that triggers intracellular signaling.

Indeed, uPA and PAI-1 are among the most highly induced proteins in several migratory or invasive tumor cell types. Even though some studies failed to show a correlation between elevated levels of PAI-1 and poor clinical prognosis [100–102], PAI-1 has been established as one of the most reliable biomarkers and prognostic markers in many cancer types, including breast [103–106], ovarian [107], bladder [108,109], colon [110], renal [111] and non-small cell lung cancers [112].

4.5. PAI-1 in Fibrosis

The plasminogen activator/plasmin system is also important for tissue remodeling and promoting wound healing, which is characterized by inflammation, cellular migration to the wound area, and the activation and differentiation of fibroblasts, and the synthesis of ECM proteins to heal the wound [113]. As PAI-1 inhibits PA-mediated plasmin formation and thus protects ECM proteins from degradation, it facilitates wound healing. However, excessive and sustained PAI-1 activity promotes excess fibrin accumulation and leads to low ECM degradation which results in excessive collagen accumulation. Several studies showed that PAI-1 contributes to tissue fibrosis which can affect multiple organs including the skin [114,115], lungs [116,117], kidneys [118], and liver [119]. In agreement with these reports, pharmacological inhibition of PAI-1 or deficiency of host PAI-1 accelerate wound healing [120] and attenuate fibrosis [121–125]. In contrast, complete PAI-1 deficiency in mice leads to spontaneous development of cardiac fibrosis in older animals. The observed accelerated fibrosis has been shown to be the result of an increased vascular permeability, local inflammation, and excessive ECM remodeling, caused by the lack of PAI-1-mediated regulation of integrin $\alpha_v\beta_3$ and a consequently increased signaling of transforming growth factor-β, a potent profibrotic molecule [126,127].

4.6. PAI-1 in the Central Nervous System

PAI-1 is also produced in brain tissue where it has an anti-apoptotic role in neurons by acting as an inhibitor to tPA. Alternatively, PAI-1 protects neurons by preventing disintegration of neuronal networks by maintaining or promoting neuroprotective signaling, independent from its function as a proteinase inhibitor [128]. PAI-1 has also been shown to promote the migration of microglial cells, which are the resident macrophages in the central nervous system, through an LRP1-dependent mechanism, as well as to modulate their phagocytic activity via a vitronectin-dependent mechanism [129]. However, when chronically activated, microglia also contribute to neurodegenerative diseases by maintaining neuroinflammation. Several reports have indicated a role of PAI-1 in central nervous system pathology, including multiple sclerosis [130,131], Alzheimer's disease [132,133], and Parkinson's disease [134,135]. In demyelinated axons in inflammatory multiple sclerosis lesions, increased PAI-1 levels impair the capacity of the tPA/plasmin system to clear fibrin(ogen) deposits and therefore contribute to axonal damage in multiple sclerosis [131,136]. Likewise, elevated levels of PAI-1 have been shown to interfere with the plasmin-mediated clearance and degradation of amyloid-β (Aβ), thereby contributing to the deposition of

Aβ into neurotoxic amyloid plaques and dementia in Alzheimer's disease. Furthermore, inhibition of PAI-1 in transgenic Aβ-producing mice significantly lowered plasma and brain Aβ levels and reversed the cognitive deficits [133,137]. Recently, it was hypothesized that the synergistic relationship between α-synuclein, aggregation and neuroinflammation up-regulate the expression of PAI-1, suggesting a pathological amplification loop, i.e., increased α-synuclein aggregation results in an inflammatory response from microglia, a subsequent increase in PAI-1 levels and thus a decrease in plasmin formation, leading to the accumulation of α-synuclein and a further amplified inflammatory response [135].

4.7. PAI-1 in Aging

Age is the largest risk factor for most chronic diseases, including CVD, metabolic syndrome, and type 2 diabetes. On a cellular level, senescence is a process characterized by the loss of normal physiological function and permanent growth arrest, which accelerates organ and systemic aging when induced by, e.g., oxidative stress. Several molecular drivers of aging have been proposed, including shortening of the telomere length, genomic instability, loss of proteostasis, and altered intercellular communication [138]. Senescent cells have been shown to secrete bioactive molecules called the senescence-associated secretory phenotype (SASP) [139]. These factors have been shown to modulate not only the functions of the secreting cells but also those of adjacent cells. Importantly, PAI-1 levels have been reported to increase with age in various tissues and PAI-1 has been identified as a fundamental component of the SASP, being part of the signaling circuit that induces senescence in neighboring cells [140]. Recently, in the Berne Amish kindred, carriers of the null *SERPINE1* allele, i.e., a rare loss-of-function mutation in the *SERPINE1* gene that encodes PAI-1 which is associated with a lifelong reduction in PAI-1 levels, were shown to have a longer life span [141]. The same study identified an association between heterozygosity of the null *SERPINE1* and a longer leukocyte telomere length, a better metabolic profile and a lower prevalence of diabetes. Therefore, PAI-1 may act not only as a marker but also as a mediator of cellular senescence associated with aging and aging-related pathologies [142].

5. Diverse Approaches to Inhibit PAI-1

From the various biological roles of PAI-1 and its contribution to a wide variety of pathological processes it is clear that targeting PAI-1 may have significant beneficial effects. Therefore, many efforts have been devoted to the development of selective PAI-1 inhibitors, in particular for the prevention or treatment of cardiovascular disease. Some marketed drugs, including insulin-sensitizing agents [143] and angiotensin-converting enzyme inhibitors [144], and antisense oligonucleotides have been shown to attenuate PAI-1 synthesis or secretion [145]. In contrast, the majority of PAI-1 inhibitors currently in development (extensively reviewed elsewhere [37,146,147] can influence PAI-1 functionality in at least four possible ways, i.e., (I) by blocking the interaction between PAI-1 and PAs, (II) by inducing substrate behavior of PAI-1, (III) by accelerating the active-to-latent transition or converting active PAI-1 to an otherwise inert form, or (IV) by interfering with interactions between PAI-1 and other biological ligands such as LRP1. These inhibitors include small molecules, peptides, antibodies (Abs), and antibody fragments such as nanobodies. A link between the mechanisms by which these inhibitors modulate PAI-1 functionality and their binding site has been provided by using a broad range of biochemical and biophysical methods, including mutagenesis studies, competitive binding experiments, computational docking, and X-ray crystallography.

PAI-1 inhibitory peptides have been shown to either induce substrate behavior of PAI-1 or to accelerate the conversion to an inert form of PAI-1. Synthetic peptides that were derived from the sequence of the RCL were shown to insert into the core of the PAI-1 protein in between strand 3 and strand 5 of the central β-sheet A. It was suggested that, depending on their position within the cleft, i.e., occupying the same space as the N-terminal part or the C-terminal part of the RCL in latent or cleaved PAI-1, they act by inducing substrate behavior of PAI-1 or by accelerating the irreversible transition to inert PAI-1, respectively [148]. In contrast, a peptide that was isolated from a phage-display peptide library, paionin-4, was shown to accelerate the active-to-latent conversion by binding to a different region in PAI-1, located at the loop between hD and s2A [149]. From the same library, the peptide paionin-1 did not affect PAI-1 activity; however, it was able to prevent the binding of the PAI-1/uPA complex to LRP1 by binding hD and hE in the flexible joint region of PAI-1, which may impair the signaling function of uPA/uPAR/LRP1 [150].

Another large category of PAI-1 inhibitors includes small organochemical molecules that are very diverse in their chemical structure. Many of these compounds have been shown to bind a common binding pocket within the area of the flexible joint region of PAI-1 [151–153], or to link structural elements within this region through interactions at the PAI-1 surface [154] (Figure 2A). By interfering with the flexible joint region, these compounds were shown to inhibit PAI-1 through a dual mechanism of action, i.e., by inducing substrate behavior of PAI-1 and converting PAI-1 to an inert form which can be latent or unreactive PAI-1 or PAI-1 in the capacity of polymers. By binding this otherwise flexible region in PAI-1, these compounds can induce substrate behavior possibly by attenuating or preventing the conformational rearrangements within this region that are required for a successful inhibitory reaction between PAI-1 and PA's or by affecting regions outside the flexible joint region through allosteric modulation. In contrast to the aforementioned compounds, compounds that bind the sheet B/sheet C (sB/sC) pocket (Figure 2A), i.e., an interface composed of residues from the s3A/s4C loop, β-sheets B and C, and hH, were shown to block initial PAI-1/PA Michaelis complex formation, possibly by a reversible allosteric modulation of the RCL [155].

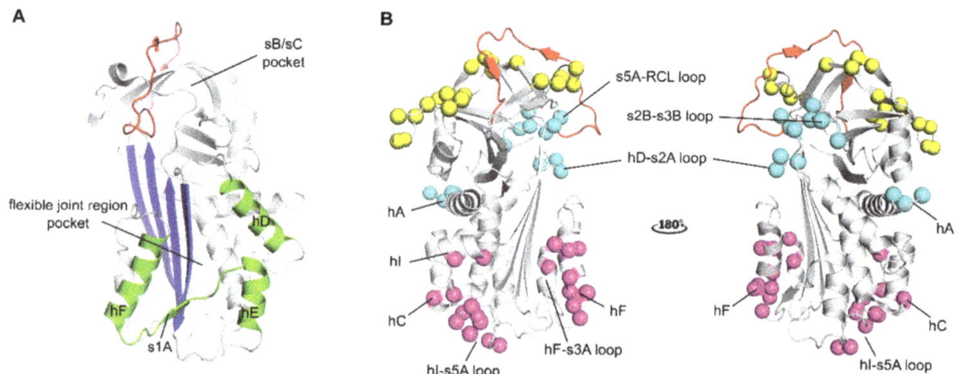

Figure 2. Localization of binding regions for PAI-1 inhibitors in the structure of active PAI-1. (**A**) Localization of the binding regions for small molecule PAI-1 inhibitors. The binding pocket in the flexible joint region is aligned by hD, hE, hF and strand 1 (shown in green). The sB/sC pocket is aligned by β-sheet B and C (**B**), localization of different epitopes of antibodies and antibody fragments as determined by mutagenesis and X-ray crystallographic studies. The epitopes of Abs that prevent the interaction between PAI-1 and PAs comprise residues that are indicated by yellow spheres (exosites for PAs on PAI-1) and residues in the reactive center loop (RCL) (shown in red). The epitopes of switching Abs are indicated by magenta spheres and comprise either residues located in hF and the loop connecting hF and s3A or residues located in the loop connecting hI and s5A, hC and hI. The epitopes of latency-inducing Abs are indicated by cyan spheres and comprise in hA or residues at the top part of the PAI-1 molecule in the hD-s2A loop, the s2B-s3B loop, and the s5A-RCL loop. All panels have been generated using the structure of active PAI-1 (PDB ID 1DB2).

In contrast to inhibitory peptides and small molecules, the binding sites of antibody-based PAI-1 inhibitors have been mapped to different regions of the PAI-1 molecule (extensively reviewed in [37]). Overall, Abs that interfere with PAI-1 activity can be divided into three categories. The first category of Abs or antibody fragments acts by interfering with the formation of the initial Michaelis complex. It was recently shown that PAI-1/PA complex formation can be prevented by destabilizing the Michaelis complex merely by hampering exosite interactions between PAI-1 and PAs or in combination with shielding the P1-P1′ reactive center in the RCL of PAI-1 (Figure 2B). The second category comprises Abs and antibody fragments that switch the PAI-1/PA reaction towards the substrate pathway. These Abs, referred to as "switching antibodies", can bind different epitope regions in the lower half of the PAI-1 molecule. Within this category, Abs that bind hF or the loop connecting hF to s3A of the central PAI-1 β-sheet A (Figure 2B) were shown to slow down the rate of cleaved RCL insertion, resulting in hydrolysis of the PAI-1/PA complex. On the other hand, Abs that bind the loop between hI and s5A at the bottom of the PAI-1 molecule, a region that is buried by the PA in the final inhibitory PAI-1/PA complex, hinder full translocation of the PA and thus prevent distortion of the catalytic triad of the PA. The third category of Abs have the ability to accelerate the active-to-latent transition of PAI-1 and were shown to bind different epitopes that are spread more across the PAI-1 surface (Figure 2B). In most cases, these epitopes comprise regions at the top of the PAI-1 molecule that are less accessible in the active form of PAI-1. Binding is therefore believed to occur to a prelatent state of PAI-1, in which the RCL is already partially inserted. On the other hand, acceleration of the active-to-latent transition was also observed for an Ab binding to the N-terminal part of hA in rat PAI-1. However, an Ab targeting as similar region in human PAI-1 was shown to be non-inhibitory.

Even though several of these molecules were proven to be efficient PAI-1 inhibitors both in vitro and in vivo, no PAI-1 inhibitor is currently approved for therapeutic use in humans. However, it should be noted that a few PAI-1 antagonists are currently proceeding through clinical trials, which underscores the clinical interest in safe and efficient modulators of PAI-1 activity for the treatment of PAI-1-related diseases.

Author Contributions: Conceptualization, M.S. and P.J.D.; writing—original draft preparation, M.S.; writing—review and editing, M.S. and P.J.D.; visualization, M.S. All authors have read and agreed to the published version of the manuscript.

Funding: This research received no external funding.

Conflicts of Interest: The authors declare no conflict of interest.

Abbreviations

Aβ	Amyloid-β
Abs	Antibodies
ARDS	Acute respiratory distress syndrome
COVID-19	Coronavirus disease 2019
CVD	Cardiovascular disease
ECM	Extracellular matrix
IHD	Ischemic heart disease
IL-6	Interleukin-6
LDL	Low-density lipoprotein
LDLR	Low-density lipoprotein receptor
LRP1	LDLR-related protein 1
MMP	Matrix metalloprotease
PA	Plasminogen activator
PAI-1	Plasminogen activator inhibitor-1
RCL	Reactive center loop
SASP	Senescence-associated secretory phenotype
SMB	Somatomedin B
TNF-α	Tumor necrosis factor-α [72]
tPA	Tissue-type plasminogen activator
uPA	Urokinase-type plasminogen activator
uPAR	uPA receptor

50. Hamsten, A.; Wiman, B.; De Faire, U.; Blombäck, M. Increased Plasma Levels of a Rapid Inhibitor of Tissue Plasminogen Activator in Young Survivors of Myocardial Infarction. *N. Engl. J. Med.* **1985**, *313*, 1557–1563. [CrossRef] [PubMed]

51. Tofler, G.; Massaro, J.; O'Donnell, C.; Wilson, P.; Vasan, R.; Sutherland, P.; Meigs, J.; Levy, D.; D'Agostino, R. Plasminogen activator inhibitor and the risk of cardiovascular disease: The Framingham Heart Study. *Thromb. Res.* **2016**, *140*, 30–35. [CrossRef] [PubMed]

52. Jung, R.G.; Motazedian, P.; Ramirez, F.D.; Simard, T.; Di Santo, P.; Visintini, S.; Faraz, M.A.; Labinaz, A.; Jung, Y.; Hibbert, B. Association between plasminogen activator inhibitor-1 and cardiovascular events: A systematic review and meta-analysis. *Thromb. J.* **2018**, *16*, 1–12. [CrossRef]

53. Song, C.; Burgess, S.; Eicher, J.D.; O'Donnell, C.J.; Johnson, A.D.; Huang, J.; Sabater-Lleal, M.; Asselbergs, F.W.; Tregouet, D.; Shin, S.; et al. Causal Effect of Plasminogen Activator Inhibitor Type 1 on Coronary Heart Disease. *J. Am. Heart Assoc.* **2017**, *6*. [CrossRef] [PubMed]

54. Meltzer, M.E.; Lisman, T.; De Groot, P.G.; Meijers, J.C.M.; Le Cessie, S.; Doggen, C.J.M.; Rosendaal, F.R. Venous thrombosis risk associated with plasma hypofibrinolysis is explained by elevated plasma levels of TAFI and PAI-1. *Blood* **2010**, *116*, 113–121. [CrossRef]

55. Chomiki, N.; Henry, M.; Alessi, M.C.; Anfosso, F.; Juhan-Vague, I. Plasminogen Activator Inhibitor-1 Expression in Human Liver and Healthy or Atherosclerotic Vessel Walls. *Thromb. Haemost.* **1994**, *72*, 44–53. [CrossRef]

56. Juhan-Vague, I.; Alessi, M.C.; Vague, P. Increased plasma plasminogen activator inhibitor 1 levels. A possible link between insulin resistance and atherothrombosis. *Diabetology* **1991**, *34*, 457–462. [CrossRef] [PubMed]

57. Carratala, A.; Martinez-Hervas, S.; Rodriguez-Borja, E.; Benito, E.; Real, J.T.; Saez, G.T.; Carmena, R.; Ascaso, J.F. PAI-1 levels are related to insulin resistance and carotid atherosclerosis in subjects with familial combined hyperlipidemia. *J. Investig. Med.* **2017**, *66*, 17–21. [CrossRef]

58. Ploplis, V.A. Effects of altered plasminogen activator inhibitor-1 expression on cardiovascular disease. *Curr. Drug Targets* **2011**, *12*, 1782–1789. [CrossRef]

59. Festa, A.; D'Agostino, R.; Tracy, R.P.; Haffner, S.M. Elevated Levels of Acute-Phase Proteins and Plasminogen Activator Inhibitor-1 Predict the Development of Type 2 Diabetes: The Insulin Resistance Atherosclerosis Study. *Diabetes* **2002**, *51*, 1131–1137. [CrossRef] [PubMed]

60. Yarmolinsky, J.; Barbieri, N.B.; Weinmann, T.; Ziegelmann, P.K.; Duncan, B.B.; Schmidt, M.I. Plasminogen activator inhibitor-1 and type 2 diabetes: A systematic review and meta-analysis of observational studies. *Sci. Rep.* **2016**, *6*, 17714. [CrossRef] [PubMed]

61. Alessi, M.-C.; Poggi, M.; Juhan-Vague, I. Plasminogen activator inhibitor-1, adipose tissue and insulin resistance. *Curr. Opin. Lipidol.* **2007**, *18*, 240–245. [CrossRef]

62. Aguilar-Salinas, C.A.; Viveros-Ruiz, T. Recent advances in managing/understanding the metabolic syndrome. *F1000Research* **2019**, *8*, 370. [CrossRef] [PubMed]

63. Ninomiya, T.; Kubo, M.; Doi, Y.; Yonemoto, K.; Tanizaki, Y.; Rahman, M.; Arima, H.; Tsuryuya, K.; Iida, M.; Kiyohara, Y. Impact of Metabolic Syndrome on the Development of Cardiovascular Disease in a General Japanese Population. *Stroke* **2007**, *38*, 2063–2069. [CrossRef] [PubMed]

64. Iwasaki, H.; Okamoto, R.; Kato, S.; Konishi, K.; Mizutani, H.; Yamada, N.; Isaka, N.; Nakano, T.; Ito, M. High glucose induces plasminogen activator inhibitor-1 expression through Rho/Rho-kinase-mediated NF-κB activation in bovine aortic endothelial cells. *Atherosclerosis* **2008**, *196*, 22–28. [CrossRef] [PubMed]

65. Alessi, M.C.; Juhan-Vague, I.; Kooistra, T.; Declerck, P.J.; Collen, D. Insulin Stimulates the Synthesis of Plasminogen Activator Inhibitor 1 by the Human Hepatocellular Cell Line Hep G2. *Thromb. Haemost.* **1988**, *60*, 491–494. [CrossRef] [PubMed]

66. Schneider, D.J.; Sobel, B.E. Synergistic augmentation of expression of plasminogen activator inhibitor type-1 induced by insulin, very-low-density lipoproteins, and fatty acids. *Coron. Artery Dis.* **1996**, *7*, 813–818. [CrossRef]

67. Nordt, T.K.; Schneider, D.J.; Sobel, B.E. Augmentation of the synthesis of plasminogen activator inhibitor type-1 by precursors of insulin. A potential risk factor for vascular disease. *Circulation* **1994**, *89*, 321–330. [CrossRef]

68. Chen, Y.; Billadello, J.J.; Schneider, D.J. Identification and localization of a fatty acid response region in the human plasminogen activator inhibitor-1 gene. *Arterioscler. Thromb. Vasc. Biol.* **2000**, *20*, 2696–2701. [CrossRef] [PubMed]

69. Fattal, P.; Schneider, D.; Sobel, B.; Billadello, J. Post-transcriptional regulation of expression of plasminogen activator inhibitor type 1 mRNA by insulin and insulin-like growth factor 1. *J. Biol. Chem.* **1992**, *267*, 12412–12415. [CrossRef]

70. Sawdey, M.S.; Loskutoff, D.J. Regulation of murine type 1 plasminogen activator inhibitor gene expression in vivo. Tissue specificity and induction by lipopolysaccharide, tumor necrosis factor-alpha, and transforming growth factor-beta. *J. Clin. Investig.* **1991**, *88*, 1346–1353. [CrossRef]

71. Alessi, M.C.; Peiretti, F.; Morange, P.; Henry, M.; Nalbone, G.; Juhan-Vague, I. Production of Plasminogen Activator Inhibitor 1 by Human Adipose Tissue: Possible Link Between Visceral Fat Accumulation and Vascular Disease. *Diabetes* **1997**, *46*, 860–867. [CrossRef] [PubMed]

72. Ellulu, M.S.; Patimah, I.; Khaza'Ai, H.; Rahmat, A.; Abed, Y. Obesity and inflammation: The linking mechanism and the complications. *Arch. Med. Sci.* **2017**, *4*, 851–863. [CrossRef] [PubMed]

73. Pandey, M.; Loskutoff, D.J.; Samad, F. Molecular mechanisms of tumor necrosis factor-α-mediated plasminogen activator inhibitor-1 expression in adipocytes. *FASEB J.* **2005**, *19*, 1317–1319. [CrossRef] [PubMed]

74. Rega, G.; Kaun, C.; Weiss, T.; Demyanets, S.; Zorn, G.; Kastl, S.; Steiner, S.; Seidinger, D.; Kopp, C.; Frey, M.; et al. Inflammatory Cytokines Interleukin-6 and Oncostatin M Induce Plasminogen Activator Inhibitor-1 in Human Adipose Tissue. *Circulation* **2005**, *111*, 1938–1945. [CrossRef] [PubMed]

75. Wang, L.; Chen, L.; Liu, Z.; Liu, Y.; Luo, M.; Chen, N.; Deng, X.; Luo, Y.; He, J.; Zhang, L.; et al. PAI-1 Exacerbates White Adipose Tissue Dysfunction and Metabolic Dysregulation in High Fat Diet-Induced Obesity. *Front. Pharmacol.* **2018**, *9*, 1087. [CrossRef] [PubMed]

76. Somodi, S.; Seres, I.; Lőrincz, H.; Harangi, M.; Fülöp, P.; Paragh, G. Plasminogen Activator Inhibitor-1 Level Correlates with Lipoprotein Subfractions in Obese Nondiabetic Subjects. *Int. J. Endocrinol.* **2018**, *2018*, 1–9. [CrossRef]

77. Levine, J.A.; Oleaga, C.; Eren, M.; Amaral, A.P.; Shang, M.; Lux, E.; Khan, S.S.; Shah, S.J.; Omura, Y.; Pamir, N.; et al. Role of PAI-1 in hepatic steatosis and dyslipidemia. *Sci. Rep.* **2021**, *11*, 1–13. [CrossRef] [PubMed]

78. Morange, P.E.; Lijnen, H.R.; Alessi, M.C.; Kopp, F.; Collen, D.; Juhan-Vague, I. Influence of PAI-1 on Adipose Tissue Growth and Metabolic Parameters in a Murine Model of Diet-Induced Obesity. *Arterioscler. Thromb. Vasc. Biol.* **2000**, *20*, 1150–1154. [CrossRef]

79. Schäfer, K.; Fujisawa, K.; Konstantinides, S.; Loskutoff, D.J. Disruption of the plasminogen activator inhibitor-1 gene reduces the adiposity and improves the metabolic profile of genetically obese and diabetic ob/ob mice. *FASEB J.* **2001**, *15*, 1840–1842. [CrossRef]

80. Ma, L.-J.; Mao, S.-L.; Taylor, K.L.; Kanjanabuch, T.; Guan, Y.; Zhang, Y.; Brown, N.J.; Swift, L.L.; McGuinness, O.P.; Wasserman, D.H.; et al. Prevention of Obesity and Insulin Resistance in Mice Lacking Plasminogen Activator Inhibitor 1. *Diabetes* **2004**, *53*, 336–346. [CrossRef]

81. Henkel, A.S.; Khan, S.S.; Olivares, S.; Miyata, T.; Vaughan, D.E. Inhibition of Plasminogen Activator Inhibitor 1 Attenuates Hepatic Steatosis but Does Not Prevent Progressive Nonalcoholic Steatohepatitis in Mice. *Hepatol. Commun.* **2018**, *2*, 1479–1492. [CrossRef] [PubMed]

82. Jensen, L.; Sloth, B.; Krog-Mikkelsen, I.; Flint, A.; Raben, A.; Tholstrup, T.; Brünner, N.; Astrup, A. A low-glycemic-index diet reduces plasma plasminogen activator inhibitor-1 activity, but not tissue inhibitor of proteinases-1 or plasminogen activator inhibitor-1 protein, in overweight women. *Am. J. Clin. Nutr.* **2008**, *87*, 97–105. [CrossRef] [PubMed]

83. Lim, J.H.; Woo, C.-H.; Li, J.-D. Critical role of type 1 plasminogen activator inhibitor (PAI-1) in early host defense against nontypeable Haemophilus influenzae (NTHi) infection. *Biochem. Biophys. Res. Commun.* **2011**, *414*, 67–72. [CrossRef]

84. Goolaerts, A.; LaFargue, M.; Song, Y.; Miyazawa, B.; Arjomandi, M.; Carlès, M.; Roux, J.; Howard, M.; Parks, D.A.; Iles, K.E.; et al. PAI-1 is an essential component of the pulmonary host response during Pseudomonas aeruginosa pneumonia in mice. *Thorax* **2011**, *66*, 788–796. [CrossRef] [PubMed]

85. Kager, L.M.; Van Der Windt, G.J.; Wieland, C.W.; Florquin, S.; van't Veer, C.; Van Der Poll, T. Plasminogen activator inhibitor type I may contribute to transient, non-specific changes in immunity in the subacute phase of murine tuberculosis. *Microbes Infect.* **2012**, *14*, 748–755. [CrossRef] [PubMed]

86. Matthay, M.A.; Ware, L.B.; Zimmerman, G.A. The acute respiratory distress syndrome. *J. Clin. Investig.* **2012**, *122*, 2731–2740. [CrossRef]

87. Ozolina, A.; Sarkele, M.; Sabelnikovs, O.; Skesters, A.; Jaunalksne, I.; Serova, J.; Ievins, T.; Bjertnaes, L.J.; Vanags, I. Activation of Coagulation and Fibrinolysis in Acute Respiratory Distress Syndrome: A Prospective Pilot Study. *Front. Med.* **2016**, *3*, 64. [CrossRef]

88. Kang, S.; Tanaka, T.; Inoue, H.; Ono, C.; Hashimoto, S.; Kioi, Y.; Matsumoto, H.; Matsuura, H.; Matsubara, T.; Shimizu, K.; et al. IL-6 trans-signaling induces plasminogen activator inhibitor-1 from vascular endothelial cells in cytokine release syndrome. *Proc. Natl. Acad. Sci. USA* **2020**, *117*, 22351–22356. [CrossRef] [PubMed]

89. Del Valle, D.M.; Kim-Schulze, S.; Huang, H.-H.; Beckmann, N.D.; Nirenberg, S.; Wang, B.; Lavin, Y.; Swartz, T.H.; Madduri, D.; Stock, A.; et al. An inflammatory cytokine signature predicts COVID-19 severity and survival. *Nat. Med.* **2020**, *26*, 1636–1643. [CrossRef] [PubMed]

90. Tsantes, A.E.; Frantzeskaki, F.; Tsantes, A.G.; Rapti, E.; Rizos, M.; Kokoris, S.I.; Paramythiotou, E.; Katsadiotis, G.; Karali, V.; Flevari, A.; et al. The haemostatic profile in critically ill COVID-19 patients receiving therapeutic anticoagulant therapy. *Medicine* **2020**, *99*, e23365. [CrossRef]

91. Kwaan, H.; Lindholm, P. The Central Role of Fibrinolytic Response in COVID-19—A Hematologist's Perspective. *Int. J. Mol. Sci.* **2021**, *22*, 1283. [CrossRef] [PubMed]

92. Mahmood, N.; Mihalcioiu, C.; Rabbani, S.A. Multifaceted Role of the Urokinase-Type Plasminogen Activator (uPA) and Its Receptor (uPAR): Diagnostic, Prognostic, and Therapeutic Applications. *Front. Oncol.* **2018**, *8*, 24. [CrossRef] [PubMed]

93. Devy, L.; Blacher, S.; Grignet-Debrus, C.; Bajou, K.; Masson, V.; Gerard, R.D.; Gils, A.; Carmeliet, G.; Carmeliet, P.; Declerck, P.J.; et al. The pro- or antiangiogenic effect of plasminogen activator inhibitor 1 is dose dependent. *FASEB J.* **2001**, *16*, 147–154. [CrossRef]

94. Fang, H.; Placencio, V.R.; Declerck, Y.A. Protumorigenic Activity of Plasminogen Activator Inhibitor-1 Through an Antiapoptotic Function. *J. Natl. Cancer Inst.* **2012**, *104*, 1470–1484. [CrossRef] [PubMed]

95. Balsara, R.D.; Ploplis, V.A. Plasminogen activator inhibitor-1: The double-edged sword in apoptosis. *Thromb. Haemost.* **2008**, *100*, 1029–1036. [CrossRef]

96. Bajou, K.; Noel, A.; Gerard, R.D.; Masson, V.; Brunner, N.; Holsthansen, C.; Skobe, M.; Fusenig, N.E.; Carmeliet, P.; Collen, D.; et al. Absence of host plasminogen activator inhibitor 1 prevents cancer invasion and vascularization. *Nat. Med.* **1998**, *4*, 923–928. [CrossRef]

97. Kubala, M.H.; Declerck, Y.A. The plasminogen activator inhibitor-1 paradox in cancer: A mechanistic understanding. *Cancer Metastasis Rev.* **2019**, *38*, 483–492. [CrossRef] [PubMed]

98. Bajou, K.; Maillard, C.; Jost, M.; Lijnen, R.H.; Gils, A.; Declerck, P.; Carmeliet, P.; Foidart, J.-M.; Noël, A. Host-derived plasminogen activator inhibitor-1 (PAI-1) concentration is critical for in vivo tumoral angiogenesis and growth. *Oncogene* **2004**, *23*, 6986–6990. [CrossRef] [PubMed]

99. Isogai, C.; Laug, W.E.; Shimada, H.; Declerck, P.J.; Stins, M.F.; Durden, D.L.; Erdreich-Epstein, A.; Declerck, Y.A. Plasminogen activator inhibitor-1 promotes angiogenesis by stimulating endothelial cell migration toward fibronectin. *Cancer Res.* **2001**, *61*, 5587–5594.

100. Li, H.; Shinohara, E.; Cai, Q.; Chen, H.; Courtney, R.; Cao, C.; Wang, Z.; Teng, M.; Zheng, W.; Lu, B. Plasminogen Activator Inhibitor-1 Promoter Polymorphism is Not Associated with the Aggressiveness of Disease in Prostate Cancer. *Clin. Oncol.* **2006**, *18*, 333–337. [CrossRef] [PubMed]

101. Błasiak, J.; Smolarz, B. Plasminogen activator inhibitor-1 (PAI-1) gene 4G/5G promoter polymorphism is not associated with breast cancer. *Acta Biochim. Pol.* **2000**, *47*, 191–199. [CrossRef] [PubMed]

102. Sternlicht, M.D.; Dunning, A.M.; Moore, D.H.; Pharoah, P.D.; Ginzinger, D.G.; Chin, K.; Gray, J.W.; Waldman, F.M.; Ponder, B.A.; Werb, Z. Prognostic Value of PAI1 in Invasive Breast Cancer: Evidence That Tumor-Specific Factors Are More Important Than Genetic Variation in Regulating PAI1 Expression. *Cancer Epidemiol. Biomark. Prev.* **2006**, *15*, 2107–2114. [CrossRef] [PubMed]

103. Jevrić, M.; Matić, I.Z.; Krivokuća, A.; Đorđić Crnogorac, M.; Besu, I.; Damjanović, A.; Branković-Magić, M.; Milovanović, Z.; Gavrilović, D.; Susnjar, S.; et al. Association of uPA and PAI-1 tumor levels and 4G/5G variants of PAI-1 gene with disease outcome in luminal HER2-negative node-negative breast cancer patients treated with adjuvant endocrine therapy. *BMC Cancer* **2019**, *19*, 71. [CrossRef]

104. Duffy, M.J.; McGowan, P.M.; Harbeck, N.; Thomssen, C.; Schmitt, M. uPA and PAI-1 as biomarkers in breast cancer: Validated for clinical use in level-of-evidence-1 studies. *Breast Cancer Res.* **2014**, *16*, 1–10. [CrossRef]

105. Duffy, M.J.; O'Donovan, N.; McDermott, E.; Crown, J. Validated biomarkers: The key to precision treatment in patients with breast cancer. *Breast* **2016**, *29*, 192–201. [CrossRef] [PubMed]

106. Kates, R.E.; Gauger, K.; Willems, A.; Kiechle, M.; Magdolen, V.; Schmitt, M.; Harbeck, N. Urokinase-type plasminogen activator (uPA) and its inhibitor PAI-1: Novel tumor-derived factors with a high prognostic and predictive impact in breast cancer. *Thromb. Haemost.* **2004**, *91*, 450–456. [CrossRef]

107. Nakatsuka, E.; Sawada, K.; Nakamura, K.; Yoshimura, A.; Kinose, Y.; Kodama, M.; Hashimoto, K.; Mabuchi, S.; Makino, H.; Morii, E.; et al. Plasminogen activator inhibitor-1 is an independent prognostic factor of ovarian cancer and IMD-4482, a novel plasminogen activator inhibitor-1 inhibitor, inhibits ovarian cancer peritoneal dissemination. *Oncotarget* **2017**, *8*, 89887–89902. [CrossRef]

108. Chan, O.T.; Furuya, H.; Pagano, I.; Shimizu, Y.; Hokutan, K.; Dyrskjøt, L.; Jensen, J.B.; Malmstrom, P.-U.; Segersten, U.; Janku, F.; et al. Association of MMP-2, RB and PAI-1 with decreased recurrence-free survival and overall survival in bladder cancer patients. *Oncotarget* **2017**, *8*, 99707–99721. [CrossRef]

109. Becker, M.; Szarvas, T.; Wittschier, M.; Dorp, F.V.; Tötsch, M.; Schmid, K.W.; Rübben, H.; Ergün, S. Prognostic impact of plasminogen activator inhibitor type 1 expression in bladder cancer. *Cancer* **2010**, *116*, 4502–4512. [CrossRef]

110. Sakakibara, T.; Hibi, K.; Koike, M.; Fujiwara, M.; Kodera, Y.; Ito, K.; Nakao, A. Plasminogen activator inhibitor-1 as a potential marker for the malignancy of colorectal cancer. *Br. J. Cancer* **2005**, *93*, 799–803. [CrossRef] [PubMed]

111. Zubac, D.P.; Wentzel-Larsen, T.; Seidal, T.; Bostad, L. Type 1 plasminogen activator inhibitor (PAI-1) in clear cell renal cell carcinoma (CCRCC) and its impact on angiogenesis, progression and patient survival after radical nephrectomy. *BMC Urol.* **2010**, *10*, 20. [CrossRef] [PubMed]

112. Sotiropoulos, G.; Kotopouli, M.; Karampela, I.; Christodoulatos, G.S.; Antonakos, G.; Marinou, I.; Vogiatzakis, E.; Lekka, A.; Papavassiliou, A.; Dalamaga, M. Circulating plasminogen activator inhibitor-1 activity: A biomarker for resectable non-small cell lung cancer? *J. BU ON Off. J. Balk. Union Oncol.* **2019**, *24*, 943–954.

113. Ghosh, A.K.; Vaughan, D.E. PAI-1 in tissue fibrosis. *J. Cell. Physiol.* **2011**, *227*, 493–507. [CrossRef]

114. Lemaire, R.; Burwell, T.; Sun, H.; Delaney, T.; Bakken, J.; Cheng, L.; Rebelatto, M.C.; Czapiga, M.; De-Mendez, I.; Coyle, A.J.; et al. Resolution of Skin Fibrosis by Neutralization of the Antifibrinolytic Function of Plasminogen Activator Inhibitor 1. *Arthritis Rheumatol.* **2015**, *68*, 473–483. [CrossRef] [PubMed]

115. Sun, Y.; Wang, Y.; Long, J.; Wang, X. Association of the Plasminogen Activator Inhibitor-1 (PAI-1) Gene -675 4G/5G and -844 A/G Promoter Polymorphism with Risk of Keloid in a Chinese Han Population. *Med. Sci. Monit. Int. Med. J. Exp. Clin. Res.* **2014**, *20*, 2069–2073. [CrossRef] [PubMed]

116. Liu, R.-M. Oxidative Stress, Plasminogen Activator Inhibitor 1, and Lung Fibrosis. *Antioxid. Redox Signal.* **2008**, *10*, 303–320. [CrossRef]

117. Courey, A.J.; Horowitz, J.C.; Kim, K.K.; Koh, T.J.; Novak, M.L.; Subbotina, N.; Warnock, M.; Xue, B.; Cunningham, A.K.; Lin, Y.; et al. The vitronectin-binding function of PAI-1 exacerbates lung fibrosis in mice. *Blood* **2011**, *118*, 2313–2321. [CrossRef]

118. Małgorzewicz, S.; Skrzypczak-Jankun, E.; Jankun, J. Plasminogen activator inhibitor-1 in kidney pathology. *Int. J. Mol. Med.* **2013**, *31*, 503–510. [CrossRef] [PubMed]

119. Bergheim, I.; Guo, L.; Davis, M.A.; Duveau, I.; Arteel, G.E. Critical Role of Plasminogen Activator Inhibitor-1 in Cholestatic Liver Injury and Fibrosis. *J. Pharmacol. Exp. Ther.* **2005**, *316*, 592–600. [CrossRef]
120. Chan, J.C.; Duszczyszyn, D.A.; Castellino, F.J.; Ploplis, V.A. Accelerated Skin Wound Healing in Plasminogen Activator Inhibitor-1-Deficient Mice. *Am. J. Pathol.* **2001**, *159*, 1681–1688. [CrossRef]
121. Shioya, S.; Masuda, T.; Senoo, T.; Horimasu, Y.; Miyamoto, S.; Nakashima, T.; Iwamoto, H.; Fujitaka, K.; Hamada, H.; Hattori, N. Plasminogen activator inhibitor-1 serves an important role in radiation-induced pulmonary fibrosis. *Exp. Ther. Med.* **2018**, *16*, 3070–3076. [CrossRef]
122. Huang, W.-T.; Vayalil, P.K.; Miyata, T.; Hagood, J.; Liu, R.-M. Therapeutic Value of Small Molecule Inhibitor to Plasminogen Activator Inhibitor–1 for Lung Fibrosis. *Am. J. Respir. Cell Mol. Biol.* **2012**, *46*, 87–95. [CrossRef]
123. Senoo, T.; Hattori, N.; Tanimoto, T.; Furonaka, M.; Ishikawa, N.; Fujitaka, K.; Haruta, Y.; Murai, H.; Yokoyama, A.; Kohno, N. Suppression of plasminogen activator inhibitor-1 by RNA interference attenuates pulmonary fibrosis. *Thorax* **2010**, *65*, 334–340. [CrossRef] [PubMed]
124. Lassila, M.; Fukami, K.; Jandeleit-Dahm, K.; Semple, T.; Carmeliet, P.; Cooper, M.E.; Kitching, A.R. Plasminogen activator inhibitor-1 production is pathogenetic in experimental murine diabetic renal disease. *Diabetologia* **2007**, *50*, 1315–1326. [CrossRef] [PubMed]
125. Noguchi, R.; Kaji, K.; Namisaki, T.; Moriya, K.; Kawaratani, H.; Kitade, M.; Takaya, H.; Aihara, Y.; Douhara, A.; Asada, K.; et al. Novel oral plasminogen activator inhibitor-1 inhibitor TM5275 attenuates hepatic fibrosis under metabolic syndrome via suppression of activated hepatic stellate cells in rats. *Mol. Med. Rep.* **2020**, *22*, 2948–2956. [CrossRef] [PubMed]
126. Xu, Z.; Castellino, F.J.; Ploplis, V.A. Plasminogen activator inhibitor-1 (PAI-1) is cardioprotective in mice by maintaining microvascular integrity and cardiac architecture. *Blood* **2010**, *115*, 2038–2047. [CrossRef]
127. Pedroja, B.S.; Kang, L.E.; Imas, A.O.; Carmeliet, P.; Bernstein, A.M. Plasminogen Activator Inhibitor-1 Regulates Integrin $\alpha v \beta 3$ Expression and Autocrine Transforming Growth Factor β Signaling. *J. Biol. Chem.* **2009**, *284*, 20708–20717. [CrossRef]
128. Soeda, S.; Koyanagi, S.; Kuramoto, Y.; Kimura, M.; Oda, M.; Kozako, T.; Hayashida, S.; Shimeno, H. Anti-apoptotic roles of plasminogen activator inhibitor-1 as a neurotrophic factor in the central nervous system. *Thromb. Haemost.* **2008**, *100*, 1014–1020. [CrossRef]
129. Jeon, H.; Kim, J.-H.; Kim, J.-H.; Lee, W.-H.; Lee, M.-S.; Suk, K. Plasminogen activator inhibitor type 1 regulates microglial motility and phagocytic activity. *J. Neuroinflamm.* **2012**, *9*, 149. [CrossRef]
130. Rodríguez-Lorenzo, S.; Francisco, D.M.F.; Vos, R.; Hof, B.V.H.; Rijnsburger, M.; Schroten, H.; Ishikawa, H.; Beaino, W.; Bruggmann, R.; Kooij, G.; et al. Altered secretory and neuroprotective function of the choroid plexus in progressive multiple sclerosis. *Acta Neuropathol. Commun.* **2020**, *8*, 1–13. [CrossRef] [PubMed]
131. Gverić, D.; Herrera, B.; Petzold, A.; Lawrence, D.A.; Cuzner, M.L. Impaired fibrinolysis in multiple sclerosis: A role for tissue plasminogen activator inhibitors. *Brain* **2003**, *126*, 1590–1598. [CrossRef]
132. Oh, J.; Lee, H.-J.; Song, J.-H.; Park, S.I.; Kim, H. Plasminogen activator inhibitor-1 as an early potential diagnostic marker for Alzheimer's disease. *Exp. Gerontol.* **2014**, *60*, 87–91. [CrossRef]
133. Jacobsen, J.S.; Comery, T.A.; Martone, R.L.; Elokdah, H.; Crandall, D.L.; Oganesian, A.; Aschmies, S.; Kirksey, Y.; Gonzales, C.; Xu, J.; et al. Enhanced clearance of A in brain by sustaining the plasmin proteolysis cascade. *Proc. Natl. Acad. Sci. USA* **2008**, *105*, 8754–8759. [CrossRef] [PubMed]
134. Pan, H.; Zhao, Y.; Zhai, Z.; Zheng, J.; Zhou, Y.; Zhai, Q.; Cao, X.; Tian, J.; Zhao, L. Role of plasminogen activator inhibitor-1 in the diagnosis and prognosis of patients with Parkinson's disease. *Exp. Ther. Med.* **2018**, *15*, 5517–5522. [CrossRef]
135. Reuland, C.J.; Church, F.C. Synergy between plasminogen activator inhibitor-1, α-synuclein, and neuroinflammation in Parkinson's disease. *Med. Hypotheses* **2020**, *138*, 109602. [CrossRef]
136. Yates, R.L.; Esiri, M.M.; Palace, J.; Jacobs, B.; Perera, R.; DeLuca, G.C. Fibrin(ogen) and neurodegeneration in the progressive multiple sclerosis cortex. *Ann. Neurol.* **2017**, *82*, 259–270. [CrossRef] [PubMed]
137. Angelucci, F.; Čechová, K.; Průša, R.; Hort, J. Amyloid beta soluble forms and plasminogen activation system in Alzheimer's disease: Consequences on extracellular maturation of brain-derived neurotrophic factor and therapeutic implications. *CNS Neurosci. Ther.* **2018**, *25*, 303–313. [CrossRef] [PubMed]
138. López-Otín, C.; Blasco, M.A.; Partridge, L.; Serrano, M.; Kroemer, G. The Hallmarks of Aging. *Cell* **2013**, *153*, 1194–1217. [CrossRef]
139. Tchkonia, T.; Zhu, Y.; Van Deursen, J.; Campisi, J.; Kirkland, J.L. Cellular senescence and the senescent secretory phenotype: Therapeutic opportunities. *J. Clin. Investig.* **2013**, *123*, 966–972. [CrossRef] [PubMed]
140. Özcan, S.; Alessio, N.; Acar, M.B.; Mert, E.; Omerli, F.; Peluso, G.; Galderisi, U. Unbiased analysis of senescence associated secretory phenotype (SASP) to identify common components following different genotoxic stresses. *Aging* **2016**, *8*, 1316–1329. [CrossRef]
141. Khan, S.S.; Shah, S.J.; Klyachko, E.; Baldridge, A.S.; Eren, M.; Place, A.T.; Aviv, A.; Puterman, E.; Lloyd-Jones, D.M.; Heiman, M.; et al. A null mutation in *SERPINE1* protects against biological aging in humans. *Sci. Adv.* **2017**, *3*, eaao1617. [CrossRef]
142. Vaughan, D.E.; Rai, R.; Khan, S.S.; Eren, M.; Ghosh, A.K. Plasminogen Activator Inhibitor-1 Is a Marker and a Mediator of Senescence. *Arterioscler. Thromb. Vasc. Biol.* **2017**, *37*, 1446–1452. [CrossRef] [PubMed]
143. Ersoy, C.; Kiyici, S.; Budak, F.; Oral, B.; Guclu, M.; Duran, C.; Selimoglu, H.; Erturk, E.; Tuncel, E.; Imamoglu, S. The effect of metformin treatment on VEGF and PAI-1 levels in obese type 2 diabetic patients. *Diabetes Res. Clin. Pract.* **2008**, *81*, 56–60. [CrossRef]
144. Brown, N.J.; Kumar, S.; Painter, C.A.; Vaughan, D.E. ACE Inhibition Versus Angiotensin Type 1 Receptor Antagonism. *Hypertension* **2002**, *40*, 859–865. [CrossRef]

145. Baluta, M.M.; Vintila, M.M. PAI-1 Inhibition—Another Therapeutic Option for Cardiovascular Protection. *Maedica* **2015**, *10*, 147–152.

146. Fortenberry, Y.M. Plasminogen activator inhibitor-1 inhibitors: A patent review (2006–present). *Expert Opin. Ther. Pat.* **2013**, *23*, 801–815. [CrossRef] [PubMed]

147. Rouch, A.; Vanucci-Bacqué, C.; Bedos-Belval, F.; Baltas, M. Small molecules inhibitors of plasminogen activator inhibitor-1—An overview. *Eur. J. Med. Chem.* **2015**, *92*, 619–636. [CrossRef]

148. D'Amico, S.; Martial, J.A.; Struman, I. A peptide mimicking the C-terminal part of the reactive center loop induces the transition to the latent form of plasminogen activator inhibitor type-1. *FEBS Lett.* **2012**, *586*, 686–692. [CrossRef] [PubMed]

149. Mathiasen, L.; Dupont, D.M.; Christensen, A.; Blouse, G.E.; Jensen, J.K.; Gils, A.; Declerck, P.J.; Wind, T.; Andreasen, P.A. A Peptide Accelerating the Conversion of Plasminogen Activator Inhibitor-1 to an Inactive Latent State. *Mol. Pharmacol.* **2008**, *74*, 641–653. [CrossRef]

150. Jensen, J.K.; Malmendal, A.; Schiøtt, B.; Skeldal, S.; Pedersen, K.E.; Celik, L.; Nielsen, N.C.; Andreasen, P.A.; Wind, T. Inhibition of plasminogen activator inhibitor-1 binding to endocytosis receptors of the low-density-lipoprotein receptor family by a peptide isolated from a phage display library. *Biochem. J.* **2006**, *399*, 387–396. [CrossRef] [PubMed]

151. Fjellström, O.; Deinum, J.; Sjögren, T.; Johansson, C.; Geschwindner, S.; Nerme, V.; Legnehed, A.; McPheat, J.; Olsson, K.; Bodin, C.; et al. Characterization of a Small Molecule Inhibitor of Plasminogen Activator Inhibitor Type 1 That Accelerates the Transition into the Latent Conformation. *J. Biol. Chem.* **2013**, *288*, 873–885. [CrossRef]

152. Lin, Z.; Jensen, J.K.; Hong, Z.; Shi, X.; Hu, L.; Andreasen, P.A.; Huang, M. Structural Insight into Inactivation of Plasminogen Activator Inhibitor-1 by a Small-Molecule Antagonist. *Chem. Biol.* **2013**, *20*, 253–261. [CrossRef] [PubMed]

153. Egelund, R.; Einholm, A.P.; Pedersen, K.E.; Nielsen, R.W.; Christensen, A.; Deinum, J.; Andreasen, P.A. A Regulatory Hydrophobic Area in the Flexible Joint Region of Plasminogen Activator Inhibitor-1, Defined with Fluorescent Activity-neutralizing Ligands. *J. Biol. Chem.* **2001**, *276*, 13077–13086. [CrossRef] [PubMed]

154. Sillen, M.; Miyata, T.; Vaughan, D.; Strelkov, S.; Declerck, P. Structural Insight into the Two-Step Mechanism of PAI-1 Inhibition by Small Molecule TM5484. *Int. J. Mol. Sci.* **2021**, *22*, 1482. [CrossRef] [PubMed]

155. Li, S.-H.; Reinke, A.A.; Sanders, K.L.; Emal, C.D.; Whisstock, J.C.; Stuckey, J.A.; Lawrence, D.A. Mechanistic characterization and crystal structure of a small molecule inactivator bound to plasminogen activator inhibitor-1. *Proc. Natl. Acad. Sci. USA* **2013**, *110*, E4941–E4949. [CrossRef] [PubMed]

International Journal of
Molecular Sciences

Review

The Contribution of the Urokinase Plasminogen Activator and the Urokinase Receptor to Pleural and Parenchymal Lung Injury and Repair: A Narrative Review

Torry A. Tucker and Steven Idell *

The Department of Cellular and Molecular Biology, The University of Texas Health Science Center at Tyler, Tyler, TX 75708, USA; torry.tucker@uthct.edu
* Correspondence: Steven.Idell@uthct.edu; Tel.: +1-903-877-7556; Fax: +1-903-877-7316

Abstract: Pleural and parenchymal lung injury have long been characterized by acute inflammation and pathologic tissue reorganization, when severe. Although transitional matrix deposition is a normal part of the injury response, unresolved fibrin deposition can lead to pleural loculation and scarification of affected areas. Within this review, we present a brief discussion of the fibrinolytic pathway, its components, and their contribution to injury progression. We review how local derangements of fibrinolysis, resulting from increased coagulation and reduced plasminogen activator activity, promote extravascular fibrin deposition. Further, we describe how pleural mesothelial cells contribute to lung scarring via the acquisition of a profibrotic phenotype. We also discuss soluble uPAR, a recently identified biomarker of pleural injury, and its diagnostic value in the grading of pleural effusions. Finally, we provide an in-depth discussion on the clinical importance of single-chain urokinase plasminogen activator (uPA) for the treatment of loculated pleural collections.

Keywords: urokinase plasminogen activator; urokinase plasminogen activator receptor; fibrinolysis; plasminogen activator inhibitor-1; acute lung injury and repair and pleural injury and pleural organization

Citation: Tucker, T.A.; Idell, S. The Contribution of the Urokinase Plasminogen Activator and the Urokinase Receptor to Pleural and Parenchymal Lung Injury and Repair: A Narrative Review. *Int. J. Mol. Sci.* **2021**, *22*, 1437. https://doi.org/10.3390/ijms22031437

Academic Editor: Hau C. Kwaan

Received: 7 December 2020
Accepted: 26 January 2021
Published: 1 February 2021

Publisher's Note: MDPI stays neutral with regard to jurisdictional claims in published maps and institutional affiliations.

1. Introduction

Derangements of pathways of fibrin turnover, including the fibrinolytic system, have long been associated with the pathogenesis of lung and pleural injury and repair [1,2]. In acute and chronic parenchymal lung injury, accelerated coagulation and a fibrinolytic defect favor the formation of extravascular fibrinous transition neomatrices, which can rapidly organize and scar [3–6]. These events occur in the setting of acute and chronic inflammation, and organization of the fibrinous neomatrices in the alveolar and interstitial lung compartments resemble those associated with microvascular leakage and organization occurring in the context of neoplasia and other inflammatory conditions [7–10]. In the pleural space, local inflammation likewise leads to increased microvascular permeability, intrapleural egress of coagulation substrates, and inhibitors of fibrinolysis and rapid organization with the formation of fibrinous intrapleural collections or loculae that are capable of impeding pleural drainage [1].

Plasminogen activators (PAs) regulate local fibrinolysis and include tissue plasminogen activator (tPA) and urokinase plasminogen activator (uPA). Both are capable of cleaving plasminogen to generate the active protease plasmin, which, in turn, degrades and remodels fibrin. While tPA is expressed in lung epithelial cells, fibroblasts, and mesothelial cells [2], it is mainly involved in intravascular fibrinolysis [11]. However, tPA has a greater affinity for fibrin than uPA, and its binding to fibrin increases its ability to cleave plasminogen [11]. uPA appears to play a greater role in pericellular proteolysis by virtue of its ability to bind to its receptor, uPA receptor (uPAR) [12]. Plasmin generated by tPA and urokinase plasminogen activator (uPA) reciprocally generates more active two chain forms of the PAs.

In both lung and pleural injury, plasminogen activator inhibitor-1 (PAI-1) appears to play a critical role in outcomes and the process of accelerated organization and scarification. PAI-1 is capable of inhibiting PAs, tPA, and uPA (two chain or active urokinase) within the injured lung and pleural space [1,2]. The activity of these PAs decreases in relation to the severity [13,14] of injury in both compartments [2,15,16]. PAI-1 thereby impedes local fibrinolysis and promotes extravascular fibrin deposition in the injured lung and pleural space [2]. Tissue factor expression by lung epithelial and/or mesothelial cells and lung fibroblasts triggers activation of the coagulation system in these injuries, while fibrinolytic activity is suppressed in resident cells mainly by increased expression of PAI-1 driven by proinflammatory mediators released into the local microenvironment [2,17–24]. These observations and those of other laboratories [13,25–47] support the concept that increments of local coagulation and concurrent decrements of fibrinolysis occur in inflammation and favor the formation and retention of fibrinous extracellular fibrin. Protracted collections of extravascular fibrin can organize with scarification in the injured lung or pleural space [1].

Our purpose in this narrative review is to provide the readers with an overview of the contributions of the urokinase plasminogen activator (uPA) and its receptor; uPAR, in the pathogenesis of lung and pleural injury with particular emphasis on pleural disease. New strategies and approaches to the development of new therapeutic interventions are also reviewed with a review of recent contributions to these fields, building on prior key foundational studies.

2. Similar Modes of Regulation Govern the Organization of the Fibrinous Transitional Neomatrix in the Settings of Lung and Pleural Injury

Fluid phase and cellular derangements in pathways of fibrin turnover promote fibrin deposition in lung and/or pleural injury. Based on studies of bronchoalveolar lavage and immunohistochemical analyses, uPA is readily detectable in normal lung lining fluids and is the major plasminogen activator represented there [4,5,21,48,49]. While studies of the very small amounts of normal pleural fluid have not, to our knowledge, been conducted, PA activity attributable to uPA and tPA occurs in pleural fluids of patients with congestive heart failure and is generally undetectable after the induction of pleural injury [19]. If unresolved, the inflammatory process suppresses local PA and fibrinolytic activities and perpetuates extravascular fibrin deposition that may rapidly organize over a few days (Figure 1) [3,50]. In the injured lung, early organization promotes accelerated fibroproliferation that can often be detected by lung imaging and can eventuate in lung restriction with long-term, severe morbidity, including respiratory compromise [51]. In the pleural compartment, early organization occurs in the setting of empyema, complicated parapneumonic pleural effusions, hemothoraces, or pleural malignancy and can likewise result in remodeling with loculation or pleurodesis, scarification, pulmonary restriction, and dyspnea [52].

The derangements in pathways of local fibrinolysis have led to the testing of interventions that target fibrin dissolution after lung or pleural injury. While anticoagulant strategies are of conceptual appeal, they have not gained traction for the treatment of patients with pleural or acute lung injury. This may be based largely on clinical trial testing that has failed to show the efficacy in severe sepsis, which is often associated with lung dysfunction [53]. Interestingly, there has been recent consideration of the targeting early pulmonary organization associated with COVID-19 [54]. Whether that approach will be of clinical benefit remains unclear but provocative [55]. Whether lung protection derives from anticoagulants otherwise administered to prevent thrombosis in COVID-19 patients is likewise unclear. On the other hand, in organizing pleural injury associated with loculation and failed drainage, intrapleural fibrinolytic therapy (IPFT) is commonly used and is particularly effective in pediatric patients, as recently reviewed [2].

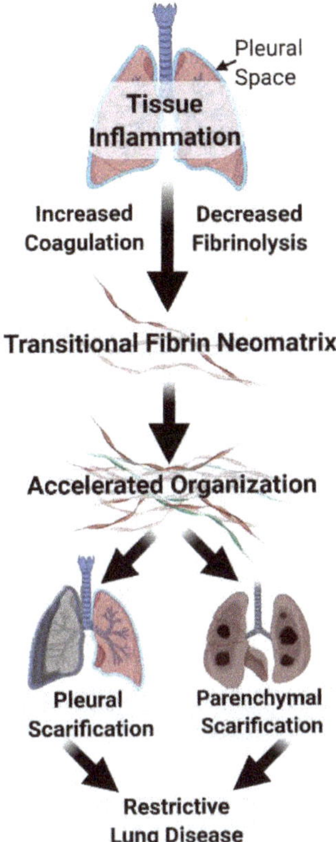

Figure 1. Aberrant fibrin turnover leads to accelerated scarification in the lung or pleural space. The accelerated organization encompasses the invasion of the fibrinous neomatrix by fibroblasts and myofibroblasts, which secrete collagen and initiate fibrotic repair. The neomatrix undergoes continuing remodeling with infiltration by inflammatory cells, including macrophages. Parenchymal lung scarification can be accelerated after acute lung injury or occur more slowly in interstitial lung disease. Pleural scarification can occur with organization, leading to sequestration of pockets of inflammatory pleural fluid; loculation, which can lead to pleural scarification. Pleural fibrosis may also occur after intrapleural bleeding or particulate exposures, such as that due to asbestos. Brown strands indicate fibrin. Green strands indicate collagen intercalated within the fibrinous neomatrix.

3. The Urokinase/Urokinase Receptor Interaction and Derangements Associated with Lung or Pleural Injury

In the injured lung, pleural space, and virtually all forms of tissue injury, the active two chain form of urokinase or the zymogen single chain urokinase binds to uPAR at the surface of resident cells bearing this Glycosylphosphatidylinositol (GPI)-linked receptor to regulate and localize pericellular proteolysis [56]. When two chain uPA cleaves plasminogen to generate plasmin, plasmin-mediated conversion of single to two chain uPA also occurs to increase the efficiency of pericellular fibrinolysis [57]. Signaling through uPAR can also occur through the binding of uPA or independent of the binding of uPA to uPAR to support the cellular invasion, migration, and cellular viability [56,58]. As uPAR lacks a cytoplasmic domain through which cellular signaling can occur, studies to identify other surface receptors that interact with uPAR are ongoing (Table 1).

Table 1. Components of the urokinase plasminogen activator (uPA)/uPA receptor (uPAR) system.

Single Chain uPA; scuPA	A Proenzyme that Can Bind uPAR and Localize PA Activity to the Cell Surface
Two-chain uPA; tcuPA	Conversion of scuPA to this two-chain form generates a much more active PA, can likewise bind uPAR and is mainly involved in pericellular proteolysis.
Tissue type plasminogen activator; tPA	Has a greater affinity for fibrin than tcuPA, relatively more involved in intravascular fibrinolysis, and binding to fibrin increases its PA activity.
Plasminogen activator inhibitor 1; PAI-1	Main PA inhibitor in extravascular fluids in lung and pleural injury, where it can exist in active, cleaved, and inactivated or latent forms. Active PAI-1 can inhibit both tcuPA and tPA.
uPA receptor; uPAR	Multidomain surface glycoprotein responsible for cellular localization of uPA and can be cleaved by uPA. Capable of mediating signaling through interactions with other surface receptors.

Cellular and extravascular derangements of the uPA/uPAR system occur in the setting of acute organizing lung injury. In the injured lung, the profile of alveolar lining fluids assumes more procoagulant and less fibrinolytic potential with increased levels of PAI-1, thereby favoring alveolar fibrin deposition. The alveolar epithelium contributes substantively to these derangements and is capable of regulating its own expression of uPA, uPAR, and PAI-1, which involve unique posttranscriptional mechanisms [59]. The posttranscriptional regulatory mechanisms involve the participation of p53 to, in turn, control the viability of the lung epithelium [59]. The viability of lung epithelial cells is increased when uPA and uPAR are relatively increased. Conversely, apoptosis of these cells is favored by relatively increased expression of PAI-1 with reciprocally decreased uPA and uPAR. In lung fibroblasts harvested from patients with idiopathic pulmonary fibrosis, uPAR expression is likewise increased compared with fibroblasts harvested from the lungs of individuals without lung disease [60]. Regulation of uPAR by lung fibroblasts is, in part, controlled via posttranscriptional regulation, as are pleural mesothelial cells [61].

Preclinical information further supports the critical involvement of the uPA/uPAR system in the pathogenesis of organizing lung injury. Increased uPA expression has been found to mitigate accelerated, fibrosing lung injury induced by bleomycin [41,44,62]. In a related vein, overexpression of PAI-1 aggravated bleomycin-induced lung injury, while PAI-1 deficiency was salutary [63]. Interestingly, uPA or uPAR deficiency did not alter lung hydroxyproline levels in bleomycin-treated mice, but areas of hemorrhage seen in wild-type mice were abridged [42]. On the other hand, uPAR deficiency has been reported to attenuate hypoxia-associated lung injury and reduce lung inflammation, while it likewise impairs inflammation but limits containment of pneumococcal pneumonia with worsened outcomes in mice [64,65].

A similar situation exists in the context of pleural injury, where greatly increased pleural fluid PAI-1 and its activity effectively limits local fibrinolysis and predisposes to a rapid, intrapleural organization that can occur over days [16,66–68]. In normalcy, the pleural compartment is actually a potential space that expands to form a defined anatomic compartment occupied by inflammatory pleural fluids in pleural infections and other organizing processes [2]. Pleural mesothelial cells express uPA, uPAR, and tPA, which may contribute to fibrinolysis at pleural surfaces. Fibrinolysis is limited by overexpression of PAI-1 in pleural fluids that characterize virtually all forms of pleural injury (Figure 2) [2,23,69]. These pleural collections may undergo rapid organization to loculate, impair pleural drainage and thereby increase morbidity in pleural infection, hemothoraces, or neoplasia. In a murine model of carbon black/bleomycin-induced pleural injury, pleural neomatrix organization was reduced in PAI-1 deficiency but significantly increased by PAI-1 overexpression [70]. Overexpression of PAI-1 has been shown to worsen tetracycline-induced pleural injury [16], and PAI-1-targeted IPFT has been shown to be beneficial [71]. In this model and an empyema model in rabbits, intrapleural administration of PAs has been shown to clear fibrinous pleural collections [2,50,72–75].

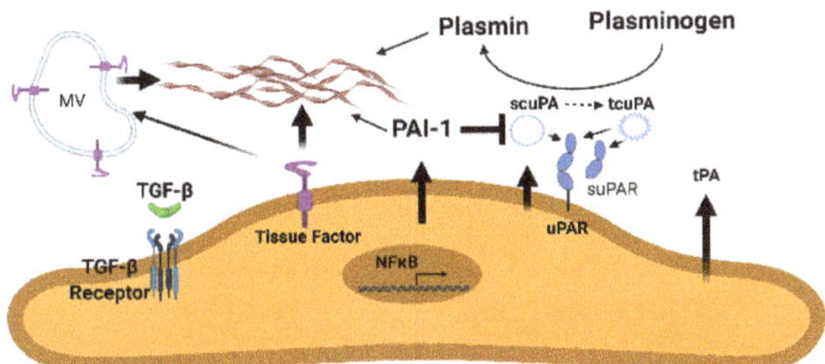

Figure 2. The mesothelium and disordered fibrin turnover. Increased TGF-β expression in response to injury increases plasminogen activator inhibitor-1 (PAI-1) expression. In the aggregate, PAI-1 expression and activity are overexpressed in exudative pleural effusions. This increase in PAI-1 reduces plasminogen conversion to active plasmin by single chain urokinase plasminogen activator (scuPA), two chain urokinase plasminogen activator (tcuPA), and tissue plasminogen activator (tPA). These effects decrease local expression of fibrinolytic activity, which decreases fibrin degradation, leading to aberrant extravascular fibrin deposition. scuPA or the high molecular weight form of tcuPA can bind uPA receptor (uPAR) and be localized to the cell surface or be present unbound in pleural fluids. NFκB is NF kappa B, a signaling mediator that is in particular implicated in the transcriptional regulation of PAI-1 expression by pleural mesothelial cells. MV is microvesicles, which have been shown to contain TF. Brown strands indicate fibrin.

The use of IPFT in clinical practice is likewise predicated on the concept that increments of pleural fluid PA activity can increase intrapleural plasmin generation and fibrinolysis sufficient to improve pleural drainage and clinical outcomes [1,2]. IPFT is generally accepted as effective in pediatric patients and can improve outcomes in adult patients, although adult patient dosing is empiric, and no PA is currently approved for this indication [2,76–78].

4. The uPA/uPAR System and the Contribution of Mesenchymal Differentiation of Pleural Mesothelial Cells to Pleural Thickening and Scarification

uPAR has been implicated in the pathogenesis of pleural injury and builds upon and extends studies linking uPAR to pleural neoplasia. Interestingly, uPAR has been shown to be involved in the pathogenesis of neoplasia and a target for the development of new therapeutics for several different forms of cancer [79]. In the setting of neoplasia, uPAR expression may be primarily upregulated in the tumor cells themselves or in endothelial or in infiltrating myeloid cells. In malignant pleural mesothelioma cells, uPAR expression correlated with tumor burden, aggressiveness and mortality in mice [58]. Cellular proliferation, migration and invasion were increased in REN malignant mesothelioma cells that have increased uPAR expression. uPAR silencing in these cells decreased cellular indices of aggressiveness, while exposure to uPA and bovine fetal serum enhanced these effects. Conversely, increasing uPAR expression in a less aggressive mesothelioma line significantly increased tumor virulence in vitro and in vivo. The responses suggest the possibility that uPAR-targeted therapeutics now in development may be of value for the treatment of pleural malignant mesothelioma.

Extending this work, uPAR expression by pleural mesothelial cells has been linked to the regulation of pleural remodeling. uPAR internalization was reduced by treatment of human pleural mesothelial cells with TNF-α or IL-1β, mediators which have been implicated in inflammation and the pathogenesis of pleural injury [1,2]. This effect was attributable to decreased expression of the lipoprotein receptor related protein-1; LRP-1, which stabilized uPAR at the cell surface. This stabilization augmented uPA-mediated proteolytic activity at the cell surface, as well as cellular migration [80]. In a murine model of pleural organization induced by carbon black and bleomycin, it was found that PAI-1 deficiency increased pleural thickness and lung restriction likely as a consequence of sustained mesomesenchymal transtion (MesoMT) [70]. The effects involved crosstalk with coagulation proteases and plasmin generation was likewise augmented in pleural lavage of PAI-1-/- mice. Increased plasmin activity was likewise detected in the pleural lavage of mice with empyema induced by intrapleural administration of *Streptoccus pneumoniae* [81], indicating that pleural fibrinolysis is enhanced in pleural injury induced by local instillation of noxious chemicals or bacterial infection.

The expansion of the pool of subpleural myofibroblasts contributes to the pleural thickening and neomatrix deposition that characterizes pleural injury and predisposes to the development of pleural fibrosis [69,82,83]. Pleural mesothelial cells have been shown to contribute to that response and undergo a process of mesenchymal phenotypic change called mesomesenchymal transition (MesoMT, Figure 3). The uPA–uPAR system plays a major role in the regulation of this process [1]. In studies to evaluate the impact of coagulation and fibrinolytic proteases on induction of MesoMT, uPA, as well as plasmin, factor Xa and thrombin were found to induce this phenotypic change in human pleural mesothelial cells [70]. Plasmin, thrombin and TGF-b commonly induce MesoMT through phosphatidyl inositol-kinase/AKT/NF kappa (κ) B signaling [84]. The therapeutic targeting of GSK-3β likewise blocked induction of MesoMT and the progression of pleural fibrosis in a novel model of pleural injury [85]. At present, the role of uPA and uPAR in the induction of MesoMT remains to be further elucidated. The effects of proinflammatory cytokines on uPAR stabilization at the surfaces of pleural mesothelial cells via regulation of LRP1 [80], uPA mediated induction of MesoMT [70] and uPA-mediated induction of collagen-1 in these cells suggest a critical role for the uPA/uPAR system in MesoMT and pleural remodeling. Future studies will be needed to define the role of uPAR in the pathogenesis of pleural organization and scarification.

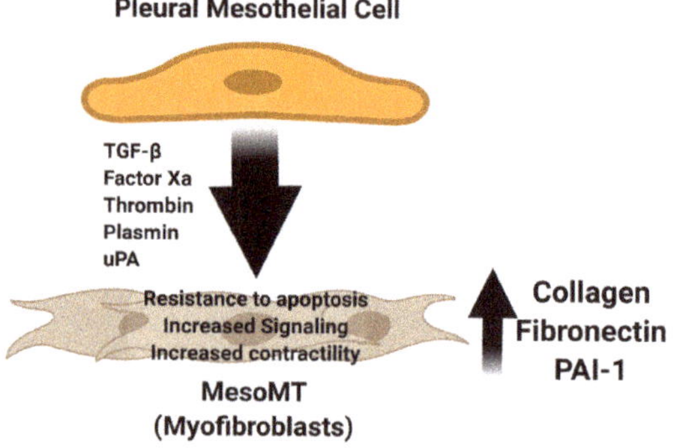

Figure 3. Induction and consequences of MesoMT. Upon stimulation by TGF-β, Factor Xa, thrombin, uPA, or plasmin, pleural mesothelial cells assume a mesenchymal phenotype by undergoing MesoMT. The population of cells undergoing MesoMT expands in the setting of organizing pleural inflammation. Cells undergoing MesoMT elongate compared to unstimulated pleural mesothelial cells and assume the functionality of myofibroblasts and increase expression of neomatrix components including collagen, fibronectin, and PAI-1.

5. The Role of Soluble uPAR in Pleural Injury: Biomarker, Effector, or Both?

The structure of uPAR incorporates three domains: DI-III. High affinity binding to its ligand, uPA, primarily involves domain I, while requiring the structural integrity of all three domains [86]. Although a splice variant of the PLAUR gene can generate a soluble uPAR construct [87], soluble uPAR is also generated by cleavage of cell-surface uPAR and occurs in serum and other biologic fluids, including pleural effusions [88,89]. suPAR can be generated by cleavage of the GPI-anchor of uPAR at the cell surface by enzymes such as phospholipase C [90]. Cleavage of a protease sensitive linker region between domains I and domains II-III can also occur via uPA or plasmin among other proteases. Thus, soluble suPAR fragments include domains I-III, I or II-III, as previously reviewed [88]. suPAR, incorporating all domains, retains the ability to bind uPA, via its growth factor domain [91], supporting the concept that this form of suPAR could be a scavenger capable of binding uPA and thereby altering effects of the interaction of uPA with cell surface uPAR such as migration or invasion.

Interestingly, Higazi and colleagues reported that binding of the proenzyme single chain uPA to suPAR increased its catalytic activity [92]. The authors speculated that scuPA exists in a latent and more active state and that binding of scuPA to suPAR favored the more active conformation, which was more susceptible to inhibition by PAI-1. On the other hand, these observations were disputed by Behrendt and colleagues, who found that the binding of scuPA to suPAR did not accelerate the PA activity of scuPA and in fact had an inhibitory effect on the activation of scuPA to two chain more active uPA by plasmin [93]. The basis for the disparities between these studies remain unclear but are possibly technical. Liberation of domain I from suPAR exposes a chemotactic sequence on the remaining domains (II-III), most of which may be shed from neutrophil membranes, which is a chemotaxin for monocytic cells [94]. Conversely, suPAR domain I has poor affinity for scuPA or uPA and is thereby unable to serve as a scavenger binding either protein [88].

In a recent study, pleural fluid suPAR levels were found to correlate with the requirement for invasive management in parapneumonic pleural effusions [89]. This work extends prior work showing that elevated levels of suPAR occur in a range of biologic fluids in infectious, autoimmune and neoplastic diseases, as previously reviewed [88]. Given the lack of reliable, validated predictors of the need for use of IPFT or surgery for relief of complicated, organizing pleural infections that loculate, Arnold and colleagues sought to determine if pleural fluid or serum suPAR levels predict the need for such management and how suPAR determinations compared with traditional biomarkers such as pleural fluid pH, glucose and lactate dehydrogenase levels. Pleural fluid and serum suPAR levels were determined in 93 subjects with parapneumonic pleural effusions and 47 controls that had either benign transudative or malignant causes of their pleural effusions. A commercially available ELISA assay was used and was able to detect intact suPAR or the domain II-III fragment. The main findings of this important study were that pleural fluid suPAR levels were greater in patients with loculated parapneumonic pleural effusions. Pleural fluid suPAR levels were also superior to pleural fluid pH, glucose, and lactate dehydrogenase (LDH) in combination in terms of predicting the need for IPFT or surgical intervention. Serum suPAR correlated with levels of C-reactive protein, a biomarker of inflammation often used to assess clinical trends in patients with parapneumonic pleural effusions [68]. Pleural fluid suPAR was significantly increased in patients with parapneumonic pleural effusions that were loculated. At a cutoff level of 35 ng/mL, pleural fluid suPAR demonstrated a 100 percent sensitivity, 91 percent specificity for predicting pleural loculation in patients with parapneumonic pleural effusions. Positive and negative likelihood ratios were likewise favorable, and pleural fluid suPAR better predicted loculation than pH. In nine patients with parapneumonic pleural effusions that were initially free-flowing and subsequently loculated, baseline pleural fluid suPAR was elevated to levels equivalent to those found in loculated pleural effusions. Pleural fluid suPAR at the same cutoff was the most accurate biomarker predicting insertion of a chest tube. Lastly, pleural fluid suPAR was superior to the conventional biomarkers in predicting the use of IPFT or surgical intervention and was the only significant baseline parameter that did so. Similar trends were seen in the smaller number of patients with malignancy, in that pleural fluid suPAR was significantly increased in loculated malignant effusions and was non significantly elevated in patients with delayed loculation. The encouraging results suggest that pleural fluid suPAR could improve clinical decision making for the management of organizing pleural infection. The authors properly point out that while pleural fluid suPAR appears to address an unmet clinical need, validation would require a future large, multicenter clinical trial of suPAR-directed clinical management versus current standard management. Whether pleural fluid suPAR can add a precision medicine dimension to such management decisions remains to be proven, but this work underscores the relevance of derangements of the uPA/uPAR system to the field.

6. The Use of scuPA for Treatment of Pleural Loculation

For over seventy years, IPFT has been used to treat pleural loculation with failed drainage [2,78]. To date, several agents have been used, now mainly including tPA with or without DNase and urokinase. Many centers have adopted the use of tPA/DNase based on the efficacy demonstrated in adult patients in the MIST-2 clinical trial [95]. Others use tPA alone or two chain urokinase where it is available. Unfortunately, the dosing, administration schedules and agents used for IPFT are all empiric, with no agent currently approved for the indication of treatment of pleural loculation and failed drainage [77]. There are relatively small reports suggesting that two chain uPA (uPA)-based IPFT may be of advantage. uPA-based IPFT was found to be effective in a small randomized, double blinded study of patient with either complicated parapneumonic pleural effusions or empyema [96]. uPA IPFT was also found to be associated with greater efficacy in patients with complicated parapneumonic pleural effusions and less bleeding complications than tPA [97]. In a third small study, uPA was as effective as tPA/DNase for patients requiring IPFT for empyema or parapneumonic pleural effusions but bleeding complications were reduced [98].

Over twenty years ago, we initiated studies in which scuPA, a relatively PAI-1-resistant PA, was used to reverse pleural organization and adhesions in rabbits with tetracycline-induced pleural loculation [50]. Given the similarities of the fibrinolytic system between rabbits and humans, the model was amenable to the testing of human plasminogen activators. We found that scuPA effectively alleviated fibrinous intrapleural collections and adhesions in the model and that single dose treatment was well-tolerated. Single dose administration was as effective as multiple dose scuPA IPFT and generated durable intrapleural PA activity supporting local fibrinolysis [50,72]. In a comparative study, it was found that scuPA was more effective at resolving intrapleural adhesions than any clinically prorated dose of low molecular weight two chain uPA and demonstrated a trend toward better efficacy than tPA [73]. We found that scuPA was processed intrapleurally with formation of bioactive uPA/a2macroglobilin complexes, which appear to contribute to slow release of PA within pleural fluids [99]. In this study, PAI-1 resistant uPA activity was increased in rabbits with tetracycline-induced pleural injury and was related to removal of fibrinous intrapleural collections. PAI-1 resistant enzymatic activity was not found in animals treated with intrapleural low molecular weight uPA or tPA. We identified an equilibrium between an active and relatively inactive form of scuPA, which limited its inactivation by PAI-1 and favored formation of the bioactive uPA/a2macroglobilin complexes.

Based upon these findings and other studies conducted by our group [2], scuPA was manufactured with support by the National Heart Lung and Blood Institute SMARTT (Science Moving towArds Research Translation and Therapy) program, providing sufficient amounts of material for formal toxicology studies needed to enable clinical trial testing. Good Manufacturing Practice (GMP) grade scuPA was also made available for use in clinical trial testing.

7. New Strategies to Limit Pleural Organization: Clinical Trial Testing of scuPA for Treatment of Empyema or Complicated Parapneumonic Pleural Effusions

Phase 1 clinical trial testing was next conducted in a multicenter clinical trial performed in Australia Trial Registration: ANZCT ID: ACTRN12616001442493 [68]. In this first-in-kind trial in the field, the safety of scuPA IPFT at daily doses over three days of 50.000–800,000 IU scuPA was tested in patients with complicated parapneumonic pleural effusions or empyema and failed drainage. In all, 14 patients were studied. scuPA was well-tolerated with no bleeding, surgical referrals, or treatment-associated adverse events. In the pleural fluids, scuPA rapidly saturated PAI-1 activity, increased pleural fluid PA and fibrinolytic activities, and generated increased pleural fluid levels of D-dimers. As expected, complexes of uPA/$a_2$macroglobilin were generated. No systemic fibrinolysis was detectable, and D-dimer levels remained unchanged from those at baseline in the patients. While this was a safety trial, hints of efficacy were demonstrated in that all but one patient had decreased pleural fluid opacification. At the 800,000 IU dose, pleural sepsis was clinically relieved in both treated patients, and the same effects were observed in two others receiving lower doses of scuPA IPFT.

The favorable Phase 1 trial findings formed the predicate to proceed with Phase 2 efficacy testing, which is now being conducted at multiple sites in the US: ClinicalTrials.gov Identifier: NCT04159831 a phase 2, randomized, placebo-controlled, double-blind, dose-ranging study evaluating LTI-01 (single chain urokinase plasminogen activator, scuPA) in patients with infected, non-draining pleural effusions. The trial design incorporates three doses; 400,000 IU, 800,000 IU, 1,200,000 IU scuPA IPFT, or saline-vehicle placebo, which will be given daily for up to three days. The estimated enrollment is a total of 160 patients or 40 per group. The primary outcome measure is treatment failure due to ongoing pleural sepsis or impaired drainage resulting in surgical referral, and the secondary outcome measure is change in pleural opacity by chest CT scanning at day 4, 1 day after the last IPFT treatment or at the time of treatment failure. The effects of scuPA IPFT on derangements of the fibrinolytic system in pleural fluids will be assessed using the same methods deployed in the Phase 1 trial [68]. This trial is designed to confirm the efficacy of scuPA IPFT in patients with infection-related loculation and failed drainage and to identify an optimal dose of scuPA that can either be used for FDA approval or a follow-up Phase 3 study, as may be required.

8. Conclusions

The uPA/uPAR system plays an important role in the pathogenesis of pleural organization through the regulation of local proteolysis, cellular differentiation, and pleural remodeling. In a translational vein, suPAR appears to be a promising new candidate to predict outcomes of pleural injury and to help clinicians to personalize decision making and inform better management of organizing pleural injury. scuPA IPFT is an equally promising translational candidate that is currently in clinical trial evaluation that could offer the first evidence-based, potentially more effective new therapy for patients with nondraining loculated pleural effusions.

Author Contributions: S.I. wrote the initial draft and edited the figures. T.A.T. critically reviewed and added to the draft and prepared the figures. Both authors edited and approved of the final version of the manuscript. All authors have read and agreed to the published version of the manuscript.

Funding: Sources of Support: R01HL130133 and HL14285301 (TAT PI, MPI), NIH R33 HL154103 (SI, PI, Contact, MPI), R01HL130402-01A1 (SI, MPI), U54 TR002804-01 CTSA (SI, Site PI), HL14285301 (SI Co-PI, MPI) U54ES027698-01 (Site PI, subcontract), R01HL130133-01A1 (Co-I), R01HL133067-01 (Co-I), R01HL130133-01 (SI Co-I), TLL Temple Endowed Chair in Idiopathic Pulmonary Fibrosis, Texas Lung Injury Institute, NIH UO-1 HL 121841-01A1 (SI, Contact PI, MPI), NIH SMARTT (Science Moving towArds Research Translation and Therapy) Contract No. HHSN268201100014C (SI, PI).

Institutional Review Board Statement: Not Applicable.

Informed Consent Statement: Not Applicable.

Data Availability Statement: Not Applicable.

Conflicts of Interest: S.I. is founder of Lung Therapeutics Incorporated and has an equity position in the company, which is commercializing single chain urokinase for the treatment of pleural loculaion. Lung Therapeutics Inc. is sponsor of the phase 1 and 2 clinical trials of this agent for this clinical problem. S.I. is also a board member of the company and serves as a compensated Chief Scientific Officer of the company.He has an institutional Conflict of Interest Management Plan at The University of Texas Health Science Center at Tyler to address this conflict of interest.

References

1. Tucker, T.; Idell, S. Plasminogen-Plasmin System in the Pathogenesis and Treatment of Lung and Pleural Injury. *Semin. Thromb. Hemost.* **2013**, *39*, 373–381. [CrossRef] [PubMed]
2. Komissarov, A.A.; Rahman, N.M.; Lee, Y.C.G.; Florova, G.; Shetty, S.; Idell, R.; Ikebe, M.; Das, K.; Tucker, T.A.; Idell, S. Fibrin turnover and pleural organization: Bench to bedside. *Am. J. Physiol. Cell. Mol. Physiol.* **2018**, *314*, L757–L768. [CrossRef] [PubMed]
3. Idell, S.; Gonzalez, K.K.; MacArthur, C.K.; Gillies, C.; Walsh, P.N.; McLarty, J.; Thrall, R.S. Bronchoalveolar Lavage Procoagulant Activity in Bleomycin-Induced Lung Injury in Marmosets: Characterization and Relationship to Fibrin Deposition and Fibrosis. *Am. Rev. Respir. Dis.* **1987**, *136*, 124–133. [CrossRef] [PubMed]
4. Idell, S.; James, K.K.; Gillies, C.; Fair, D.S.; Thrall, R.S. Abnormalities of pathways of fibrin turnover in lung lavage of rats with oleic acid and bleomycin-induced lung injury support alveolar fibrin deposition. *Am. J. Pathol.* **1989**, *135*, 387–399. [PubMed]
5. Idell, S.; James, K.K.; Levin, E.G.; Schwartz, B.S.; Manchanda, N.; Maunder, R.J.; Martin, T.R.; McLarty, J.; Fair, D.S. Local abnormalities in coagulation and fibrinolytic pathways predispose to alveolar fibrin deposition in the adult respiratory distress syndrome. *J. Clin. Investig.* **1989**, *84*, 695–705. [CrossRef]
6. Idell, S.; Peters, J.; James, K.K.; Fair, D.S.; Coalson, J.J. Local abnormalities of coagulation and fibrinolytic pathways that promote alveolar fibrin deposition in the lungs of baboons with diffuse alveolar damage. *J. Clin. Investig.* **1989**, *84*, 181–193. [CrossRef] [PubMed]
7. Dvorak, H.F. Tumors: Wounds That Do Not Heal. Similarities between Tumor Stroma Generation and Wound Healing. *N. Engl. J. Med.* **1986**, *315*, 1650–1659.
8. Dvorak, H.F.; Senger, D.R.; Dvorak, A.M. Fibrin as a component of the tumor stroma: Origins and biological significance. *Cancer Metastasis Rev.* **1983**, *2*, 41–73. [CrossRef]
9. Dvorak, H.F.; Senger, D.R.; Dvorak, A.M.; Harvey, V.S.; McDonagh, J. Regulation of extravascular coagulation by microvascular permeability. *Science* **1985**, *227*, 1059–1061. [CrossRef]
10. Brown, L.F.; Dvorak, A.M.; Dvorak, H.F. Leaky Vessels, Fibrin Deposition, and Fibrosis: A Sequence of Events Common to Solid Tumrs and Many Other Types of Disease. *Am. Rev. Respir. Dis.* **1989**, *140*, 1104–1107. [CrossRef]
11. Cesarman-Maus, G.; Hajjar, K.A. Molecular mechanisms of fibrinolysis. *Br. J. Haematol.* **2005**, *129*, 307–321. [CrossRef] [PubMed]
12. Quax, P.H.; Grimbergen, J.M.; Lansink, M.; Bakker, A.H.; Blatter, M.C.; Belin, D.; van Hinsbergh, V.W.; Verheijen, J.H. Binding of Human Urokinase-Type Plasminogen Activator to Its Receptor: Residues Involved in Species Specificity and Binding. *Arterioscler. Thromb. Vasc. Biol.* **1998**, *18*, 693–701. [CrossRef] [PubMed]
13. Sitrin, R.G. Plasminogen Activation in the Injured Lung: Pulmonology Does Not Recapitulate Hematology. *Am. J. Respir. Cell Mol. Biol.* **1992**, *6*, 131–132. [CrossRef] [PubMed]
14. Gyetko, M.R.; Shollenberger, S.B.; Sitrin, R.G. Urokinase expression in mononuclear phagocytes: Cytokine-specific modulation by interferon-gamma and tumor necrosis factor-alpha. *J. Leukoc. Biol.* **1992**, *51*, 256–263. [CrossRef] [PubMed]
15. Prabhakaran, P.; Ware, L.B.; White, K.E.; Cross, M.T.; Matthay, M.A.; Olman, M.A. Elevated levels of plasminogen activator inhibitor-1 in pulmonary edema fluid are associated with mortality in acute lung injury. *Am. J. Physiol. Cell. Mol. Physiol.* **2003**, *285*, L20–L28. [CrossRef] [PubMed]
16. Karandashova, S.; Florova, G.; Azghani, A.O.; Komissarov, A.A.; Koenig, K.; Tucker, T.A.; Allen, T.C.; Stewart, K.; Tvinnereim, A.; Idell, S. Intrapleural Adenoviral Delivery of Human Plasminogen Activator Inhibitor–1 Exacerbates Tetracycline-Induced Pleural Injury in Rabbits. *Am. J. Respir. Cell Mol. Biol.* **2013**, *48*, 44–52. [CrossRef]
17. Bhandary, Y.P.; Shetty, S.K.; Marudamuthu, A.S.; Gyetko, M.R.; Idell, S.; Gharaee-Kermani, M.; Shetty, R.S.; Starcher, B.C.; Shetty, S. Regulation of alveolar epithelial cell apoptosis and pulmonary fibrosis by coordinate expression of components of the fibrinolytic system. *Am. J. Physiol. Cell. Mol. Physiol.* **2012**, *302*, L463–L473. [CrossRef]
18. Bhandary, Y.P.; Shetty, S.K.; Marudamuthu, A.S.; Ji, H.-L.; Neuenschwander, P.F.; Boggaram, V.; Morris, G.F.; Fu, J.; Idell, S.; Shetty, S. Regulation of Lung Injury and Fibrosis by p53-Mediated Changes in Urokinase and Plasminogen Activator Inhibitor-1. *Am. J. Pathol.* **2013**, *183*, 131–143. [CrossRef]
19. Idell, S.; Girard, W.; Koenig, K.B.; McLarty, J.; Fair, D.S. Abnormalities of Pathways of Fibrin Turnover in the Human Pleural Space. *Am. Rev. Respir. Dis.* **1991**, *144*, 187–194. [CrossRef]
20. Idell, S.; Kumar, A.; Zwieb, C.; Holiday, D.; Koenig, K.B.; Johnson, A.R. Effects of TGF-beta and TNF-alpha on procoagulant and fibrinolytic pathways of human tracheal epithelial cells. *Am. J. Physiol. Content* **1994**, *267*, 693. [CrossRef]
21. Idell, S.; Koenig, K.B.; Fair, D.S.; Martin, T.R.; McLarty, J.; Maunder, R.J. Serial abnormalities of fibrin turnover in evolving adult respiratory distress syndrome. *Am. J. Physiol. Cell. Mol. Physiol.* **1991**, *261*, L240–L248. [CrossRef] [PubMed]

22. Idell, S.; Gonzalez, K.; Bradford, H.; MacArthur, C.K.; Fein, A.M.; Maunder, R.J.; Garcia, J.G.; Griffith, D.E.; Weiland, J.; Martin, T.R. Procoagulant Activity in Bronchoalveolar Lavage in the Adult Respiratory Distress Syndrome. Contribution of Tissue Factor Associated with Factor Vii. *Am. Rev. Respir. Dis.* **1987**, *136*, 1466–1474. [CrossRef] [PubMed]

23. Idell, S.; Zwieb, C.; Kumar, A.; Koenig, K.B.; Johnson, A.R. Pathways of Fibrin Turnover of Human Pleural Mesothelial CellsIn Vitro. *Am. J. Respir. Cell Mol. Biol.* **1992**, *7*, 414–426. [CrossRef] [PubMed]

24. Idell, S.; Zwieb, C.; Boggaram, J.; Holiday, D.; Johnson, A.R.; Raghu, G. Mechanisms of fibrin formation and lysis by human lung fibroblasts: Influence of TGF-beta and TNF-alpha. *Am. J. Physiol. Content* **1992**, *263*, L487–L494. [CrossRef]

25. Bertozzi, P.; Astedt, B.; Zenzius, L.; Lynch, K.; LeMaire, F.; Zapol, W.; Chapman, H.A. Depressed Bronchoalveolar Urokinase Activity in Patients with Adult Respiratory Distress Syndrome. *N. Engl. J. Med.* **1990**, *322*, 890–897. [CrossRef]

26. Chapman, H.A. Disorders of lung matrix remodeling. *J. Clin. Investig.* **2004**, *113*, 148–157. [CrossRef]

27. Chapman, H.A. Plasminogen activators, integrins, and the coordinated regulation of cell adhesion and migration. *Curr. Opin. Cell Biol.* **1997**, *9*, 714–724. [CrossRef]

28. Chapman, H.A.; Stahl, M.; Allen, C.L.; Yee, R.; Fair, D.S. Regulation of the Procoagulant Activity within the Bronchoalveolar Compartment of Normal Human Lung. *Am. Rev. Respir. Dis.* **1988**, *137*, 1417–1425. [CrossRef]

29. Gross, T.J.; Simon, R.H.; Kelly, C.J.; Sitrin, R.G. Rat alveolar epithelial cells concomitantly express plasminogen activator inhibitor-1 and urokinase. *Am. J. Physiol. Cell. Mol. Physiol.* **1991**, *260*, L286–L295. [CrossRef]

30. Gross, T.J.; Simon, R.H.; Sitrin, R.G. Expression of Urokinase-type Plasminogen Activator by Rat Pulmonary Alveolar Epithelial Cells. *Am. J. Respir. Cell Mol. Biol.* **1990**, *3*, 449–456. [CrossRef]

31. Gross, T.J.; Simon, R.H.; Sitrin, R.G. Tissue Factor Procoagulant Expression by Rat Alveolar Epithelial Cells. *Am. J. Respir. Cell Mol. Biol.* **1992**, *6*, 397–403. [CrossRef] [PubMed]

32. Sitrin, R.G.; Pan, P.; Blackwood, R.A.; Huang, J.; Petty, H.R. Cutting edge: Evidence for a signaling partnership between urokinase receptors (CD87) and L-selectin (CD62L) in human polymorphonuclear neutrophils. *J. Immunol.* **2001**, *166*, 4822–4825. [CrossRef] [PubMed]

33. Sitrin, R.G.; Pan, P.M.; Harper, H.A.; Todd, R.F.; Harsh, D.M.; Blackwood, R.A. Clustering of Urokinase Receptors (uPAR.; CD87) Induces Proinflammatory Signaling in Human Polymorphonuclear Neutrophils. *J. Immunol.* **2000**, *165*, 3341–3349. [CrossRef] [PubMed]

34. Sitrin, R.G.; Todd, R.F.; Albrecht, E.; Gyetko, M.R. The urokinase receptor (CD87) facilitates CD11b/CD18-mediated adhesion of human monocytes. *J. Clin. Investig.* **1996**, *97*, 1942–1951. [CrossRef] [PubMed]

35. Sitrin, R.G.; Todd, R.F.; Mizukami, I.F.; Gross, T.J.; Shollenberger, S.B.; Gyetko, M.R. Cytokine-specific regulation of urokinase receptor (CD87) expression by U937 mononuclear phagocytes. *Blood* **1994**, *84*, 1268–1275.

36. Bastarache, J.A. The complex role of fibrin in acute lung injury. *Am. J. Physiol. Cell. Mol. Physiol.* **2009**, *296*, L275–L276. [CrossRef]

37. Bastarache, J.A.; Wang, L.; Geiser, T.; Wang, Z.; Albertine, K.H.; Matthay, M.A.; Ware, L.B. The alveolar epithelium can initiate the extrinsic coagulation cascade through expression of tissue factor. *Thorax* **2007**, *62*, 608–616. [CrossRef]

38. Van der Poll, T.; Levi, M.; Nick, J.A.; Abraham, E. Activated Protein C Inhibits Local Coagulation after Intrapulmonary Delivery of Endotoxin in Humans. *Am. J. Respir. Crit. Care Med.* **2005**, *171*, 1125–1128. [CrossRef]

39. Bdeir, K.; Murciano, J.C.; Tomaszewski, J.; Koniaris, L.; Martinez, J.; Cines, D.B.; Muzykantov, V.R.; Higazi, A.A. Urokinase Mediates Fibrinolysis in the Pulmonary Microvasculature. *Blood* **2000**, *96*, 1820–1826. [CrossRef]

40. Higazi, A.A.; Bdeir, K.; Hiss, E.; Arad, S.; Kuo, A.; Barghouti, I.; Cines, D.B. Lysis of Plasma Clots by Urokinase-Soluble Urokinase Receptor Complexes. *Blood* **1998**, *92*, 2075–2083. [CrossRef]

41. Sisson, T.H.; Hanson, K.E.; Subbotina, N.; Patwardhan, A.; Hattori, N.; Simon, R.H. Inducible lung-specific urokinase expression reduces fibrosis and mortality after lung injury in mice. *Am. J. Physiol. Cell. Mol. Physiol.* **2002**, *283*, L1023–L1032. [CrossRef] [PubMed]

42. Swaisgood, C.M.; French, E.L.; Noga, C.; Simon, R.H.; Ploplis, V.A. The Development of Bleomycin-Induced Pulmonary Fibrosis in Mice Deficient for Components of the Fibrinolytic System. *Am. J. Pathol.* **2000**, *157*, 177–187. [CrossRef]

43. Hattori, N.; Degen, J.L.; Sisson, T.H.; Liu, H.; Moore, B.B.; Pandrangi, R.G.; Simon, R.H.; Drew, A.F. Bleomycin-induced pulmonary fibrosis in fibrinogen-null mice. *J. Clin. Investig.* **2000**, *106*, 1341–1350. [CrossRef] [PubMed]

44. Horowitz, J.C.; Tschumperlin, D.J.; Kim, K.K.; Osterholzer, J.J.; Subbotina, N.; Ajayi, I.O.; Teitz-Tennenbaum, S.; Virk, A.; Dotson, M.; Liu, F.; et al. Urokinase Plasminogen Activator Overexpression Reverses Established Lung Fibrosis. *Thromb. Haemost.* **2019**, *119*, 1968–1980. [CrossRef] [PubMed]

45. Chambers, R.C.; Scotton, C.J. Coagulation Cascade Proteinases in Lung Injury and Fibrosis. *Proc. Am. Thorac. Soc.* **2012**, *9*, 96–101. [CrossRef] [PubMed]

46. Günther, A.; Lübke, N.; Ermert, M.; Schermuly, R.T.; Weissmann, N.; Breithecker, A.; Markart, P.; Ruppert, C.; Quanz, K.; Ermert, L.; et al. Prevention of Bleomycin-induced Lung Fibrosis by Aerosolization of Heparin or Urokinase in Rabbits. *Am. J. Respir. Crit. Care Med.* **2003**, *168*, 1358–1365. [CrossRef]

47. Günther, A.; Mosavi, P.; Heinemann, S.; Ruppert, C.; Muth, H.; Markart, P.; Grimminger, F.; Walmrath, D.; Temmesfeld-Wollbrück, B.; Seeger, W. Alveolar Fibrin Formation Caused by Enhanced Procoagulant and Depressed Fibrinolytic Capacities in Severe Pneumonia. *Am. J. Respir. Crit. Care Med.* **2000**, *161*, 454–462. [CrossRef]

48. Idell, S.; Peterson, B.T.; Gonzalez, K.K.; Gray, L.D.; Bach, R.; McLarty, J.; Fair, D.S. Local Abnormalities of Coagulation and Fibrinolysis and Alveolar Fibrin Deposition in Sheep with Oleic Acid-induced Lung Injury. *Am. Rev. Respir. Dis.* **1988**, *138*, 1282–1294. [CrossRef]

49. Idell, S.; James, K.K.; Coalson, J.J. Fibrinolytic Activity in Bronchoalveolar Lavage of Baboons with Diffuse Alveolar Damage: Trends in Two Forms of Lung Injury. *Crit. Care Med.* **1992**, *20*, 1431–1440. [CrossRef]

50. Idell, S.; Mazar, A.; Cines, D.; Kuo, A.; Parry, G.; Gawlak, S.; Juarez, J.; Koenig, K.; Azghani, A.; Hadden, W.; et al. Single-Chain Urokinase Alone or Complexed to Its Receptor in Tetracycline-induced Pleuritis in Rabbits. *Am. J. Respir. Crit. Care Med.* **2002**, *166*, 920–926. [CrossRef]

51. Burnham, E.L.; Hyzy, R.C.; Paine, R., III; Kelly, A.M.; Quint, L.E.; Lynch, D.; Curran-Everett, D.; Moss, M.; Standiford, T.J. Detection of Fibroproliferation by Chest High-Resolution Ct Scan in Resolving Ards. *Chest* **2014**, *146*, 1196–1204. [CrossRef] [PubMed]

52. Mutsaers, S.E.; Prele, C.M.; Brody, A.R.; Idell, S. Pathogenesis of pleural fibrosis. *Respirology* **2004**, *9*, 428–440. [CrossRef] [PubMed]

53. Murao, S.; Yamakawa, K. A Systematic Summary of Systematic Reviews on Anticoagulant Therapy in Sepsis. *J. Clin. Med.* **2019**, *8*, 1869. [CrossRef] [PubMed]

54. Lechowicz, K.; Drozdzal, S.; Machaj, F.; Rosik, J.; Szostak, B.; Zegan-Baranska, M.; Biernawska, J.; Dabrowski, W.; Rotter, I.; Kotfis, K. Covid-19: The Potential Treatment of Pulmonary Fibrosis Associated with Sars-Cov-2 Infection. *J. Clin. Med.* **2020**, *9*, 1917. [CrossRef]

55. Ji, H.J.; Su, Z.; Zhao, R.; Komissarov, A.A.; Yi, G.; Liu, S.L.; Idell, S.; Matthay, M.A. Insufficient Hyperfibrinolysis in Covid-19: A Systematic Review of Thrombolysis Based on Meta-Analysis and Meta-Regression. *medRxiv* **2020**. [CrossRef]

56. Smith, H.W.; Marshall, C.J. Regulation of Cell Signalling by Upar. *Nat. Rev. Mol. Cell Biol.* **2010**, *11*, 23–36. [CrossRef]

57. Ellis, V.; Behrendt, N.; Danø, K. Plasminogen activation by receptor-bound urokinase. A kinetic study with both cell-associated and isolated receptor. *J. Biol. Chem.* **1991**, *266*, 12752–12758. [CrossRef]

58. Tucker, T.A.; Dean, C.; Komissarov, A.A.; Koenig, K.; Mazar, A.P.; Pendurthi, U.; Allen, T.; Idell, S. The Urokinase Receptor Supports Tumorigenesis of Human Malignant Pleural Mesothelioma Cells. *Am. J. Respir. Cell Mol. Biol.* **2010**, *42*, 685–696. [CrossRef]

59. Shetty, S.; Padijnayayveetil, J.; Tucker, T.; Stankowska, D.; Idell, S. The fibrinolytic system and the regulation of lung epithelial cell proteolysis, signaling, and cellular viability. *Am. J. Physiol. Cell. Mol. Physiol.* **2008**, *295*, L967–L975. [CrossRef]

60. Shetty, S.; Kumar, A.; Johnson, A.R.; Pueblitz, S.; Holiday, D.; Raghu, G.; Idell, S. Differential expression of the urokinase receptor in fibroblasts from normal and fibrotic human lungs. *Am. J. Respir. Cell Mol. Biol.* **1996**, *15*, 78–87. [CrossRef]

61. Shetty, S.; Idell, S. A urokinase receptor mRNA binding protein from rabbit lung fibroblasts and mesothelial cells. *Am. J. Physiol. Cell. Mol. Physiol.* **1998**, *274*, L871–L882. [CrossRef] [PubMed]

62. Sisson, T.H.; Hattori, N.; Xu, Y.; Simon, R.H. Treatment of Bleomycin-Induced Pulmonary Fibrosis by Transfer of Urokinase-Type Plasminogen Activator Genes. *Hum. Gene Ther.* **1999**, *10*, 2315–2323. [CrossRef] [PubMed]

63. Eitzman, D.T.; McCoy, R.D.; Zheng, X.; Fay, W.P.; Shen, T.; Ginsburg, D.; Simon, R.H. Bleomycin-induced pulmonary fibrosis in transgenic mice that either lack or overexpress the murine plasminogen activator inhibitor-1 gene. *J. Clin. Investig.* **1996**, *97*, 232–237. [CrossRef] [PubMed]

64. Van Zoelen, M.A.; Florquin, S.; de Beer, R.; Pater, J.M.; Verstege, M.I.; Meijers, J.C.; van der Poll, T. Urokinase Plasminogen Activator Receptor-Deficient Mice Demonstrate Reduced Hyperoxia-Induced Lung Injury. *Am. J. Pathol.* **2009**, *174*, 2182–2189. [CrossRef] [PubMed]

65. Rijneveld, A.W.; Levi, M.; Florquin, S.; Speelman, P.; Carmeliet, P.; Van Der Poll, T. Urokinase Receptor Is Necessary for Adequate Host Defense Against Pneumococcal Pneumonia. *J. Immunol.* **2002**, *168*, 3507–3511. [CrossRef] [PubMed]

66. Philip-Joët, F.; Alessi, M.-C.; Aillaud, M.; Barriere, J.-R.; Arnaud, A.; Juhan-Vague, I. Fibrinolytic and inflammatory processes in pleural effusions. *Eur. Respir. J.* **1995**, *8*, 1352–1356. [CrossRef]

67. Chung, C.-L.; Chen, Y.-C.; Chang, S.-C. Effect of Repeated Thoracenteses on Fluid Characteristics, Cytokines, and Fibrinolytic Activity in Malignant Pleural Effusion. *Chest* **2003**, *123*, 1188–1195. [CrossRef] [PubMed]

68. Beckert, L.; Brockway, B.; Simpson, G.; Southcott, A.M.; Lee, Y.C.G.; Rahman, N.; Light, R.W.; Shoemaker, S.; Gillies, J.; Komissarov, A.A.; et al. Phase 1 Trial of Intrapleural Lti-01; Single Chain Urokinase in Complicated Parapneumonic Effusions or Empyema. *JCI Insight* **2019**, *5*, e127470. [CrossRef]

69. Idell, S. The pathogenesis of pleural space loculation and fibrosis. *Curr. Opin. Pulm. Med.* **2008**, *14*, 310–315. [CrossRef]

70. Tucker, T.A.; Jeffers, A.; Alvarez, A.; Owens, S.; Koenig, K.; Quaid, B.; Komissarov, A.A.; Florova, G.; Kothari, H.; Pendurthi, U.; et al. Plasminogen Activator Inhibitor-1 Deficiency Augments Visceral Mesothelial Organization, Intrapleural Coagulation and Lung Restriction in Mice with Carbon Black/Bleomycin-Induced Pleural Injury. *Am. J. Respir. Cell Mol. Biol.* **2013**, *50*, 316–327. [CrossRef]

71. Florova, G.; Azghani, A.; Karandashova, S.; Schaefer, C.; Koenig, K.; Stewart-Evans, K.; Declerck, P.J.; Idell, S.; Komissarov, A.A. Targeting of Plasminogen Activator Inhibitor 1 Improves Fibrinolytic Therapy for Tetracycline-Induced Pleural Injury in Rabbits. *Am. J. Respir. Cell Mol. Biol.* **2015**, *52*, 429–437. [CrossRef] [PubMed]

72. Idell, S.; Allen, T.; Chen, S.; Koenig, K.; Mazar, A.; Azghani, A. Intrapleural activation, processing, efficacy, and duration of protection of single-chain urokinase in evolving tetracycline-induced pleural injury in rabbits. *Am. J. Physiol. Cell. Mol. Physiol.* **2007**, *292*, L25–L32. [CrossRef] [PubMed]

73. Idell, S.; Azghani, A.; Chen, S.; Koenig, K.; Mazar, A.; Kodandapani, L.; Bdeir, K.; Cines, D.; Kulikovskaya, I.; Allen, T. Intrapleural Low-Molecular-Weight Urokinase or Tissue Plasminogen Activator Versus Single-Chain Urokinase in Tetracycline-Induced Pleural Loculation in Rabbits. *Exp. Lung Res.* **2007**, *33*, 419–440. [CrossRef] [PubMed]

74. Komissarov, A.A.; Florova, G.; Azghani, A.O.; Buchanan, A.; Boren, J.; Allen, T.; Rahman, N.M.; Koenig, K.; Chamiso, M.; Karandashova, S.; et al. Dose dependency of outcomes of intrapleural fibrinolytic therapy in new rabbit empyema models. *Am. J. Physiol. Cell. Mol. Physiol.* **2016**, *311*, L389–L399. [CrossRef] [PubMed]

75. Komissarov, A.A.; Florova, G.; Azghani, A.O.; Buchanan, A.; Bradley, W.M.; Schaefer, C.; Koenig, K.; Idell, S. The Time Course of Resolution of Adhesions During Fibrinolyitic Therapy in Tetracycline-Induced Pleural Injury in Rabbits. *Am. J. Physiol. Lung. Cell Mol. Physiol.* **2015**. [CrossRef] [PubMed]

76. Corcoran, J.P.; Hallifax, R.; Rahman, N.M. New therapeutic approaches to pleural infection. *Curr. Opin. Infect. Dis.* **2013**, *26*, 196–202. [CrossRef]

77. Idell, S.; Rahman, N.M. Intrapleural Fibrinolytic Therapy for Empyema and Pleural Loculation: Knowns and Unknowns. *Ann. Am. Thorac. Soc.* **2018**, *15*, 515–517. [CrossRef]

78. Idell, S. Update on the Use of Fibrinolysins in Pleural Disease. *Clin. Pulm. Med.* **2005**, *12*, 184–190. [CrossRef]

79. Mazar, A.P.; Ahn, R.W.; O'Halloran, T.V. Development of novel therapeutics targeting the urokinase plasminogen activator receptor (uPAR) and their translation toward the clinic. *Curr. Pharm. Des.* **2011**, *17*, 1970–1978. [CrossRef]

80. Tucker, T.A.; Williams, L.; Koenig, K.; Kothari, H.; Komissarov, A.A.; Florova, G.; Mazar, A.P.; Allen, T.C.; Bdeir, K.; Mohan Rao, L.V.; et al. Lipoprotein Receptor-Related Protein 1 Regulates Collagen 1 Expression, Proteolysis, and Migration in Human Pleural Mesothelial Cells. *Am. J. Respir. Cell Mol. Biol.* **2012**, *46*, 196–206. [CrossRef]

81. Tucker, T.A.; Jeffers, A.; Boren, J.; Quaid, B.; Owens, S.; Koenig, K.B.; Tsukasaki, Y.; Florova, G.; Komissarov, A.A.; Ikebe, M.; et al. Organizing empyema induced in mice by Streptococcus pneumoniae: Effects of plasminogen activator inhibitor-1 deficiency. *Clin. Transl. Med.* **2016**, *5*, 17. [CrossRef]

82. Decologne, N.; Wettstein, G.; Kolb, M.; Margetts, P.; Garrido, C.; Camus, P.; Bonniaud, P. Bleomycin induces pleural and subpleural fibrosis in the presence of carbon particles. *Eur. Respir. J.* **2009**, *35*, 176–185. [CrossRef] [PubMed]

83. Decologne, N.; Kolb, M.; Margetts, P.J.; Menetrier, F.; Artur, Y.; Garrido, C.; Gauldie, J.; Camus, P.; Bonniaud, P. Tgf-Beta1 Induces Progressive Pleural Scarring and Subpleural Fibrosis. *J. Immunol.* **2007**, *179*, 6043–6051. [CrossRef] [PubMed]

84. Owens, S.; Jeffers, A.; Boren, J.; Tsukasaki, Y.; Koenig, K.; Ikebe, M.; Idell, S.; Tucker, T.A. Mesomesenchymal Transition of Pleural Mesothelial Cells Is Pi3k and Nf-Kappab Dependent. *Am. J. Physiol. Lung. Cell Mol. Physiol.* **2015**, *308*, L1265–L1273. [CrossRef] [PubMed]

85. Boren, J.; Shryock, G.; Fergis, A.; Jeffers, A.; Owens, S.; Qin, W.; Koenig, K.B.; Tsukasaki, Y.; Komatsu, S.; Ikebe, M.; et al. Inhibition of Glycogen Synthase Kinase 3beta Blocks Mesomesenchymal Transition and Attenuates Streptococcus Pneumonia-Mediated Pleural Injury in Mice. *Am. J. Pathol.* **2017**, *187*, 2461–2472. [CrossRef]

86. Ploug, M.; Gardsvoll, H.; Jorgensen, T.J.; Lonborg, H.L.; Dano, K. Structural Analysis of the Interaction between Urokinase-Type Plasminogen Activator and Its Receptor: A Potential Target for Anti-Invasive Cancer Therapy. *Biochem. Soc. Trans.* **2002**, *30*, 177–183. [CrossRef]

87. Pyke, C.; Eriksen, J.; Solberg, H.; Nielsen, B.; Kristensen, P.; Lund, L.R.; Dano, K. An alternatively spliced variant of mRNA for the human receptor for urokinase plasminogen activator. *FEBS Lett.* **1993**, *326*, 69–74. [CrossRef]

88. Thunø, M.; Macho, B.; Eugen-Olsen, J. suPAR: The Molecular Crystal Ball. *Dis. Markers* **2009**, *27*, 157–172. [CrossRef]

89. Arnold, D.T.; Hamilton, F.W.; Elvers, K.T.; Frankland, S.W.; Zahan-Evans, N.; Patole, S.; Medford, A.; Bhatnagar, R.; Maskell, N.A. Pleural Fluid suPAR Levels Predict the Need for Invasive Management in Parapneumonic Effusions. *Am. J. Respir. Crit. Care Med.* **2020**, *201*, 1545–1553. [CrossRef]

90. Van Veen, M.; Matas-Rico, E.; van de Wetering, K.; Leyton-Puig, D.; Kedziora, K.M.; De Lorenzi, V.; Stijf-Bultsma, Y.; van den Broek, B.; Jalink, K.; Sidenius, N.; et al. Negative Regulation of Urokinase Receptor Activity by a Gpi-Specific Phospholipase C in Breast Cancer Cells. *eLife* **2017**, *6*, e23649. [CrossRef]

91. Masucci, M.; Pedersen, N.; Blasi, F. A soluble, ligand binding mutant of the human urokinase plasminogen activator receptor. *J. Biol. Chem.* **1991**, *266*, 8655–8658. [CrossRef]

92. Higazi, A.A.-R.; Cohen, R.L.; Henkin, J.; Kniss, D.; Schwartz, B.S.; Cines, D.B. Enhancement of the Enzymatic Activity of Single-chain Urokinase Plasminogen Activator by Soluble Urokinase Receptor. *J. Biol. Chem.* **1995**, *270*, 17375–17380. [CrossRef] [PubMed]

93. Behrendt, N.; Danø, K. Effect of purified, soluble urokinase receptor on the plasminogen-prourokinase activation system. *FEBS Lett.* **1996**, *393*, 31–36. [CrossRef]

94. Resnati, M.; Pallavicini, I.; Wang, J.M.; Oppenheim, J.; Serhan, C.N.; Romano, M.; Blasi, F. The fibrinolytic receptor for urokinase activates the G protein-coupled chemotactic receptor FPRL1/LXA4R. *Proc. Natl. Acad. Sci. USA* **2002**, *99*, 1359–1364. [CrossRef]

95. Rahman, N.M.; Maskell, N.A.; West, A.; Teoh, R.; Arnold, A.; Mackinlay, C.; Peckham, D.; Davies, C.W.; Ali, N.; Kinnear, W.; et al. Intrapleural Use of Tissue Plasminogen Activator and DNase in Pleural Infection. *N. Engl. J. Med.* **2011**, *365*, 518–526. [CrossRef]

96. Bouros, D.; Schiza, S.; Tzanakis, N.; Chalkiadakis, G.; Drositis, J.; Siafakas, N. Intrapleural Urokinase versus Normal Saline in the Treatment of Complicated Parapneumonic Effusions and Empyema. *Am. J. Respir. Crit. Care Med.* **1999**, *159*, 37–42. [CrossRef]

97. Alemán, C.; Porcel, J.M.; Alegre-Martin, J.; Ruiz, E.; Bielsa, S.; Andreu, J.; Deu, M.; Suñé, P.; Deu-Martín, M.; López, I.; et al. Intrapleural Fibrinolysis with Urokinase Versus Alteplase in Complicated Parapneumonic Pleural Effusions and Empyemas: A Prospective Randomized Study. *Lung* **2015**, *193*, 993–1000. [CrossRef]
98. Bédat, B.; Plojoux, J.; Noel, J.; Morel, A.; Worley, J.; Triponez, F.; Karenovics, W. Comparison of intrapleural use of urokinase and tissue plasminogen activator/DNAse in pleural infection. *ERJ Open Res.* **2019**, *5*, 00084–02019. [CrossRef]
99. Komissarov, A.A.; Mazar, A.P.; Koenig, K.; Kurdowska, A.K.; Idell, S. Regulation of Intrapleural Fibrinolysis by Urokinase-Alpha-Macroglobulin Complexes in Tetracycline-Induced Pleural Injury in Rabbits. *Am. J. Physiol. Lung Cell Mol. Physiol.* **2009**, *297*, L568–L577. [CrossRef]

International Journal of
Molecular Sciences

MDPI

Review

The Central Role of Fibrinolytic Response in COVID-19—A Hematologist's Perspective

Hau C. Kwaan [1],* and Paul F. Lindholm [2]

[1] Division of Hematology/Oncology, Department of Medicine, Feinberg School of Medicine, Northwestern University, Chicago, IL 60611, USA

[2] Department of Pathology, Feinberg School of Medicine, Northwestern University, Chicago, IL 60611, USA; p-lindholm@northwestern.edu

* Correspondence: h-kwaan@northwestern.edu

Abstract: The novel coronavirus disease (COVID-19) has many characteristics common to those in two other coronavirus acute respiratory diseases, severe acute respiratory syndrome (SARS) and Middle East respiratory syndrome (MERS). They are all highly contagious and have severe pulmonary complications. Clinically, patients with COVID-19 run a rapidly progressive course of an acute respiratory tract infection with fever, sore throat, cough, headache and fatigue, complicated by severe pneumonia often leading to acute respiratory distress syndrome (ARDS). The infection also involves other organs throughout the body. In all three viral illnesses, the fibrinolytic system plays an active role in each phase of the pathogenesis. During transmission, the renin-aldosterone-angiotensin-system (RAAS) is involved with the spike protein of SARS-CoV-2, attaching to its natural receptor angiotensin-converting enzyme 2 (ACE 2) in host cells. Both tissue plasminogen activator (tPA) and plasminogen activator inhibitor 1 (PAI-1) are closely linked to the RAAS. In lesions in the lung, kidney and other organs, the two plasminogen activators urokinase-type plasminogen activator (uPA) and tissue plasminogen activator (tPA), along with their inhibitor, plasminogen activator 1 (PAI-1), are involved. The altered fibrinolytic balance enables the development of a hypercoagulable state. In this article, evidence for the central role of fibrinolysis is reviewed, and the possible drug targets at multiple sites in the fibrinolytic pathways are discussed.

Keywords: COVID-19; fibrinolysis; renin-aldosterone-angiotensin-system (RAAS); fibrinolysis; plasminogen activator inhibitor 1 (PAI-1)

Citation: Kwaan, H.C.; Lindholm, P.F. The Central Role of Fibrinolytic Response in COVID-19—A Hematologist's Perspective. *Int. J. Mol. Sci.* **2021**, 22, 1283. https://doi.org/10.3390/ijms22031283

Academic Editor: Toshiyuki Kaji
Received: 1 January 2021
Accepted: 26 January 2021
Published: 28 January 2021

Publisher's Note: MDPI stays neutral with regard to jurisdictional claims in published maps and institutional affiliations.

1. Introduction

Infection by the highly contagious coronavirus SARS-CoV-2 has resulted in a global pandemic of coronavirus disease 2019 (COVID-19) [1–3]. This virus shares a 79.5% homology to SARS-CoV, the virus responsible for severe acute respiratory syndrome (SARS) [2]. COVID-19 has many clinical and pathologic characteristics in common with SARS and the Middle East respiratory syndrome (MERS) caused by the virus MERS-CoV [4–6]. These viruses gain entry into the host cells by attaching a spike protein on their envelope to the angiotensin-converting enzyme 2 (ACE2) expressed on cell surfaces. Clinically, type 2 alveolar cells in the lung are the predominant target, resulting in severe pneumonitis [1,3,6]. In addition, many other organs including the kidneys and cardiovascular system are affected, leading to multiorgan failure. The injury to the lungs is worse in COVID-19 than in SARS or in MERS. In the affected tissues, there is an acute inflammatory response. In the lungs, this manifests as edema, macrophage infiltration and intra-alveolar fibrin deposition, leading to acute respiratory distress syndrome (ARDS) and acute respiratory failure. In autopsied specimens, diffuse alveolar damage, microthrombi in perialveolar vessels and intra-alveolar hemorrhage have been observed [7–13]. The clinical course is associated with a hypercoagulable state, resulting in thrombotic complications in arteries, veins, catheters, arteriovenous fistulas and implantable devices such as the ECMO (extracorporeal membrane oxygenation circuit). In all these aspects, the fibrinolytic system is involved. The fibrinolytic system is activated by acute tissue injury and inflammation and is an important element in the body's response to these acute viral infections. In this article, the evidence of the participation of various components of the fibrinolytic system in the pathogenesis of these disorders is reviewed. In addition, interference with specific sites of the fibrinolytic pathways may mitigate the tissue injury and thus they are compelling drug targets.

2. The Fibrinolytic System (Aka Plasminogen-Plasmin System)

The chief component of this system is a serine protease, plasmin. It is formed by the activation of its precursor plasminogen (Figure 1) [14,15]. In humans, the two activators are tissue type plasminogen activator (tPA) and urokinase-type plasminogen activator (uPA). They bind to their respective receptors on cell surfaces. The receptor for tPA is a heterotetramer complex of annexin A2 and a surface binding protein S100A10 [16,17], whereas that for uPA is urokinase receptor (uPAR) [18]. To maintain a physiologic balance, the fibrinolytic system is regulated by serine protease inhibitors (serpins) at various activation sites. Those inhibiting the conversion of plasminogen to plasmin are plasminogen activator inhibitor 1 (PAI-1), plasminogen activator inhibitor 2 (PAI-2), activated protein C inhibitor (APC) inhibitor (PAI-3), protease nexin 1 and defensin (for tPA only). Those that inhibit plasmin are α2-antiplasmin, α2-macroglobulin, as well as several serine proteases, including antithrombin, α2-antitrypsin and protease nexin 1. In addition, thrombin activatable fibrinolytic inhibitor (TAFI), a carboxypeptidase, prevents plasminogen binding and plasmin formation by cleaving lysine residues on fibrin. Among the inhibitors of plasmin, α-antiplasmin [19] and PAI-1 [20] are the most effective ones. The components of the fibrinolytic system regulate many physiologic functions and participate in the pathogenesis of numerous pathologic disorders. tPA regulates fibrin formation in blood and takes part in many functions in the brain. On the other hand, uPA and its receptor uPAR are involved in inflammation, tissue repair, cell proliferation and a multitude of other body functions. In particular, the uPA and uPA/uPAR complex is highly expressed in the airway epithelium in the lung [21,22] and plays an important role in acute lung injury. Under physiologic conditions, the activators and inhibitors are in a state of balance and regulate hemostasis. This balance is deranged in COVID-19, as discussed below.

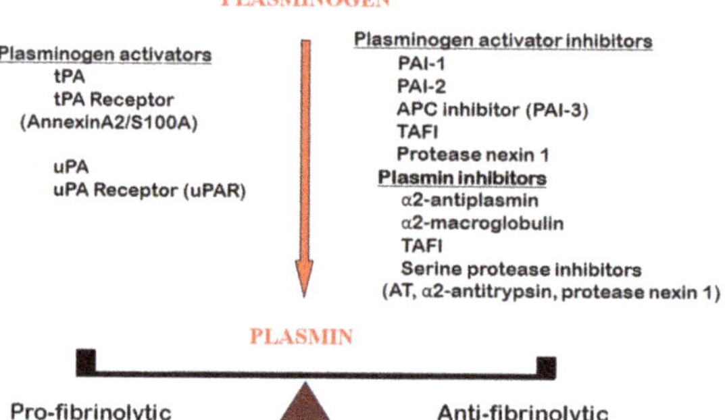

Figure 1. The fibrinolytic system, aka the plasminogen-plasmin system, consisting of the serine protease plasmin, derived from plasminogen by the activation of its activators, tissue plasminogen activator (tPA) and urokinase-type plasminogen activator (uPA). These two activators are ligands to their cellular receptors, urokinase receptor (uPAR) and a heterotetramer annexin, A2/S100A, respectively. The plasminogen activators and plasmin are inhibited by plasminogen activator inhibitors, plasminogen activator inhibitor 1 (PAI-1), plasminogen activator inhibitor 2 (PAI-2), activated protein C inhibitor (APC) inhibitor (PAI-3), thrombin activatible fibrinolysis inhibitor (TAFI) and protease nexin 1. Plasmin is inhibited by α2-antiplasmin, α2-macroglobulin, TAFI and several serine protease inhibitors (AT, α2-antitrypsin, protease nexin 1).

3. Invasion of SARS-CoV-2 into Host Cells and Subsequent Events

SARS-CoV-2 is highly contagious and is transmitted both as droplets and as aerosols. The virus gains entry to the host cells by attaching a spike protein on the viral envelop to angiotensin converting enzyme 2 (ACE2) on the cell surface. ACE2 is an integral component of the renin-aldosterone-angiotensin-system (RAAS) (Figure 2) [23,24]. In SARS-CoV-2 infection, the pathogenesis involves the following steps. First, the RAAS is an essential regulator of vascular functions including blood pressure, sodium balance and blood volume. In the RAAS, a plasma protein, angiotensinogen, is converted by a renal aspartic protease renin to angiotensin I. Angiotensin I is then metabolized by angiotensin converting enzyme (ACE) to angiotensin II. Angiotensin II is further metabolized by ACE2, a homolog of ACE [25], producing a vasodilator, angiotensin 1–7. ACE2 is expressed on cell membranes in the lung in the trachea and bronchial epithelial cells, type 2 alveolar cells and macrophages, in the kidney on the luminal surface of tubular epithelial cells, in the heart endothelium and myocytes, in the gastrointestinal epithelial cells, in the testis and in the brain [26–30]. ACE-2 acts as the receptor for SARS-CoV-2, as well as for other coronaviruses such as SARS-CoV [31,32], by binding to the spike protein on the viral envelop. This binding requires the proteolysis of the S1/S2 cleavage site of the spike protein by proteases in the pulmonary airway, including plasmin, furin, trypsin and transmembrane proteases (TMPRSS2) [33,34], for entry into host cells. Following its attachment to the spike protein, ACE2 is internalized and downregulated [35,36]. The reduction in ACE2 results in a diminished degradation of angiotensin II, leading to a buildup of angiotensin II. This produces several deleterious effects. Angiotensin II binds to a cellular receptor angiotensin II type 1a receptor in the lung, causing acute lung injury [26,37,38]. This injury is characterized by pulmonary edema, infiltration of monocytes and macrophages, diffuse alveolar damage, along with increased fibrin deposition, hyaline membrane formation and microvascular thrombosis [7,13,39,40]. Autopsy specimens of lungs from COVID-19 patients showed a high prevalence of diffuse alveolar damage, capillary congestion and capillary microthrombi [41]. There is severe capillary endothelial injury with widespread capillary fibrinous microthrombi [42]. The lungs of COVID-19 patients contain a much higher load of capillary microthrombi and less thrombi in post-capillary venules than those in influenza patients. Pulmonary capillary microthrombi are also found in association with complement components [43] and neutrophil extracellular traps (NETs) [44]. Proteolytic breakdown of the fibrin results in the generation of D-dimer [45,46]. The magnitude of D-dimer has been correlated to the severity of the infection [45]. Angiotensin II also increases the expression of PAI-1 in endothelial cells [47–49], resulting in decreased fibrinolysis and contributing to a hypercoagulable state. The lung injury occurs with the viral entry into the type II alveolar cells. As these cells are the source of surfactant, there is loss of surfactant. The decrease in surfactant leads to the induction of the p53 pathway. p53 binds to uPA/uPAR mRNAs and suppresses their expression while increasing PAI-1mRNA [50,51]. The diminished effect of ACE2 also leads to decreased formation of angiotensin 1–7 and angiotensin 1–9. Angiotensin 1–7 has been shown to impair the release of PAI-1 by cultured endothelial cells in vitro [52]; hence, a diminished angiotensin 1–7 will result in more PAI-1. On the other hand, angiotensin 1–9 increased PAI-1 and thus the thrombotic tendency [53]. In addition, angiotensin 1–9 activates platelets with the release of PAI-1 from their α-granules. The sum effect is enhanced PAI-1 activity. PAI-1 contributes to many changes in the lung. It keeps in check excessive fibrinolysis and lessens the risk of progression of the alveolar damage to intra-alveolar hemorrhage. PAI-1 further enhances epithelial–mesenchymal transition and fibrosis [50,51,54–56]. The infiltration of monocytes and macrophages evokes an acute inflammatory response, with elevated levels of proinflammatory cytokines such as interleukin (IL)-6, IL-1, tumor necrosis factor (TNF)α and IL-8. In many of the severe cases, an uncontrolled and continuous interaction between natural killer (NK) cells of the innate immune system and CD8 positive cytolytic T cells of the adaptive immune system leads to a cytokine storm with very high levels of serum pro-inflammatory cytokines and ferritin [57,58]. The inflammatory cytokines activate the intrinsic pathway of the coagulation cascade, with factor XII activation and progression to thrombin generation. A thrombo-inflammatory condition then ensues. These inflammatory cytokines significantly upregulate the expression of PAI-1 [59].

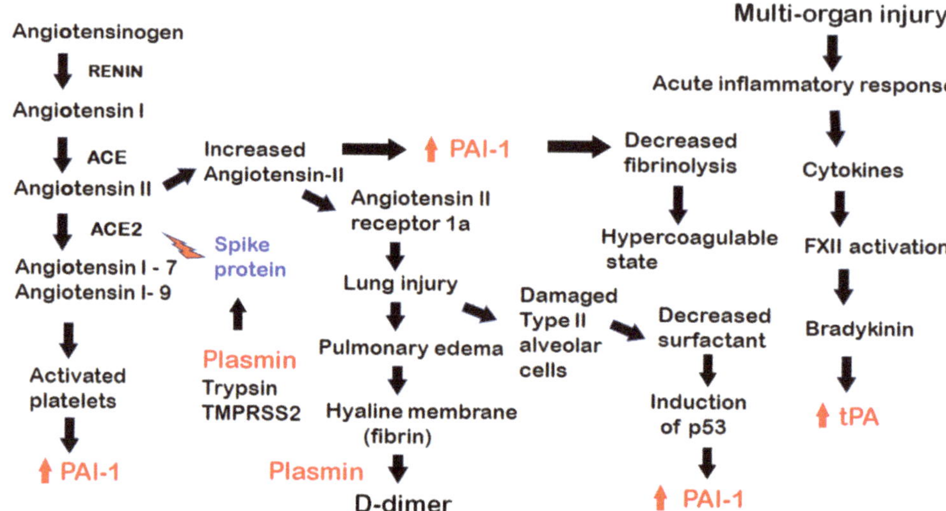

Figure 2. The pathogenesis of COVID-19, showing the involvement of components of the fibrinolytic system in various steps. From the left, the renin-aldosterone-angiotensin-system (RAAS) pathway is shown. Plasmin and other proteases, trypsin and TMPRSS2 act on the SARS-CoV-2 spike protein to facilitate its binding to ACE2 on the surface of host cells. With the binding, the virus invades the host cells while ACE2 is internalized and unable to process the breakdown of angiotensin II, leading to its excess. The excess of angiotensin II leads to an increase in PAI-1 and decreased fibrinolysis, creating a hypercoagulable state, while the excess of angiotensin II binds to its receptor angiotensin II receptor 1a, causing lung injury and leading to pulmonary edema with the formation of a hyaline membrane with fibrin in the alveoli. This is broken down by plasmin with the formation of D-dimer. The diffuse alveolar damage with damaged type II alveolar cells leads to decreased surfactant, which results in induction of the p53 pathway and increased in PAI-1. Of note (on the left), platelets are activated by angiotensin 1–9 and by other pathways. They then release PAI-1 into the circulation. There is also an acute inflammatory response with multi-organ injury. Inflammatory cytokines activate factor XII leading to bradykinin formation and a subsequent increase in tPA.

The role of PAI-1 has been extensively studied in many pathological conditions. In health, PAI-1 originates from endothelial cells, platelets, liver, adipose tissues and macrophages [60]. Though PAI-1 is present in an active conformation, especially in platelets, it can be readily converted to the inactive latent form. PAI-1 is well recognized to play an important role in the pathogenesis of a wide variety of conditions, including aging, cellular senescence, obesity, cardiovascular disease, hypertension, diabetes, fibrosis and thrombosis [20,56,61–65], It is notable that people with many of these conditions are more susceptible to COVID-19 and have worse outcomes [66].

4. Hypercoagulability

A marked increase in thrombotic complications due to hypercoagulability has been observed in patients with COVID-19. Thrombosis had been observed in both superficial and deep veins, in arteries and in the microvasculature. One remarkable example is acute ischemic strokes in young patients with no previous arterial disease [67]. The acute lung injury in COVID-19 patients shows fibrin deposits in the pulmonary microcirculation, forming microthrombi, which are thought to be due to viral injury to the endothelium [68].

Thrombosis is more common in the severely ill patients in the ICU, with a notably greater incidence of pulmonary embolism [69–71].

Microthrombi have been observed clinically in multiple organs: in the heart, resulting in acute myocardial infarction [72], within tumors [73], in the lungs [74,75] and in the kidneys, resulting in acute renal failure [76,77], whereas autopsy findings have revealed more widespread microthrombosis elsewhere [9,12,13,39,77,78]. Heart failure has been found in almost a third of hospitalized COVID patients and is the second leading cause of death following ARDS [79]. Notably, the microthrombi were found to be rich in platelets and megakaryocytes [80].

The coagulation profile in blood shows mild prolongations in prothrombin time (PT) and activated partial thromboplastin time (APTT), as well as mild thrombocytopenia, and does not reflect the in vivo state of hypercoagulability. On the other hand, a high fibrinogen and a high D-dimer are characteristic of patients with COVID-19 [45,81–84]. The acute inflammatory response in the infection accounts for some of the increase in the fibrinogen levels. In response to acute inflammation, the hepatic synthesis of fibrinogen has been shown to increase two- to ten-fold as an acute phase reactant [85]. The increase in fibrinogen was found to be proportional to the severity of the disease [45,86]. The D-dimer increase is observed in most patients and found to be correlated not only with the severity of the infection but also with thrombotic events [45,86]. Since D-dimer is the product of plasmin degradation of cross-linked fibrin [46] and since fibrinolytic activity is low [87,88], the enigma of the high D-dimer has not been fully elucidated [89]. Based on observations in acute lung injury and acute respiratory distress syndrome in SARS-CoV, it is generally believed that the major portion of the circulating D-dimer originates from the pulmonary lesion [90,91]. Continuous fibrin deposition into the alveoli with its breakdown by plasmin leads to the production of D-dimer. D-dimer has been recovered in the bronchoalveolar lavage (BAL) fluid of patients with acute respiratory distress syndrome (ARDS), indicating that intra-alveolar coagulation and fibrinolysis occur in this syndrome [92]. However, in this study, plasma and BAL D-dimer levels did not correlate. The generation of D-dimer occurs despite the antifibrinolytic effects of thrombin activatable fibrinolysis inhibitor (TAFI), protein C inhibitor (aka PAI-3) [93] and PAI-1 [40,94]. The APTT is often prolonged with the presence of lupus anticoagulant [95,96]. Factor XII is frequently low, partly due to loss in the exudation in pulmonary lesions, as a high level of this factor was found in BAL [97]. Activated factor XII has high homology to tPA and can activate plasminogen to plasmin [98]. Thus, a low factor XII would contribute to reduced fibrinolysis in this setting.

Although each of the individual conventional clotting tests does not give a whole picture of hemostasis, a more complete hemostatic picture can be found using whole blood testing in viscoelastography (VE). Two common techniques are thromboelastography (TEG) and rotational thromboelastometry (ROTEM). The progressive increase in viscosity of clotting whole blood and subsequent clot lysis is plotted graphically. In TEG, the time taken for a clot to form is recorded as the r value, and the velocity of clot formation as K (minutes) and K (angle). The clot strength is the maximum amplitude (MA), whereas clot lysis at 30 min is Lys30. Short r and K values, with increased MA, were found in COVID-19 patients, whereas in most of the ICU patients no clot lysis was seen at 30 min [87,88]. These findings indicate a hypercoagulable state. They were correlated with a high D-dimer and severity of the disease. The absence of fibrinolysis characterizes a pathologic hypercoagulable condition often associated with thrombosis, showing decreased viscoelastic fibrinolysis associated with elevated D-dimer and plasmin-antiplasmin (PAP) complexes [87]. Those patients with fibrinolysis shutdown had a 40% incidence of VTE compared to 5% in those without shutdown [87]. Fibrinolysis shutdown with a high D-dimer level is also correlated with renal failure [87]. Similar findings were seen using the ROTEM [99–102]. In a systemic review of ten studies of 389 COVID-19 patients, 90% of which were severe, the above-described hypercoagulability and fibrinolysis shutdown were present [103]. These patients were also on anticoagulant therapy. However, the mechanism by which these changes occur are not clearly shown. Obviously, many more studies are needed to provide a clear-cut explanation. Results of these studies would enable a better use of viscoelastography in guiding the use of anticoagulation therapy.

Both TEG and ROTEM have the advantage of being readily available at the point-of-care, with a short turnaround time for the results. It can also be used to guide anticoagulant therapy.

The fibrinolysis shutdown had previously been shown to be a poor prognostic indicator in acute sepsis and in severe trauma [104].

5. Role of Platelets

Recent observations in COVID-19 patients revealed that platelet activation takes place. Platelets form aggregates with neutrophils, monocytes and T-lymphocytes. Monocyte aggregates were found to release tissue factor [104,105], which can activate platelets. The activation of platelets is correlated to the severity and poor outcome of the disease [105]. The evidence for platelet activation were alterations in the gene transcriptome; increased P-selectin; and increased platelet aggregation to ADP, thrombin and collagen [106]. In addition, there was increased adhesion and spreading of platelets on fibrinogen and collagen. These changes contribute to the hypercoagulability not only through increased platelet activation but also through increased tissue factor release. In addition, platelets are the richest source of PAI-1 in the circulation. Activated platelets release PAI-1 from the α granules, as has been shown in trauma patients [107]. However, it is not known if PAI-1 is similarly released by platelets in COVID-19 infection.

6. Activation of the Complement System

One major innate immune response is the complement system [108,109]. It is composed of protein elements that can be activated by three different pathways—classic, alternative and lectin. Each pathway is activated respectively by the antigen/antibody complex, spontaneous hydrolysis of C3 and the mannose-binding lectin (MBL)–mannose complex. Following activation, C3 convertase cleaves C3 to C3a and C3b, resulting in the generation of C5 convertase, which cleaves C5 to C5a and C5b. C5b then forms a complex with other complement proteins to generate the membrane attack complex (MAC), consisting of C5b-C6-C7-C8-C9 (often abbreviated as C5b–9). MAC acts on cells by disrupting the cell surface, resulting in cell lysis. There are complement control proteins in the plasma that regulate the complement system. They are C1-inhibitor (C1-INH) which binds to C1 and prevents its activation, decay accelerating factor (CD55), membrane cofactor protein (CD46), protectin (CD59), complement C3b/C4b receptor 1, CR1 (CD35) and factor H.

The complement system has been shown to be activated in animal models of SARS-CoV and MERS-CoV infections [40,110,111] and was observed in COVID-19 patients [43]. Autopsy examination of COVID-19 patients using immune-histochemical staining revealed the presence of mannose-binding lectin-associated serine protease (MASP-2), C4d (lectin pathway) and C5b–9 (membrane attack complex) in the microvasculature of the lung in the alveolar septa and in the skin. These complement components colocalize with SARS-CoV-2. In both biopsy and autopsy materials, vascular lesions with microthrombi were found, along with C4d and C5b–9.

These observations are highly significant since the complement system is linked to both the coagulation and the fibrinolytic systems. They support another explanation for the high incidence of thrombotic complications in COVID-19.

There is crosstalk between members of the complement system and coagulation factors on the one hand, and components of the fibrinolytic system on the other [109], and this is illustrated in Figure 3. C5a is a procoagulant and activates tissue factor, and suppression of C3 or of C5 reduces tissue factor activity. With activation of tissue factor, the extrinsic pathway of the coagulation cascade is triggered, leading to thrombin generation. Thrombin converts a carboxy-peptidase B-like proenzyme to thrombin-activatable fibrinolysis inhibitor (TAFI), which blocks the conversion of plasminogen to plasmin [112]. TAFI can also be activated by another component of the complement system, MASP-1 [113,114]. The inhibitor of complement activation, C1-protease inhibitor(C1-inh), blocks the activation of coagulation and of complement [115]. It also inhibits plasmin [116]. In this complex setting, activation of the complement system affects fibrinolysis and vice versa. Plasmin has been shown to activate C3 and C5 directly and lead to formation of the membrane attack (MAC) [117]. C5a has been shown to increase the expression of PAI-1 in mast cells [118].

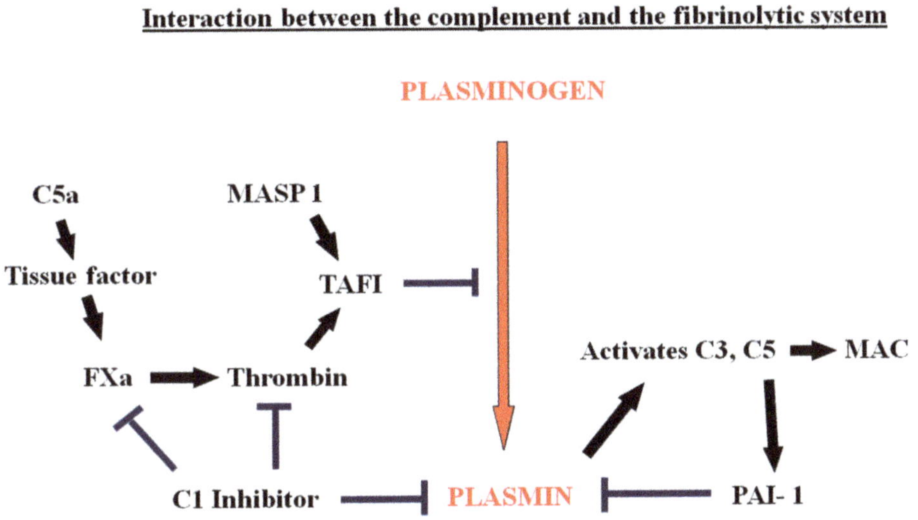

Interaction between the complement and the fibrinolytic system

Figure 3. The interactions of the complement system with coagulation and fibrinolysis. C5a activates tissue factor, initiating the coagulation cascade, leading to thrombin generation and the formation of thrombin-activatable fibrinolysis inhibitor (TAFI). TAFI blocks the conversion of plasminogen to plasmin. TAFI can also be activated by another component of the complement system, MASP-1. C1 inhibitor blocks the activation of coagulation and inhibits plasmin. In addition, plasmin activates C3 and C5 directly and this leads to formation of the membrane attack (MAC). C5a also increases the expression of PAI-1 in mast cells.

In addition, C3a and C5b–9 were found to activate platelets [119–121]. Whether or not PAI-1 is released is not known. If it does, this will be an additional pathway by which complement activation may inhibit fibrinolysis.

Complement components have been found to colocalize in microvasculature with SARS-CoV-2, activate coagulation and inhibit fibrinolysis, thus raising the question of whether eculizumab or related C5 cleavage targeted therapies would be useful to prevent detrimental thrombosis and microvascular injury [43]. However, in a case report of a patient with preexisting atypical hemolytic syndrome (aHUS) who contracted severe COVID-19, eculizumab therapy did not prevent severe endothelial injury or D-dimer elevations [122]. Nonetheless, treatment that inhibits the complement system remains an attractive drug target. Two approved drugs are eculizumab and the longer acting ravulizumab, both being monoclonal antibodies with high affinity to C5 and preventing the cleavage of C5 to C5a, thus blocking the formation of C5–9. In addition, there are other agents blocking C3 activation under development. One of these is narsoplimab, a monoclonal antibody against MASP-2 [123].

7. Fibrinolytic Balance

Under physiological conditions, the body's fibrinolytic components are kept in a state of balance between the profibrinolytic and the antifibrinolytic factors (Figure 1). This delicate balance can be perturbed under many pathological conditions. In COVID-19, an abnormal fibrinolytic balance is one of the major players in its pathogenesis. As the lung is the most commonly involved organ, we have examined the local changes in the fibrinolytic components and their effects on the fibrinolytic balance (Figure 4).

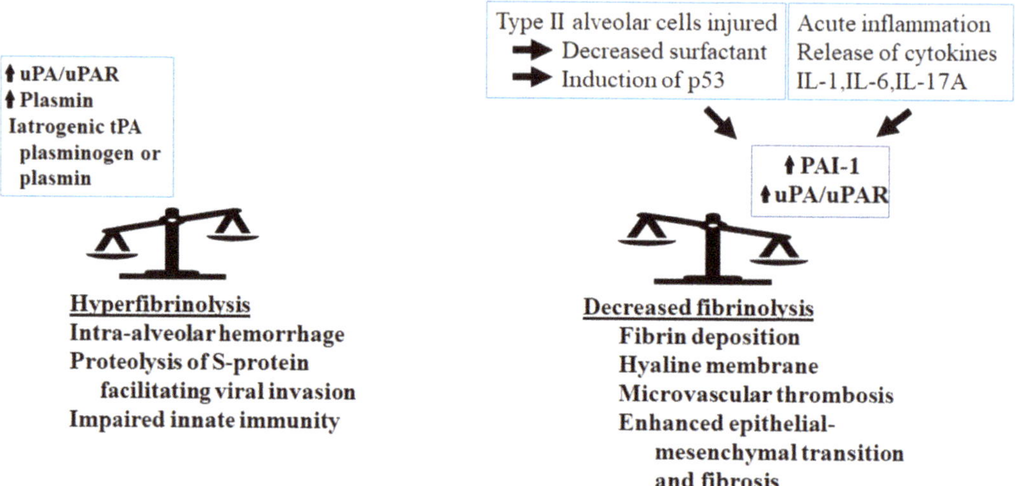

Figure 4. Changes in the fibrinolytic balance in the lung in COVID-19. On the one hand, an increase in fibrinolytic components, including uPA/UPAR, plasmin, iatrogenic tPA, plasminogen or plasmin, would tilt the balance towards hyperfibrinolysis. This enhances the proteolysis of the viral S-protein and facilitates viral invasion; increases the breakdown of fibrin in the alveoli, generating more D-dimer; enhances the risk of intra-alveolar hemorrhage and impairs the innate immunity of host cells. On the other hand, the acute inflammatory response in the lung releases inflammatory cytokines (IL-1, IL-6 and IL-17A), leading to an increase in PAI-1 and a decrease in the uPA/uPAR complexes. In addition, injury of the type II alveolar cells results in decreased surfactant and the induction of the p53 pathway that upregulates PAI-1. These will tilt the balance to hypofibrinolysis, with the consequences of increased fibrin deposition, hyaline membrane formation, microvascular thrombosis, enhanced epithelial–mesenchymal transition and fibrosis.

Studies in the epithelial cells in the normal lung have indicated that uPA, its receptor uPAR and PAI-1 are expressed [22,124,125]. This state of fibrinolytic balance keeps the airways and alveoli free from fibrin deposition. In healthy subjects, uPA and uPAR are present in the bronchoalveolar lavage fluids with no procoagulants, keeping the airways and alveoli clear. tPA is not involved. There is a temporal relationship between COVID-19 infection and fibrinolysis. In the acute phase of the infection, the inflow of inflammatory fluids containing fibrinogen and coagulation factors leads to fibrin deposition and hyaline membrane formation. The acute inflammatory cytokines consisting of IL-1, IL-6 and IL-17A upregulates PAI-1 and suppresses the expression of uPA and uPAR [50,51].

The relationship between lung injury and changes in uPA, uPAR and PAI-1 was fully demonstrated in a bleomycin-induced model in mice [50]. Following injury to type II alveolar cells by bleomycin, there is suppression of uPA and uPAR, along with an increase in PAI-1. As uPA induces alveolar epithelial cell proliferation, these changes promote apoptosis and fibrosis. The bleomycin lung injury also induces p53 expression, leading to further downregulation of uPA and uPAR and upregulation of PAI-1 [125]. Similar changes were observed in lungs injured by smoking [50].

In COVID-19 infection, there is injury to the type II alveolar cells, with the same results of increased p53 expression, suppression of uPA, uPAR and an increase in PAI-1. In addition, the type II alveolar cells are the source of surfactant. The damage by the viral infection leads to a marked reduction in surfactant. With the decreased surfactant, the p53 pathway is induced, leading to an increase in PAI-1 and a decrease in uPA and uPAR [51,126]. The fibrinolytic balance is then shifted to a hypofibrinolytic state. This enhances fibrin deposition, hyaline membrane formation and microvascular thrombosis. In the later phase of the infection, the lung epithelial cells undergo epithelial–mesenchymal transition [127] and fibrosis [128]. These changes are also enhanced by PAI-1 [54].

8. Bleeding Complications

In contrast to thrombosis, bleeding in COVID-19 patients has received less attention in the current literature. The incidence ranges from 6–21% in hospitalized patients [129–132]. The largest retrospective study, with 102 hospitalized patients anticoagulated on therapeutic doses, showed an incidence of 11%, well correlated to mortality. There was little or no increase in fibrinolysis. The etiology of the bleeding is likely endothelial damage from the viral infection. The main risk factors for bleeding are the use of antiplatelet drugs, anticoagulation and underlying vascular lesions. A recent report of major bleeding in anticoagulated patients with macrothrombosis prompted caution in using therapeutic doses of heparin and recommended close monitoring [133].

9. Therapeutic Targets

As impaired fibrinolysis due to increased PAI-1 levels was believed to be involved in the pathogenesis of acute lung injury [22], various fibrinolytic agents have been used in its management. These agents include tPA, uPA, plasminogen and plasmin. A meta-analysis of 22 studies of fibrinolytic therapy in animal models of acute lung injury revealed that the lung injury, oxygenation, local neutrophil infiltration and mortality are improved [134]. A similar approach is being employed to mitigate the clinical course of COVID-19 using various components of the fibrinolytic system. Inhalation of plasminogen was found to improve the lung lesions and hypoxemia in 13 patients with COVID 19 [135]. To augment fibrinolysis, tPA has been used and delivered intravenously in ongoing clinical trials [136,137]. Another strategy employs small molecules that inhibit PAI-1, one of which is TM5411 [62,138]. At the time of this writing, it is also is undergoing clinical trial. Results of these trials will verify the adverse role of PAI-1 in COVID-19.

10. Conclusions

This review of the pathogenesis of COVID-19 provides findings in the published literature showing that the components of the fibrinolytic system are involved in multiple steps of the viral infection. This involvement ranges from the invasion of the host by the virus to organ damage and a variety of complications, including thrombosis and fibrosis. Plasmin processes the viral S-protein for its entry into the host cells. The subsequent binding of the S-protein to ACE-2 triggers a rise in angiotensin II, which upregulates PAI-1. The lung injury with edema, hyaline membrane formation and alveolar damage are examples of the ways in which fibrinolysis is heavily involved. Other changes such as monocyte and macrophage infiltration in the lesions, evoking an acute inflammatory and cytokine response, further enhance the fibrinolytic changes, especially an increase in PAI-1. This close relationship between fibrinolysis and the disease process offers an opportunity for a therapeutic target.

The pathogenesis of this viral infection also involves the activation of the complement system. Notably, there is interaction among coagulation activation, with tissue factor initiating the coagulation cascade and thrombin generation, TAFI activation and the inhibition of fibrinolysis. Fibrinolysis is also inhibited by TAFI. Several steps in the complement activation pathways also inhibit fibrinolysis. These actions form an attractive basis to explain the hypercoagulability in COVID-19, with its increased incidence of thrombotic complications. It is also notable that the degree of fibrinolytic involvement varies with the severity of the illness and furthermore may change temporally during the course of the disease. As such, the therapeutic approach should be appropriately tailored to the specific phases of the clinical course. We expect that ongoing clinical trials will verify the role of altered fibrinolysis in the pathogenesis of COVID-19.

Author Contributions: Concept and design: H.C.K. Manuscript writing: H.C.K. and P.F.L. All authors have read and agreed to the published version of the manuscript.

Funding: This work received no external funding. H.C.K. is supported by the Hematology Research Fund, Division of Hematology/Oncology, Feinberg School of Medicine, Northwestern University.

Conflicts of Interest: The authors declare no conflict of interest.

References

1. Guan, W.J.; Ni, Z.Y.; Hu, Y.; Liang, W.H.; Ou, C.Q.; He, J.X.; Kiu, L.; Shan, H.; Lei, C.L.; Hui, D.S.C.; et al. Clinical Characteristics of Coronavirus Disease 2019 in China. *N. Engl. J. Med.* **2020**, *382*, 1708–1720. [CrossRef]
2. Zhou, P.; Yang, X.L.; Wang, X.G.; Hu, B.; Zhang, L.; Zhang, W.; Si, H.R.; Zhu, Y.; Li, B.; Huang, C.L.; et al. A pneumonia outbreak associated with a new coronavirus of probable bat origin. *Nature* **2020**, *579*, 270–273. [CrossRef] [PubMed]
3. Guo, Y.R.; Cao, Q.D.; Hong, Z.S.; Tan, Y.Y.; Chen, S.D.; Jin, H.J.; Tan, K.S.; Wang, D.Y.; Yan, Y. The origin, transmission and clinical therapies on coronavirus disease 2019 (COVID-19) outbreak—An update on the status. *Mil. Med. Res.* **2020**, *7*, 11. [CrossRef] [PubMed]
4. Lau, Y.L.; Peiris, J.S. Pathogenesis of severe acute respiratory syndrome. *Curr. Opin. Immunol.* **2005**, *17*, 404–410. [CrossRef] [PubMed]
5. Yin, Y.; Wunderink, R.G. MERS, SARS and other coronaviruses as causes of pneumonia. *Respirology* **2018**, *23*, 130–137. [CrossRef] [PubMed]
6. Zhou, M.; Zhang, X.; Qu, J. Coronavirus disease 2019 (COVID-19): A clinical update. *Front. Med.* **2020**, *14*, 126–135. [CrossRef] [PubMed]
7. Barton, L.M.; Duval, E.J.; Stroberg, E.; Ghosh, S.; Mukhopadhyay, S. COVID-19 Autopsies, Oklahoma, USA. *Am. J. Clin. Pathol.* **2020**, *153*, 725–733. [CrossRef] [PubMed]
8. Xu, Z.; Shi, L.; Wang, Y.; Zhang, J.; Huang, L.; Zhang, C.; Liu, S.; Zhao, P.; Liu, H.; Zhu, L.; et al. Pathological findings of COVID-19 associated with acute respiratory distress syndrome. *Lancet Respir Med.* **2020**, *8*, 420–422. [CrossRef]
9. Tian, S.; Xiong, Y.; Liu, H.; Niu, L.; Guo, J.; Liao, M.; Xiao, S.-Y. Pathological study of the 2019 novel coronavirus disease (COVID-19) through postmortem core biopsies. *Mod. Pathol.* **2020**, *33*, 1007–1014. [CrossRef] [PubMed]
10. Sauter, J.L.; Baine, M.K.; Butnor, K.J.; Buonocore, D.J.; Chang, J.C.; Jungbluth, A.A.; Szabolcs, M.J.; Morjaria, S.; Mount, S.L.; Rekhtman, N.; et al. Insights into pathogenesis of fatal COVID-19 pneumonia from histopathology with immunohistochemical and viral RNA studies. *Histopathology* **2020**, *77*, 915–925. [CrossRef] [PubMed]
11. Nicholls, J.M.; Poon, L.L.; Lee, K.C.; Ng, W.F.; Lai, S.T.; Leung, C.Y.; Chu, C.M.; Hui, P.K.; Mak, K.L.; Lim, W.; et al. Lung pathology of fatal severe acute respiratory syndrome. *Lancet* **2003**, *361*, 1773–1778. [CrossRef]
12. Menter, T.; Haslbauer, J.D.; Nienhold, R.; Savic, S.; Hopfer, H.; Deigendesch, N.; Frank, S.; Turek, D.; Willi, N.; Pargger, H.; et al. Postmortem examination of COVID-19 patients reveals diffuse alveolar damage with severe capillary congestion and variegated findings in lungs and other organs suggesting vascular dysfunction. *Histopathology* **2020**, *77*, 198–209. [CrossRef] [PubMed]
13. Edler, C.; Schroder, A.S.; Aepfelbacher, M.; Fitzek, A.; Heinemann, A.; Heinrich, F.; Klein, A.; Langenwalder, F.; Lutgehetmann, M.; Meisner, K.; et al. Dying with SARS-CoV-2 infection-an autopsy study of the first consecutive 80 cases in Hamburg, Germany. *Int. J. Leg. Med.* **2020**, *134*, 1275–1284. [CrossRef] [PubMed]
14. Kwaan, H.C. From fibrinolysis to the plasminogen-plasmin system and beyond: A remarkable growth of knowledge, with personal observations on the history of fibrinolysis. *Semin. Thromb. Hemost.* **2014**, *40*, 585–591. [CrossRef] [PubMed]
15. Pechet, L. Fibrinolysis. *N. Engl. J. Med.* **1965**, *273*, 1024–1034. [CrossRef]
16. Hajjar, K.A.; Menell, J.S. Annexin II: A novel mediator of cell surface plasmin generation. *Ann. N. Y. Acad. Sci.* **1997**, *811*, 337–349. [CrossRef]
17. Surette, A.P.; Madureira, P.A.; Phipps, K.D.; Miller, V.A.; Svenningsson, P.; Waisman, D.M. Regulation of fibrinolysis by S100A10 in vivo. *Blood* **2011**, *118*, 3172–3181. [CrossRef]
18. Ploug, M.; Ronne, E.; Behrendt, N.; Jensen, A.L.; Blasi, F.; Dano, K. Cellular receptor for urokinase plasminogen activator. Carboxyl-terminal processing and membrane anchoring by glycosyl-phosphatidylinositol. *J. Biol. Chem.* **1991**, *266*, 1926–1933. [CrossRef]
19. Abdul, S.; Leebeek, F.W.; Rijken, D.C.; Uitte de Willige, S. Natural heterogeneity of alpha2-antiplasmin: Functional and clinical consequences. *Blood* **2016**, *127*, 538–545. [CrossRef]
20. Van De Craen, B.; Declerck, P.J.; Gils, A. The Biochemistry, Physiology and Pathological roles of PAI-1 and the requirements for PAI-1 inhibition in vivo. *Thromb. Res.* **2012**, *130*, 576–585. [CrossRef]
21. Hattori, N.; Sisson, T.H.; Xu, Y.; Desai, T.J.; Simon, R.H. Participation of urokinase-type plasminogen activator receptor in the clearance of fibrin from the lung. *Am. J. Physiol.* **1999**, *277*, L573–L579. [CrossRef] [PubMed]
22. Idell, S. Coagulation, fibrinolysis, and fibrin deposition in acute lung injury. *Crit. Care Med.* **2003**, *31*, S213–S220. [CrossRef] [PubMed]
23. Tikellis, C.; Thomas, M.C. Angiotensin-Converting Enzyme 2 (ACE2) Is a Key Modulator of the Renin Angiotensin System in Health and Disease. *Int. J. Pept.* **2012**, *2012*, 256294. [CrossRef] [PubMed]

24. Ferrario, C.M. The renin-angiotensin system: Importance in physiology and pathology. *J. Cardiovasc. Pharmcol.* **1990**, *15* (Suppl. 3), S1–S5. [CrossRef]
25. Donoghue, M.; Hsieh, F.; Baronas, E.; Godbout, K.; Gosselin, M.; Stagliano, N.; Donovan, M.; Woolf, B.; Robison, K.; Jeyaseelan, R.; et al. A novel angiotensin-converting enzyme-related carboxypeptidase (ACE2) converts angiotensin I to angiotensin 1–9. *Circ. Res.* **2000**, *87*, E1–E9. [CrossRef] [PubMed]
26. Kuba, K.; Imai, Y.; Rao, S.; Gao, H.; Guo, F.; Guan, B.; Huan, Y.; Yang, P.; Zhang, Y.; Deng, W.; et al. A crucial role of angiotensin converting enzyme 2 (ACE2) in SARS coronavirus-induced lung injury. *Nat. Med.* **2005**, *11*, 875–879. [CrossRef]
27. Patel, A.B.; Verma, A. COVID-19 and Angiotensin-Converting Enzyme Inhibitors and Angiotensin Receptor Blockers: What Is the Evidence? *JAMA* **2020**, *323*, 1769–1770. [CrossRef]
28. Turner, A.J.; Hiscox, J.A.; Hooper, N.M. ACE2: From vasopeptidase to SARS virus receptor. *Trends Pharmcol. Sci.* **2004**, *25*, 291–294. [CrossRef]
29. Zhang, H.; Rostami, M.R.; Leopold, P.L.; Mezey, J.G.; O'Beirne, S.L.; Strulovici-Barel, Y.; Crystal, R.G. Expression of the SARS-CoV-2 ACE2 Receptor in the Human Airway Epithelium. *Am. J. Respir. Crit. Care Med.* **2020**, *202*, 219–229. [CrossRef]
30. Zhang, H.; Li, H.B.; Lyu, J.R.; Lei, X.M.; Li, W.; Wu, G.; Lyu, J.; Dai, Z.M. Specific ACE2 expression in small intestinal enterocytes may cause gastrointestinal symptoms and injury after 2019-nCoV infection. *Int. J. Infect. Dis.* **2020**, *96*, 19–24. [CrossRef]
31. Jia, H.P.; Look, D.C.; Shi, L.; Hickey, M.; Pewe, L.; Netland, J.; Farzan, M.; Wohlford-Lenane, C.; Perlman, S.; McCray, P.B., Jr. ACE2 receptor expression and severe acute respiratory syndrome coronavirus infection depend on differentiation of human airway epithelia. *J. Virol.* **2005**, *79*, 14614–14621. [CrossRef] [PubMed]
32. Wan, Y.; Shang, J.; Graham, R.; Baric, R.S.; Li, F. Receptor Recognition by the Novel Coronavirus from Wuhan: An Analysis Based on Decade-Long Structural Studies of SARS Coronavirus. *J. Virol.* **2020**, *94*. [CrossRef] [PubMed]
33. Kam, Y.W.; Okumura, Y.; Kido, H.; Ng, L.F.; Bruzzone, R.; Altmeyer, R. Cleavage of the SARS coronavirus spike glycoprotein by airway proteases enhances virus entry into human bronchial epithelial cells in vitro. *PLoS ONE* **2009**, *4*, e7870. [CrossRef] [PubMed]
34. Hoffmann, M.; Kleine-Weber, H.; Pohlmann, S. A Multibasic Cleavage Site in the Spike Protein of SARS-CoV-2 Is Essential for Infection of Human Lung Cells. *Mol. Cell* **2020**, *78*, 779–784.e5. [CrossRef] [PubMed]
35. Verdecchia, P.; Cavallini, C.; Spanevello, A.; Angeli, F. The pivotal link between ACE2 deficiency and SARS-CoV-2 infection. *Eur. J. Intern. Med.* **2020**, *76*, 14–20. [CrossRef]
36. Hamming, I.; Cooper, M.E.; Haagmans, B.L.; Hooper, N.M.; Korstanje, R.; Osterhaus, A.D.; Timens, W.; Turner, A.J.; Navis, G.; van Goor, H. The emerging role of ACE2 in physiology and disease. *J. Pathol.* **2007**, *212*, 1–11. [CrossRef] [PubMed]
37. Dijkman, R.; Jebbink, M.F.; Deijs, M.; Milewska, A.; Pyrc, K.; Buelow, E.; van der Bijl, A.; van der Hoek, L. Replication-dependent downregulation of cellular angiotensin-converting enzyme 2 protein expression by human coronavirus NL63. *J. Gen. Virol.* **2012**, *93*, 1924–1929. [CrossRef] [PubMed]
38. Abassi, Z.A.; Skorecki, K.; Heyman, S.N.; Kinaneh, S.; Armaly, Z. Covid-19 infection and mortality: A physiologist's perspective enlightening clinical features and plausible interventional strategies. *Am. J. Physiol. Lung Cell Mol. Physiol.* **2020**, *318*, L1020–L1022. [CrossRef] [PubMed]
39. Franks, T.J.; Chong, P.Y.; Chui, P.; Galvin, J.R.; Lourens, R.M.; Reid, A.H.; Selbs, E.; McEvoy, C.P.; Hayden, C.D.; Fukuoka, J.; et al. Lung pathology of severe acute respiratory syndrome (SARS): A study of 8 autopsy cases from Singapore. *Hum. Pathol.* **2003**, *34*, 743–748. [CrossRef]
40. Gralinski, L.E.; Bankhead, A., 3rd; Jeng, S.; Menachery, V.D.; Proll, S.; Belisle, S.E.; Matzke, M.; Webb-Robertson, B.J.; Luna, M.L.; Shukla, A.K.; et al. Mechanisms of severe acute respiratory syndrome coronavirus-induced acute lung injury. *mBio* **2013**, *4*. [CrossRef] [PubMed]
41. Carsana, L.; Sonzogni, A.; Nasr, A.; Rossi, R.S.; Pellegrinelli, A.; Zerbi, P.; Rech, R.; Colombo, R.; Antinori, S.; Corbellino, M.; et al. Pulmonary post-mortem findings in a series of COVID-19 cases from northern Italy: A two-centre descriptive study. *Lancet Infect. Dis.* **2020**, *20*, 1135–1140. [CrossRef]
42. Ackermann, M.; Verleden, S.E.; Kuehnel, M.; Haverich, A.; Welte, T.; Laenger, F.; Vanstapel, A.; Werlein, C.; Stark, H.; Tzankov, A.; et al. Pulmonary Vascular Endothelialitis, Thrombosis, and Angiogenesis in Covid-19. *N. Engl. J. Med.* **2020**, *383*, 120–128. [CrossRef] [PubMed]
43. Mulvey, J.J. Complement associated microvascular injury and thrombosis in the pathogenesis of severe COVID-19 infection: A report of five cases. *Transl. Res.* **2020**, *220*, 1–13.
44. Leppkes, M.; Knopf, J.; Naschberger, E.; Lindemann, A.; Singh, J.; Herrmann, I.; Sturzl, M.; Staats, L.; Mahajan, A.; Schauer, C.; et al. Vascular occlusion by neutrophil extracellular traps in COVID-19. *EBioMedicine* **2020**, *58*, 102925. [CrossRef] [PubMed]
45. Hayiroglu, M.I.; Cinar, T.; Tekkesin, A.I. Fibrinogen and D-dimer variances and anticoagulation recommendations in Covid-19: Current literature review. *Rev. Assoc. Med. Bras.* **2020**, *66*, 842–848. [CrossRef]
46. Weitz, J.I.; Fredenburgh, J.C.; Eikelboom, J.W. A Test in Context: D-Dimer. *J. Am. Coll Cardiol.* **2017**, *70*, 2411–2420. [CrossRef]
47. Vaughan, D.E.; Lazos, S.A.; Tong, K. Angiotensin II regulates the expression of plasminogen activator inhibitor-1 in cultured endothelial cells. A potential link between the renin-angiotensin system and thrombosis. *J. Clin. Investig.* **1995**, *95*, 995–1001. [CrossRef]
48. Vaughan, D.E. Endothelial function, fibrinolysis, and angiotensin-converting enzyme inhibition. *Clin. Cardiol.* **1997**, *20*, II-34–II-37. [CrossRef]

49. Nakamura, S.; Nakamura, I.; Ma, L.; Vaughan, D.E.; Fogo, A.B. Plasminogen activator inhibitor-1 expression is regulated by the angiotensin type 1 receptor in vivo. *Kidney Int.* **2000**, *58*, 251–259. [CrossRef]
50. Bhandary, Y.P.; Shetty, S.K.; Marudamuthu, A.S.; Ji, H.L.; Neuenschwander, P.F.; Boggaram, V.; Morris, G.F.; Fu, J.; Idell, S.; Shetty, S. Regulation of lung injury and fibrosis by p53-mediated changes in urokinase and plasminogen activator inhibitor-1. *Am. J. Pathol.* **2013**, *183*, 131–143. [CrossRef]
51. Puthusseri, B.; Marudamuthu, A.; Tiwari, N.; Fu, J.; Idell, S.; Shetty, S. Regulation of p53-mediated changes in the uPA-fibrinolytic system and in lung injury by loss of surfactant protein C expression in alveolar epithelial cells. *Am. J. Physiol. Lung Cell Mol. Physiol.* **2017**, *312*, L783–L796. [CrossRef] [PubMed]
52. Yoshida, M.; Naito, Y.; Urano, T.; Takada, A.; Takada, Y. L-158,809 and (D-Ala(7))-angiotensin I/II (1-7) decrease PAI-1 release from human umbilical vein endothelial cells. *Thromb. Res.* **2002**, *105*, 531–536. [CrossRef]
53. Mogielnicki, A.; Kramkowski, K.; Hermanowicz, J.M.; Leszczynska, A.; Przyborowski, K.; Buczko, W. Angiotensin-(1-9) enhances stasis-induced venous thrombosis in the rat because of the impairment of fibrinolysis. *J. Renin Angiotensin Aldosterone Syst.* **2014**, *15*, 13–21. [CrossRef] [PubMed]
54. Flevaris, P.; Vaughan, D. The Role of Plasminogen Activator Inhibitor Type-1 in Fibrosis. *Semin. Thromb. Hemost.* **2017**, *43*, 169–177. [CrossRef] [PubMed]
55. Marudamuthu, A.S.; Bhandary, Y.P.; Shetty, S.K.; Fu, J.; Sathish, V.; Prakash, Y.; Shetty, S. Role of the urokinase-fibrinolytic system in epithelial-mesenchymal transition during lung injury. *Am. J. Pathol.* **2015**, *185*, 55–68. [CrossRef]
56. Ghosh, A.K.; Vaughan, D.E. PAI-1 in tissue fibrosis. *J. Cell Physiol.* **2012**, *227*, 493–507. [CrossRef]
57. Ye, Q.; Wang, B.; Mao, J. The pathogenesis and treatment of the 'Cytokine Storm' in COVID-19. *J. Infect.* **2020**, *80*, 607–613. [CrossRef]
58. Soy, M.; Keser, G.; Atagunduz, P.; Tabak, F.; Atagunduz, I.; Kayhan, S. Cytokine storm in COVID-19: Pathogenesis and overview of anti-inflammatory agents used in treatment. *Clin. Rheumatol.* **2020**, *39*, 2085–2094. [CrossRef]
59. Rega, G.; Kaun, C.; Weiss, T.W.; Demyanets, S.; Zorn, G.; Kastl, S.P.; Steiner, S.; Seidinger, D.; Kopp, C.W.; Frey, M.; et al. Inflammatory cytokines interleukin-6 and oncostatin m induce plasminogen activator inhibitor-1 in human adipose tissue. *Circulation* **2005**, *111*, 1938–1945. [CrossRef]
60. Declerck, P.J.; Gils, A. Three decades of research on plasminogen activator inhibitor-1: A multifaceted serpin. *Semin. Thromb. Hemost.* **2013**, *39*, 356–364.
61. Vaughan, D.E. PAI-1 and atherothrombosis. *J. Thromb. Haemost.* **2005**, *3*, 1879–1883. [CrossRef] [PubMed]
62. Sun, T.; Ghosh, A.K.; Eren, M.; Miyata, T.; Vaughan, D.E. PAI-1 contributes to homocysteine-induced cellular senescence. *Cell Signal.* **2019**, *64*, 109394. [CrossRef] [PubMed]
63. Yamamoto, K.; Kojima, T.; Adachi, T.; Hayashi, M.; Matsushita, T.; Takamatsu, J.; Loskutoff, D.J.; Saito, H. Obesity enhances the induction of plasminogen activator inhibitor-1 by restraint stress: A possible mechanism of stress-induced renal fibrin deposition in obese mice. *J. Thromb. Haemost.* **2005**, *3*, 1495–1502. [CrossRef] [PubMed]
64. Yamamoto, K.; Saito, H. A pathological role of increased expression of plasminogen activator inhibitor-1 in human or animal disorders. *Int. J. Hematol.* **1998**, *68*, 371–385. [CrossRef]
65. Yamamoto, K.; Takeshita, K.; Kojima, T.; Takamatsu, J.; Saito, H. Aging and plasminogen activator inhibitor-1 (PAI-1) regulation: Implication in the pathogenesis of thrombotic disorders in the elderly. *Cardiovasc. Res.* **2005**, *66*, 276–285. [CrossRef]
66. Guan, W.J.; Liang, W.H.; Zhao, Y.; Liang, H.R.; Chen, Z.S.; Li, Y.M.; Liu, X.Q.; Chen, R.C.; Tang, C.L.; Wang, T.; et al. Comorbidity and its impact on 1590 patients with COVID-19 in China: A nationwide analysis. *Eur. Respir. J.* **2020**, *55*, 2000547. [CrossRef]
67. Oxley, T.J.; Mocco, J.; Majidi, S.; Kellner, C.P.; Shoirah, H.; Singh, I.P.; De Leacy, R.A.; Shigematsu, T.; Ladner, T.R.; Yaeger, K.A.; et al. Large-Vessel Stroke as a Presenting Feature of Covid-19 in the Young. *N. Engl. J. Med.* **2020**, *382*, e60. [CrossRef]
68. Paramo, J.A. Pulmonary Embolism, Pulmonary Microvascular Thrombosis, or Both in COVID-19? *Clin. Appl. Thromb. Hemost.* **2020**, *26*. [CrossRef]
69. Klok, F.A.; Kruip, M.J.H.A.; van der Meer, N.J.M.; Arbous, M.S.; Gommers, D.; Kant, K.M.; Kaptein, F.H.J.; van Paassen, J.; Stals, M.A.M.; Huisman, M.V.; et al. Confirmation of the high cumulative incidence of thrombotic complications in critically ill ICU patients with COVID-19: An updated analysis. *Thromb. Res.* **2020**, *191*, 148–150. [CrossRef]
70. Middeldorp, S.; Coppens, M.; van Haaps, T.F.; Foppen, M.; Vlaar, A.P.; Muller, M.C.A.; Bouman, C.C.S.; Beenen, L.F.M.; Kootte, R.S.; Heijmans, J.; et al. Incidence of venous thromboembolism in hospitalized patients with COVID-19. *J. Thromb. Haemost.* **2020**, *18*, 1995–2002. [CrossRef]
71. Poissy, J.; Goutay, J.; Caplan, M.; Parmentier, E.; Duburcq, T.; Lassalle, F.; Jeanpierre, E.; Rauch, A.; Labreuche, J.; Susen, S.; et al. Pulmonary Embolism in COVID-19 Patients: Awareness of an Increased Prevalence. *Circulation* **2020**, *142*, 184–186. [CrossRef]
72. Guagliumi, G.; Sonzogni, A.; Pescetelli, I.; Pellegrini, D.; Finn, A.V. Microthrombi and ST-Segment Elevation Myocardial Infarction in COVID-19. *Circulation* **2020**, *142*, 804–809. [CrossRef]
73. Pun, M.; Haggerty-Skeans, J.; Pratt, D.; Fudym, Y.; Al-Holou, W.N.; Camelo-Piragua, S.; Venneti, S. H3K27M-mutant diffuse midline glioma with extensive intratumoral microthrombi in a young adult with COVID-19-associated coagulopathy. *Acta Neuropathol.* **2020**, *140*, 227–229. [CrossRef]
74. Tee, A.; Wong, A.; Yusuf, G.T.; Rao, D.; Sidhu, P.S. Contrast-enhanced ultrasound (CEUS) of the lung reveals multiple areas of microthrombi in a COVID-19 patient. *Intensive Care Med.* **2020**, *46*, 1660–1662. [CrossRef]

75. Tutiya, C.T.; Siaulys, M.M.; Kondo, M.M.; Miglioli-Galvao, L.C.A.; Galvao, E.; Pinheiro, C.C.; Torloni, M.R.; de Mello, F.B. Possible formation of pulmonary microthrombi in the early puerperium of pregnant women critically ill with COVID-19: Two case reports. *Case Rep. Womens Health* **2020**, *27*, e00237. [CrossRef]
76. Batlle, D.; Soler, M.J.; Sparks, M.A.; Hiremath, S.; South, A.M.; Welling, P.A.; Swaminathan, S. COVID-19 and ACE2 in Cardiovascular, Lung, and Kidney Working Group. Acute Kidney Injury in COVID-19: Emerging Evidence of a Distinct Pathophysiology. *J. Am. Soc. Nephrol.* **2020**, *31*, 1380–1383. [CrossRef]
77. Su, H.; Yang, M.; Wan, C.; Yi, L.X.; Tang, F.; Zhu, H.Y.; Yi, F.; Yang, H.C.; Fogo, A.B.; Nie, X.; et al. Renal histopathological analysis of 26 postmortem findings of patients with COVID-19 in China. *Kidney Int.* **2020**, *98*, 219–227. [CrossRef]
78. Hanley, B.; Lucas, S.B.; Youd, E.; Swift, B.; Osborn, M. Autopsy in suspected COVID-19 cases. *J. Clin. Pathol.* **2020**, *73*, 239–242. [CrossRef]
79. Gencer, S.; Lacy, M.; Atzler, D.; van der Vorst, E.P.C.; Doring, Y.; Weber, C. Immunoinflammatory, Thrombohaemostatic, and Cardiovascular Mechanisms in COVID-19. *Thromb. Haemost.* **2020**, *120*, 1629–1641. [CrossRef]
80. Rapkiewicz, A.V.; Mai, X.; Carsons, S.E.; Pittaluga, S.; Kleiner, D.E.; Berger, J.S.; Thomas, S.; Adler, N.M.; Charytan, D.M.; Gasmi, B.; et al. Megakaryocytes and platelet-fibrin thrombi characterize multi-organ thrombosis at autopsy in COVID-19: A case series. *EClinicalMedicine* **2020**, *24*, 100434. [CrossRef]
81. Tang, N.; Li, D.; Wang, X.; Sun, Z. Abnormal coagulation parameters are associated with poor prognosis in patients with novel coronavirus pneumonia. *J. Thromb. Haemost.* **2020**, *18*, 844–847. [CrossRef]
82. Levi, M.; Thachil, J.; Iba, T.; Levy, J.H. Coagulation abnormalities and thrombosis in patients with COVID-19. *Lancet Haematol.* **2020**, *7*, e438–e440. [CrossRef]
83. Blasi, A.; von Meijenfeldt, F.A.; Adelmeijer, J.; Calvo, A.; Ibanez, C.; Perdomo, J.; Reverter, J.C.; Lisman, T. In vitro hypercoagulability and ongoing in vivo activation of coagulation and fibrinolysis in COVID-19 patients on anticoagulation. *J. Thromb. Haemost.* **2020**, *18*, 2646–2653. [CrossRef]
84. Han, H.; Yang, L.; Liu, R.; Liu, F.; Wu, K.L.; Li, J.; Liu, X.H.; Zhu, C.L. Prominent changes in blood coagulation of patients with SARS-CoV-2 infection. *Clin. Chem. Lab. Med.* **2020**, *58*, 1116–1120. [CrossRef]
85. Dowton, S.B.; Colten, H.R. Acute phase reactants in inflammation and infection. *Semin. Hematol.* **1988**, *25*, 84–90.
86. Bi, X.; Su, Z.; Yan, H.; Du, J.; Wang, J.; Chen, L.; Peng, M.; Chen, S.; Shen, B.; Li, J. Prediction of severe illness due to COVID-19 based on an analysis of initial Fibrinogen to Albumin Ratio and Platelet count. *Platelets* **2020**, *31*, 674–679. [CrossRef]
87. Wright, F.L.; Vogler, T.O.; Moore, E.E.; Moore, H.B.; Wohlauer, M.V.; Urban, S.; Nydam, T.L.; Moore, P.K.; McIntyre, R.C., Jr. Fibrinolysis Shutdown Correlation with Thromboembolic Events in Severe COVID-19 Infection. *J. Am. Coll. Surg.* **2020**, *231*, 193–203. [CrossRef]
88. Panigada, M.; Bottino, N.; Tagliabue, P.; Grasselli, G.; Novembrino, C.; Chantarangkul, V.; Pesenti, A.; Peyvandi, F.; Tripodi, A. Hypercoagulability of COVID-19 patients in Intensive Care Unit. A Report of Thromboelastography Findings and other Parameters of Hemostasis. *J. Thromb. Haemost.* **2020**, *18*, 1738–1742. [CrossRef]
89. Ibanez, C.; Perdomo, J.; Calvo, A.; Ferrando, C.; Reverter, J.C.; Tassies, D.; Blasi, A. High D dimers and low global fibrinolysis coexist in COVID19 patients: What is going on in there? *J. Thromb. Thrombolysis* **2020**. [CrossRef]
90. Medcalf, R.L.; Keragala, C.B.; Myles, P.S. Fibrinolysis and COVID-19: A plasmin paradox. *J. Thromb. Haemost.* **2020**, *18*, 2118–2122. [CrossRef]
91. Prabhakaran, P.; Ware, L.B.; White, K.E.; Cross, M.T.; Matthay, M.A.; Olman, M.A. Elevated levels of plasminogen activator inhibitor-1 in pulmonary edema fluid are associated with mortality in acute lung injury. *Am. J. Physiol. Lung Cell Mol. Physiol.* **2003**, *285*, L20–L28. [CrossRef] [PubMed]
92. Fuchs-Buder, T.; de Moerloose, P.; Ricou, B.; Reber, G.; Vifian, C.; Nicod, L.; Romand, J.A.; Suter, P.M. Time course of procoagulant activity and D dimer in bronchoalveolar fluid of patients at risk for or with acute respiratory distress syndrome. *Am. J. Respir. Crit. Care Med.* **1996**, *153*, 163–167. [CrossRef] [PubMed]
93. Fujimoto, H.; Gabazza, E.C.; Hataji, O.; Yuda, H.; D'Alessandro-Gabazza, C.N.; Nakano, M.; Franco, O.E.; Hayashi, T.; Suzuki, K.; Adachi, Y.; et al. Thrombin-activatable fibrinolysis inhibitor and protein C inhibitor in interstitial lung disease. *Am. J. Respir. Crit. Care Med.* **2003**, *167*, 1687–1694. [CrossRef] [PubMed]
94. Bertozzi, P.; Astedt, B.; Zenzius, L.; Lynch, K.; LeMaire, F.; Zapol, W.; Chapman, H.A., Jr. Depressed bronchoalveolar urokinase activity in patients with adult respiratory distress syndrome. *N. Engl. J. Med.* **1990**, *322*, 890–897. [CrossRef] [PubMed]
95. Zhang, Y.; Cao, W.; Jiang, W.; Xiao, M.; Li, Y.; Tang, N.; Liu, Z.; Yan, X.; Zhao, Y.; Li, T.; et al. Profile of natural anticoagulant, coagulant factor and anti-phospholipid antibody in critically ill COVID-19 patients. *J. Thromb. Thromb.* **2020**, *50*, 580–586. [CrossRef] [PubMed]
96. Bowles, L.; Platton, S.; Yartey, N.; Dave, M.; Lee, K.; Hart, D.P.; MacDonald, V.; Green, L.; Sivapalaratnam, S.; Pasi, K.J.; et al. Lupus Anticoagulant and Abnormal Coagulation Tests in Patients with Covid-19. *N. Engl. J. Med.* **2020**, *383*, 288–290. [CrossRef]
97. Idell, S.; Kueppers, F.; Lippmann, M.; Rosen, H.; Niederman, M.; Fein, A. Angiotensin converting enzyme in bronchoalveolar lavage in ARDS. *Chest* **1987**, *91*, 52–56. [CrossRef]
98. Konings, J.; Hoving, L.R.; Ariens, R.S.; Hethershaw, E.L.; Ninivaggi, M.; Hardy, L.J.; de Laat, B.; Ten Cate, H.; Philippou, H.; Govers-Riemslag, J.W. The role of activated coagulation factor XII in overall clot stability and fibrinolysis. *Thromb. Res.* **2015**, *136*, 474–480. [CrossRef]

99. Kruse, J.M.; Magomedov, A.; Kurreck, A.; Munch, F.H.; Koerner, R.; Kamhieh-Milz, J.; Kahl, A.; Gotthardt, I.; Piper, S.K.; Eckardt, K.U.; et al. Thromboembolic complications in critically ill COVID-19 patients are associated with impaired fibrinolysis. *Crit. Care* **2020**, *24*, 676. [CrossRef]

100. Pavoni, V.; Gianesello, L.; Pazzi, M.; Stera, C.; Meconi, T.; Frigieri, F.C. Evaluation of coagulation function by rotation thromboelastometry in critically ill patients with severe COVID-19 pneumonia. *J. Thromb. Thrombolysis* **2020**, *50*, 281–286. [CrossRef]

101. van Veenendaal, N.; Scheeren, T.W.L.; Meijer, K.; van der Voort, P.H.J. Rotational thromboelastometry to assess hypercoagulability in COVID-19 patients. *Thromb. Res.* **2020**, *196*, 379–381. [CrossRef] [PubMed]

102. Spiezia, L.; Boscolo, A.; Poletto, F.; Cerruti, L.; Tiberio, I.; Campello, E.; Navalesi, P.; Simioni, P. COVID-19-Related Severe Hypercoagulability in Patients Admitted to Intensive Care Unit for Acute Respiratory Failure. *Thromb. Haemost.* **2020**, *120*, 998–1000. [CrossRef] [PubMed]

103. Slomka, A.; Kowalewski, M.; Zekanowska, E. Haemostasis in coronavirus disease 2019—Lesson from viscoelastic methods—A systemic review. *Thromb. Haemost.* **2021**. [CrossRef]

104. Moore, H.B.; Moore, E.E.; Neal, M.D.; Sheppard, F.R.; Kornblith, L.Z.; Draxler, D.F.; Walsh, M.; Medcalf, R.L.; Cohen, M.J.; Cotton, B.A.; et al. Fibrinolysis Shutdown in Trauma: Historical Review and Clinical Implications. *Anesth. Analg.* **2019**, *129*, 762–773. [CrossRef] [PubMed]

105. Hottz, E.D.; Azevedo-Quintanilha, I.G.; Palhinha, L.; Teixeira, L.; Barreto, E.A.; Pao, C.R.R.; Righy, C.; Franco, S.; Souza, T.M.L.; Kurtz, P.; et al. Platelet activation and platelet-monocyte aggregates formation trigger tissue factor expression in severe COVID-19 patients. *Blood* **2020**, *136*, 1330–1341. [CrossRef]

106. Manne, B.K.; Denorme, F.; Middleton, E.A.; Portier, I.; Rowley, J.W.; Stubben, C.; Petrey, A.C.; Tolley, N.D.; Guo, L.; Cody, M.; et al. Platelet Gene Expression and Function in COVID-19 Patients. *Blood* **2020**, *136*, 1317–1329. [CrossRef]

107. Huebner, B.R.; Moore, E.E.; Moore, H.B.; Stettler, G.R.; Nunns, G.R.; Lawson, P.; Sauaia, A.; Kelher, M.; Banerjee, A.; Silliman, C.C. Thrombin Provokes Degranulation of Platelet alpha-Granules Leading to the Release of Active Plasminogen Activator Inhibitor-1 (PAI-1). *Shock* **2018**, *50*, 671–676. [CrossRef]

108. Merle, N.S.; Noe, R.; Halbwachs-Mecarelli, L.; Fremeaux-Bacchi, V.; Roumenina, L.T. Complement System Part II: Role in Immunity. *Front. Immunol.* **2015**, *6*, 257. [CrossRef]

109. Merle, N.S.; Church, S.E.; Fremeaux-Bacchi, V.; Roumenina, L.T. Complement System Part I—Molecular Mechanisms of Activation and Regulation. *Front. Immunol.* **2015**, *6*, 262. [CrossRef]

110. Fletcher-Sandersjoo, A.; Bellander, B.M. Is COVID-19 associated thrombosis caused by overactivation of the complement cascade? A literature review. *Thromb. Res.* **2020**, *194*, 36–41. [CrossRef]

111. Jiang, Y.; Zhao, G.; Song, N.; Li, P.; Chen, Y.; Guo, Y.; Li, J.; Du, L.; Jiang, S.; Guo, R.; et al. Blockade of the C5a-C5aR axis alleviates lung damage in hDPP4-transgenic mice infected with MERS-CoV. *Emerg. Microbes Infect.* **2018**, *7*, 77. [CrossRef] [PubMed]

112. Mosnier, L.O.; Bouma, B.N. Regulation of fibrinolysis by thrombin activatable fibrinolysis inhibitor, an unstable carboxypeptidase B that unites the pathways of coagulation and fibrinolysis. *Arter. Thromb. Vasc. Biol.* **2006**, *26*, 2445–2453. [CrossRef] [PubMed]

113. Kozarcanin, H.; Lood, C.; Munthe-Fog, L.; Sandholm, K.; Hamad, O.A.; Bengtsson, A.A.; Skjoedt, M.O.; Huber-Lang, M.; Garred, P.; Ekdahl, K.N.; et al. The lectin complement pathway serine proteases (MASPs) represent a possible crossroad between the coagulation and complement systems in thromboinflammation. *J. Thromb. Haemost.* **2016**, *14*, 531–545. [CrossRef] [PubMed]

114. Hess, K.; Ajjan, R.; Phoenix, F.; Dobo, J.; Gal, P.; Schroeder, V. Effects of MASP-1 of the complement system on activation of coagulation factors and plasma clot formation. *PLoS ONE* **2012**, *7*, e35690. [CrossRef]

115. Davis, A.E., 3rd; Lu, F.; Mejia, P. C1 inhibitor, a multi-functional serine protease inhibitor. *Thromb. Haemost.* **2010**, *104*, 886–893.

116. Brown, E.W.; Ravindran, S.; Patston, P.A. The reaction between plasmin and C1-inhibitor results in plasmin inhibition by the serpin mechanism. *Blood Coagul. Fibrinolysis* **2002**, *13*, 711–714. [CrossRef]

117. Foley, J.H.; Walton, B.L.; Aleman, M.M.; O'Byrne, A.M.; Lei, V.; Harrasser, M.; Foley, K.A.; Wolberg, A.S.; Conway, E.M. Complement Activation in Arterial and Venous Thrombosis is Mediated by Plasmin. *EBioMedicine* **2016**, *5*, 175–182. [CrossRef]

118. Wojta, J.; Huber, K.; Valent, P. New aspects in thrombotic research: Complement induced switch in mast cells from a profibrinolytic to a prothrombotic phenotype. *Pathophysiol. Haemost. Thromb.* **2003**, *33*, 438–441. [CrossRef]

119. Fletcher-Sandersjoo, A.; Maegele, M.; Bellander, B.M. Does Complement-Mediated Hemostatic Disturbance Occur in Traumatic Brain Injury? A Literature Review and Observational Study Protocol. *Int. J. Mol. Sci.* **2020**, *21*, 1596. [CrossRef]

120. Sauter, R.J.; Sauter, M.; Obrich, M.; Emschermann, F.N.; Nording, H.; Patzelt, J.; Wendel, H.P.; Reil, J.C.; Edlich, F.; Langer, H.F. Anaphylatoxin Receptor C3aR Contributes to Platelet Function, Thrombus Formation and In Vivo Haemostasis. *Thromb. Haemost.* **2019**, *119*, 179–182. [CrossRef]

121. Sauter, R.J.; Sauter, M.; Reis, E.S.; Emschermann, F.N.; Nording, H.; Ebenhoch, S.; Kraft, P.; Munzer, P.; Mauler, M.; Rheinlaender, J.; et al. Functional Relevance of the Anaphylatoxin Receptor C3aR for Platelet Function and Arterial Thrombus Formation Marks an Intersection Point Between Innate Immunity and Thrombosis. *Circulation* **2018**, *138*, 1720–1735. [CrossRef]

122. Trimarchi, H.; Gianserra, R.; Lampo, M.; Monkowski, M.; Lodolo, J. Eculizumab, SARS-CoV-2 and atypical hemolytic uremic syndrome. *Clin. Kidney J.* **2020**, *13*, 739–741. [CrossRef] [PubMed]

123. Zhou, Y.; Lu, K.; Pfefferle, S.; Bertram, S.; Glowacka, I.; Drosten, C.; Pohlmann, S.; Simmons, G. A single asparagine-linked glycosylation site of the severe acute respiratory syndrome coronavirus spike glycoprotein facilitates inhibition by mannose-binding lectin through multiple mechanisms. *J. Virol.* **2010**, *84*, 8753–8764. [CrossRef] [PubMed]

124. Sisson, T.H.; Simon, R.H. The plasminogen activation system in lung disease. *Curr. Drug Targets* **2007**, *8*, 1016–1029. [CrossRef] [PubMed]
125. Shetty, S.; Padijnayayveetil, J.; Tucker, T.; Stankowska, D.; Idell, S. The fibrinolytic system and the regulation of lung epithelial cell proteolysis, signaling, and cellular viability. *Am. J. Physiol. Lung Cell Mol. Physiol.* **2008**, *295*, L967–L975. [CrossRef] [PubMed]
126. Raghavendran, K.; Willson, D.; Notter, R.H. Surfactant therapy for acute lung injury and acute respiratory distress syndrome. *Crit. Care Clin.* **2011**, *27*, 525–559. [CrossRef] [PubMed]
127. Gouda, M.M.; Shaikh, S.B.; Bhandary, Y.P. Inflammatory and Fibrinolytic System in Acute Respiratory Distress Syndrome. *Lung* **2018**, *196*, 609–616. [CrossRef]
128. Kalluri, R.; Neilson, E.G. Epithelial-mesenchymal transition and its implications for fibrosis. *J. Clin. Investig.* **2003**, *112*, 1776–1784. [CrossRef]
129. Al-Samkari, H.; Karp Leaf, R.S.; Dzik, W.H.; Carlson, J.C.T.; Fogerty, A.E.; Waheed, A.; Goodarzi, K.; Bendapudi, P.K.; Bornikova, L.; Gupta, S.; et al. COVID-19 and coagulation: Bleeding and thrombotic manifestations of SARS-CoV-2 infection. *Blood* **2020**, *136*, 489–500. [CrossRef]
130. Desborough, M.J.R.; Doyle, A.J.; Griffiths, A.; Retter, A.; Breen, K.A.; Hunt, B.J. Image-proven thromboembolism in patients with severe COVID-19 in a tertiary critical care unit in the United Kingdom. *Thromb. Res.* **2020**, *193*, 1–4. [CrossRef]
131. Fraisse, M.; Logre, E.; Pajot, O.; Mentec, H.; Plantefeve, G.; Contou, D. Thrombotic and hemorrhagic events in critically ill COVID-19 patients: A French monocenter retrospective study. *Crit. Care* **2020**, *24*, 275. [CrossRef] [PubMed]
132. Musoke, N.; Lo, K.B.; Albano, J.; Peterson, E.; Bhargav, R.; Gul, F.; DeJoy, R., 3rd; Salacup, G.; Pelayo, J.; Tipparaju, P.; et al. Anticoagulation and bleeding risk in patients with COVID-19. *Thromb. Res.* **2020**, *196*, 227–230. [CrossRef] [PubMed]
133. Stillson, J.E.; Bunch, C.M.; Gillespie, L.; Khan, R.; Wierman, J.; Pulvirenti, J.; Phyu, H.; Anderson, S.; Al-Fadhl, M.; Thomas, A.V.; et al. Thromboelastography guided management of the anticoagulated COVID-19 patient. *Semin. Thromb. Hematol.* **2021**, *47*. in press.
134. Liu, C.; Ma, Y.; Su, Z.; Zhao, R.; Zhao, X.; Nie, H.-G.; Ping, X.; Zhu, L.; Zhang, M.; Li, X.; et al. Meta-Analysis of Preclinical Studies of Fibrinolytic Therapy for Acute Lung Injury. *Front. Immunol.* **2018**, *9*, 1898. [CrossRef] [PubMed]
135. Wu, Y.; Wang, T.; Guo, C.; Zhang, D.; Ge, X.; Huang, Z.; Zhou, X.; Li, Y.; Peng, Q.; Li, J. Plasminogen improves lung lesions and hypoxemia in patients with COVID-19. *QJM* **2020**, *113*, 539–545. [CrossRef]
136. Choudhury, R.; Barrett, C.D.; Moore, H.B.; Moore, E.E.; McIntyre, R.C.; Moore, P.K.; Talmor, D.S.; Nydam, T.L.; Yaffe, M.B. Salvage use of tissue plasminogen activator (tPA) in the setting of acute respiratory distress syndrome (ARDS) due to COVID-19 in the USA: A Markov decision analysis. *World J. Emerg. Surg.* **2020**, *15*, 29. [CrossRef]
137. Wang, J.; Hajizadeh, N.; Moore, E.E.; McIntyre, R.C.; Moore, P.K.; Veress, L.A.; Yaffe, M.B.; Moore, H.B.; Barrett, C.D. Tissue plasminogen activator (tPA) treatment for COVID-19 associated acute respiratory distress syndrome (ARDS): A case series. *J. Thromb. Haemost.* **2020**, *18*, 1752–1755. [CrossRef]
138. Ghosh, A.K.; Rai, R.; Park, K.E.; Eren, M.; Miyata, T.; Wilsbacher, L.D.; Vaughan, D.E. A small molecule inhibitor of PAI-1 protects against doxorubicin-induced cellular senescence. *Oncotarget* **2016**, *7*, 72443–72457. [CrossRef]

MDPI
St. Alban-Anlage 66
4052 Basel
Switzerland
Tel. +41 61 683 77 34
Fax +41 61 302 89 18
www.mdpi.com

International Journal of Molecular Sciences Editorial Office
E-mail: ijms@mdpi.com
www.mdpi.com/journal/ijms